SYMPOSIA OF THE ZOOLOGICAL SOCIETY OF LONDON NUMBER 56

Immune Mechanisms in Invertebrate Vectors

SYMPOSIA OF THE ZOOLOGICAL SOCIETY OF LONDON NUMBER 56

Immune Mechanisms in Invertebrate Vectors

The Proceedings of a Symposium
held at the Zoological Society of London
on 14th and 15th November 1985

Edited by A. M. LACKIE
Department of Zoology, University of Glasgow

Published for THE ZOOLOGICAL SOCIETY OF LONDON
by CLARENDON PRESS · OXFORD
1986

Oxford University Press, Walton Street, Oxford OX2 6DP

Oxford New York Toronto
Delhi Bombay Calcutta Madras Karachi
Petaling Jaya Singapore Hong Kong Tokyo
Nairobi Dar es Salaam Cape Town
Melbourne Auckland

and associated companies in
Beirut Berlin Ibadan Nicosia

Oxford is a trade mark of Oxford University Press

Published in the United States
by Oxford University Press, New York

British Library Cataloguing in Publication Data
Immune mechanisms in invertebrate vectors:
the proceedings of a symposium held at the
Zoological Society of London on 14th and
15th November 1985. — (Symposia of the
Zoological Society of London; no. 56)
1. Invertebrates — Physiology
2. Parasites 3. Immune response
I. Lackie, A.M. II. Zoological Society of
London III. Series
592'.02'95 QL364
ISBN 0–19–854004–3

Library of Congress Cataloging in Publication Data
Immune mechanisms in invertebrate vectors.
(Symposia of the Zoological Society of London; no. 56)
Includes bibliographies and index.
1. Immunity—Congresses. 2. Invertebrates—Physiology
—Congresses. I. Lackie, A. M. (Anne M.) II. Zoological
Society of London. III. Series.
QL1.Z733 no. 56 591 s 86–12755
[QR185.3] [592'.029]
ISBN 0–19–854004–3

Set by Promenade Graphics Ltd., Cheltenham
Printed in Great Britain by
St Edmundsbury Press,
Bury St Edmunds, Suffolk.

Preface

The aim of this Symposium, which was held at the Zoological Society in November 1985, was to bring together information on the mechanisms of immunity, and the efficacy of these mechanisms in controlling parasitic infection, in invertebrate vectors of diseases of humans and their domestic animals. The main vectors are the arthropods and molluscs, the blood-feeding insects and ticks being prominent among the former group and the gastropod snails in the latter. In this Symposium, we were concerned predominantly with the insects and the molluscs, since a considerable amount is known not only about the recognition and effector arms of their immune responses but also about the responses of these hosts to protozoan and helminth parasites. Much less is known about the immunoparasitology of crustacean arthropods, although certain aspects of their immunology are well-researched; Dr Smith's chapter should thus provide a useful basis for future work on the responses of Crustacea to the helminth larvae they carry. Since surprisingly little research appears to have been carried out on the immune responses of arachnid hosts, discussion of this group was not included in the Symposium.

Many of the insect vectors are biting flies which, owing to their small size and haemolymph volume, make difficult experimental animals. For this reason, much of our information on cellular and 'humoral' components of the insect immune response has been obtained from larger animals. As will be seen from Professor Boman's chapter, there appear to be similarities in the 'humoral' antibacterial responses of lepidopteran and dipteran hosts, so extrapolation from one Order to another would seem to be permissible in some cases; that there are similarities in the mechanisms of cellular immunity between different Orders is readily apparent. Despite the technical difficulties of working with small vector species, work on mosquitoes as vectors of filarial nematode larvae and of arboviruses is described by Professor Christensen and Dr Bishop, and we are provided with a review of insect immune responses, including those of the triatomine bug *Rhodnius* to trypanosomatid Protozoa, by Professor Molyneux.

In the course of investigations into the evasion, by schistosome larvae, of gastropod snail immunity, many interesting insights into the development and efficacy of the host's cellular immune mechanisms have been provided, and were discussed in the papers presented by Dr van der Knaap, Dr Loker and Dr Yoshino. The availability of genetic strains of snail with differing degrees of susceptibility to *Schistosoma mansoni* means that much relevant and useful information on vector resistance to this important parasite has been gained.

The final chapter of this volume, by Professor Anderson, provides a theoretical treatment of the underlying causes of invertebrate immune systems, and suggests how host resistance affects both host and parasite numbers and distributions; the present lack of data to insert into the mathematical models presented should point out future areas of research for those investigators concerned with the effects of invertebrate host resistance on parasite populations and thus on the transmission of vectored parasites.

The relaxed and informal air of the symposium and its discussion sessions was due both to the smooth and efficient organization provided by Miss Unity McDonnell, Administrative Assistant of the Zoological Society, and to the efficiency and promptness with which the participants produced their manuscripts. I am also very grateful to the Company of Biologists for providing financial assistance.

Glasgow
March 1986

A. M. L.

Contents

Antibacterial immune proteins in insects

H. G. BOMAN

Cellular immune mechanisms in the Crustacea

VALERIE J. SMITH & KENNETH SÖDERHÄLL

Lectins in molluscs and arthropods: their occurrence, origin and roles in immunity

LOTHAR RENWRANTZ

Interference—immunity of mosquitoes to bunyavirus superinfection

DAVID H. L. BISHOP & BARRY J. BEATY

Insect immunity to Trypanosomatidae

D. H. MOLYNEUX, G. TAKLE, E. A. IBRAHIM & G. A. INGRAM

Immune mechanisms and mosquito-filarial worm relationships

BRUCE M. CHRISTENSEN

Evasion of insect immunity by helminth larvae

ANN M. LACKIE

Interaction between the immune system of lymnaeid snails and trematode parasites

W. P. W. van der KNAAP & ELISABETH A. MEULEMAN

Immunity to trematode larvae in the snail *Biomphalaria*

ERIC S. LOKER & CHRISTOPHER J. BAYNE

Antigen sharing between larval trematodes and their snail hosts: how real a phenomenon in immune evasion?

TIMOTHY P. YOSHINO & CARL A. BOSWELL

Genetic variability in resistance to parasitic invasion: population implications for invertebrate host species

ROY M. ANDERSON

Contributors

ANDERSON, R. M., Department of Pure and Applied Biology, Imperial College of Science and Technology, University of London, Prince Consort Road, London SW7 2BB, UK.

BAYNE, C. J., Department of Zoology, Oregon State University, Corvallis, Oregon 97331, USA.

BEATY, B. J., Department of Microbiology, College of Veterinary Medicine and Biomedical Studies, Colorado State University, Fort Collins, Colorado 80523, USA.

BISHOP, D. H. L., N.E.R.C. Institute of Virology, Mansfield Road, Oxford OX1 3SR, UK.

BOMAN, H. G., Department of Microbiology, University of Stockholm, S–106 91 Stockholm, Sweden.

BOSWELL, C. A., Department of Zoology, University of Oklahoma, Norman, Oklahoma 73019, USA.

CHRISTENSEN, B., Department of Veterinary Science, University of Wisconsin, 1655 Linden Drive, Madison, Wisconsin 53706, USA.

GÖTZ, P., Institut für Allgemeine Zoologie der Freien Universität Berlin, Königin-Luise-Strasse 1–3, D-1000 Berlin 33, West Germany.

IBRAHIM, E. A., Department of Biological Sciences, University of Salford, Salford M5 4WT, UK.

INGRAM, G. A., Department of Biological Sciences, University of Salford, Salford M5 4WT, UK.

VAN DER KNAAP, W. P. W., Laboratory of Medical Parasitology, Faculty of Medicine, Free University, PO Box 7161, 1007 MC Amsterdam, The Netherlands.

LACKIE, A. M., Department of Zoology, The University, Glasgow G12 8QQ, UK.

LOKER, E. S., Department of Biology, The University of New Mexico, Albuquerque, New Mexico 87131, USA.

MEULEMAN, E. A., Laboratory of Medical Parasitology, Faculty of Medicine, Free University, PO Box 7161, 1007 MC Amsterdam, The Netherlands.

MOLYNEUX, D. H., Department of Biological Sciences, University of Salford, Salford M5 4WT, UK.

RATCLIFFE, N. A., Biomedicine and Physiology Research Group, School of Biological Sciences, University College, Swansea SA2 8PP, Wales, UK.

RENWRANTZ, L., Zoologisches Institut und Museum, Universität Hamburg, Martin-Luther-King-Platz 3, 2000 Hamburg 13, West Germany.

SMITH, V. J., University Marine Biological Station, Millport, Isle of Cumbrae, Scotland KA28 0EG, UK.

Söderhäll, K. Institute of Physiological Botany, University of Uppsala, Box 540, S751–21, Uppsala, Sweden.

Takle, G., Department of Zoology, The University, Glasgow G12 8QQ, UK.

Yoshino, T. P., Department of Zoology, University of Oklahoma, Norman, Oklahoma, 73109, USA.

Organizer of symposium

A. M. LACKIE, Department of Zoology, The University, Glasgow G12 8QQ.

Chairmen of sessions

R. P. DALES, Department of Biological Sciences, Royal Holloway and Bedford New College, Egham, Surrey TW20 9TY, UK.

P. GÖTZ, Institut für Allgemeine Zoologie der Freien Universität Berlin, Königin-Luise-Strasse 1–3, D-1000 Berlin 33, West Germany.

A. M. LACKIE, Department of Zoology, The University, Glasgow G12 8QQ, UK.

T. P. YOSHINO, Department of Zoology, University of Oklahoma, Norman, Oklahoma 73019, USA.

Symp. zool. Soc. Lond. (1986) No. 56 1–19

Mechanisms of encapsulation in dipteran hosts

Peter GÖTZ[1]

Institute of General Zoology, Free University of Berlin, Königin-Luise-Str. 1–3, D-1000 Berlin (West) 33

Synopsis

Diptera, like other insects, react against haemocoelic parasites by cellular and humoral immune mechanisms. Cellular reactions include phagocytosis, nodule formation and cellular encapsulation, whilst humoral reactions consist of humoral encapsulation and the formation of antibacterial proteins and other antibiotic factors. This contribution concentrates on humoral encapsulation, a unique phenomenon that, in insects, is only found in certain dipteran families, such as Culicidae or Chironomidae. Dipteran species which exhibit humoral encapsulation have fewer haemocytes than those which react by cellular encapsulation. Humoral encapsulation, which consists of the deposition of a quickly hardening material on foreign surfaces without the visible participation of blood cells, offers direct access to biochemical reactions that are also responsible for cellular defence reactions. Humoral encapsulation was mainly investigated in *Chironomus* larvae; it occurs, *in vivo* or *in vitro*, when haemolymph comes into contact with foreign organisms, heterologous tissues or certain organic compounds. The freshly deposited capsule material contains active phenoloxidase; the consequent hardening of the capsule is tyrosine- or DOPA-dependent, and is comparable to the quinone-sclerotization of the arthropod cuticle. The activation of *Chironomus* haemolymph pro-phenoloxidase follows a biochemical pathway that has already been shown to be responsible for initiating cellular encapsulation. It can be triggered by β-1,3-glucans, which are common in fungal cell walls. A lack of divalent cations, or the presence of a trypsin inhibitor, prevents the deposition of capsule material. This inhibition can be reversed by the addition of calcium ions or trypsin respectively. Competitive inhibitors of phenoloxidase activity, e.g. phenylthiourea, significantly delay the deposition of capsule material and its further hardening and pigmentation.

Diptera as vectors of diseases

Amongst the Diptera, there are numerous important vectors of pathogens and parasites responsible for initiating severe diseases in man and animals. Infective stages of protozoans and nematodes are transferred by blood-

[1] This paper is dedicated to Prof. Dr Wolfgang Wülker on the occasion of his 60th birthday.

ZOOLOGICAL SYMPOSIUM No. 56
ISBN 0-19-854004-3

sucking mosquitoes, blackflies, tabanids, phlebotomids and tsetse-flies. In general, these dipteran vectors are more than just transport organisms that transfer parasites from infected to non-infected animals; most of them are also intermediate hosts of the parasites transmitted, and part of the life cycle of these parasites takes place within them. The parasites must, therefore, be able to live and develop in two different host organisms and consequently deal with two different defence systems. During transmission, they must be able to withstand the physiological alterations that are brought about by changing from a vertebrate host to an insect vector and vice versa. Finally, host-parasite interactions occur between the vector and parasite, i.e. the defence reactions of the vector organism attack the parasite, and the parasite exhibits pathogenic effects towards the vector (Maier 1976; Ratcliffe 1982; Kaaya & Ratcliffe 1982). An established equilibrium between the vector and parasite is, therefore, the result of a long period of co-evolution (Anderson & May 1982). Resistant or sensitive organisms are not particularly suited to the role of a permanent vector, but tolerant species that are not seriously damaged by the parasite, and that do not seriously react against it, are optimal vectors. The evolution of new strains of vectors may always occur, and this essentially complicates the control of vector-borne diseases.

Encapsulation in Diptera

Defence reactions in Diptera against haemocoelic parasites have been repeatedly reported (e.g. Maier 1976; Kaaya & Ratcliffe 1982; Götz & Boman 1985). The most obvious reaction is encapsulation, since it leaves melanized remnants of the encapsulated organisms, which can be observed by dissecting insect bodies. In some rather translucent larvae or adults, such traces of capsule or nodule formation can even be seen from outside; they are known as black spots or black bodies. In general, encapsulation is a cellular reaction, leading to an aggregation of haemocytes around the foreign organism, and disintegration and melanization of the innermost cells of the cellular envelope. It is interesting to note that the segregation of foreign organisms from the haemocoel may also occur as a non-cellular reaction, which was named humoral encapsulation (Götz 1969), melanotic encapsulation (Poinar & Leutenegger 1971) or melanization response (Christensen, Sutherland & Gleason 1984). This contribution concentrates on humoral encapsulation, whereas cellular immunity is discussed in the following chapter by Ratcliffe.

Occurrence of humoral encapsulation in Diptera

Humoral encapsulation has been found in some species of the families of Chironomidae, Culicidae, Psychodidae, Syrphidae and Stratiomyidae,

Table 1. Type of encapsulation (humoral or cellular) found in different insects.

Insect Order	Genus or species	Number insects tested	Humoral encapsulation	Cellular encapsulation	Mean number of haemocytes/mm^3 (+/− s.d.)	Remarks
Diptera	*Chironomus riparius*	30	x		1,800±300	L (a)
	Endochironomus tendens	25	x		2,600±300	L (a)
	Glyptotendipes sp.	10	x		1,300±200	L (a)
	Psectrotanypus varius	15		o	8,600±700	L (a)
	Prodiamesa olivacea	30		o	7,100±600	L (a)
	Chaoborus sp.	30	x		500±100	L (a)
	Culex pipiens	30	x		4,100±400	L (a)
	Anopheles sp.	10	x		5,100±800	L (a)
	Simulium sp.	30		o	49,000±8,000	L (a)
	Liponeura sp.	20		o	15,000±900	L (a)
	Psychoda sp.	30	x		3,800±800	L (a)
	Eristalomya tenax	25	x		(1)	L,I (a)
	Tubifera sp.	3	x		3,600±800	L (a)
	Microchrysa sp.	15	x		1,200±300	L (a)
	Exechia separata	20		o	12,800±2,000	L (a)
	Tabanus sp.	4		o	8,000±900	L (a,b)
	Tipula sp.	6		o	38,400±5,000	L (a,b)
	Phormia sp.	20		o	25,000±6,300	L,I (a,b)
	Drosophila melanogaster	10		o	2,300±300	L (a,b)
Lepidoptera	*Galleria mellonella*	30		o	54,000±4,500	L (a,b)
Coleoptera	*Dytiscus marginatus*	3		o	11,400±1,500	L (a)
	Tenebrio mollitor	30		o	26,400±3,100	L (a,b)
Mecoptera	*Panorpa* sp.	15		o	33,000±1,400	L,I (a,b)
Megaloptera	*Sialis* sp.	5		o	34,500±2,500	L (a,b)
Heteroptera	*Ranatra linearis*	2		o	5,500±800	L,I (a,b)
	Corixa punctata	5		o	(1)	I (a)
	Notonecta glauca	2		o	5,000±750	I (a)
Saltatoria	*Gryllotalpa gryllotalpa*	3		o	(1)	I (a,b)
Blattaria	*Periplaneta americana*	10		o	34,400±2,900	I (a,b)
Trichoptera	*Hydropsyche* sp.	10		o	6,900±900	L (a,b)
	Rhyacophila sp.	3		o	35,000±2,300	L (a,b)
	Stenophylax sp.	3		o	23,000±1,000	L (a,b)
Plecoptera	*Perla* sp.	1		o	(1)	L (a)
Odonata	*Aeschna* sp.	3		o	15,300±800	L (a,b)
Ephemeroptera	*Ephemera* sp.	4		o	(1)	L (a)
	Epeorus sp.	5		o	16,900±900	L (a)

Haemolymph from larvae of the indicated insects was tested *in vivo* or *in vitro*. The materials used as provokers (Sephadex, nematodes or fungi) were suited to inducing either cellular or humoral encapsulation, depending on the insect taxa investigated. Larvae rather than adults were tested, since they were easier to collect, handle and bleed. However, taxonomic determination with larval stages is difficult, and therefore often only genera were identified. L, I = test performed with haemolymph from larvae and imagos respectively, a = *in vitro* test; b = *in vivo* test; number of haemocytes +/− standard deviation; (1) = blood cell number not determined. (From Lingg 1976.)

where it replaces nodule formation and cellular encapsulation (Table 1). In Diptera, a correlation was found between low blood cell counts (less than 6,000 cells/mm^3) and the occurrence of humoral encapsulation, whereas higher blood cell numbers coincided with cellular encapsulation (Lingg 1976; Götz, Roettgen & Lingg 1977).

In Chironomidae, all investigated species of the subfamily of Chironominae (*Chironomus annularius*, *C. anthracinus*, *C. luridus*, *C. melanotus*, *C. plumosus*, *C. riparius*, *C. tentans*, *Endochironomus tendens* and *Glypto-*

tendipes sp.), and an undetermined genus and species of the subfamily of Orthocladiinae, reacted by humoral encapsulation. A comparison of these different species revealed only slight differences in their defence reactions, with regard to the time when encapsulation started, and the colour of the resulting capsule (Table 2). Larvae of *Glyptotendipes* exhibited the fastest encapsulation reaction; capsule material was visible 30 s after incubating isolated haemolymph with a provoking agent, and 3 min later a complete cover had already formed (Lingg 1976). Species from Diamesinae and Tanypodinae showed a cellular type of encapsulation against the provoking materials tested (Sephadex, nematodes and fungi); these species possess higher blood cell numbers (7,100 and 8,600/mm^3 respectively) than those chironomids which react by humoral encapsulation (see Table 1).

Humoral encapsulation in Culicidae against parasitic nematodes was first mentioned by Bronskill (1962) and Esslinger (1962). These authors observed the deposition of a cell-free layer of pigmented material, which was then followed by an aggregation of blood cells. The defence reaction of *Culex pipiens* against the larvae of the insect pathogenic nematode *Neoaplectana carpocapsae* was investigated in detail by Poinar & Leutenegger (1971), who described an homogenous matrix of capsule material on the surface of the nematodes, 25 min after they had invaded the haemocoel. Within the next few hours, the capsule changed to a strongly pigmented, rigid sheath, which tightly surrounded the nematode parasite. Electron micrographs revealed the aggregation of electron-dense granules in the matrix adjacent to the surface of the parasite. The capsule material finely outlined the cuticular striations and lateral lines of the nematode. Since these investigations, humoral encapsulation has now been observed in a number of other species of Culicidae (Götz & Vey 1974; Lingg 1976; Poinar, Hess & Petersen 1979; Chen & Laurence 1985; Christensen *et al.* 1984). In his comparative study of encapsulation in insects, Lingg (1976) observed the formation of abundant amounts of capsule material in haemolymph of *C. pipiens* that was incubated with fungi or Sephadex beads. The reaction was especially strong on the fungal spores, but weaker on the filaments. In Culicidae, some haemocyte involvement in the encapsulation reaction was repeatedly reported (Bronskill 1962; Poinar & Leutenegger 1971; Christensen *et al.* 1984). The haemocytes attached to the surface of the parasite and disintegrated. Nevertheless, capsules were formed within a few minutes and consisted predominantly of an homogenous material with some enclosed cell debris. In contrast, typical cellular encapsulation is characterized by the formation of a multicellular envelope, which becomes melanized only after several hours. Such true cellular encapsulation is not the rule in Culicidae, and has so far only been found in the larvae of *C. territans*, a species which differs from other Culicidae in having an unusually high number of plasmatocytes (Poinar, Hess, Hansen & Hansen 1979).

Larvae of *Psychoda* sp. (Fam.: Psychodidae) also reacted by humoral encapsulation, but their reaction was weaker than humoral encapsulation observed in Chironomidae and Culicidae. The deposition of capsule material only started after 9 min of incubation, and it took as long as 2 h until a complete sheath of black-pigmented material had formed. Nematodes (*Hydromermis contorta*) did not provoke encapsulation in the haemolymph of *Psychoda* larvae.

Among the suborder Brachycera, some species of Syrphidae and Stratiomyidae exhibited humoral encapsulation, which is in accordance with their low haemocyte concentration (Table 1). In *Eristalomya tenax* (Fam.: Syrphidae), the defence reactions of different developmental stages were found to be similar, as was shown by comparing the reactions of first-stage larvae, prepupal larvae and adults. The haemolymph of Syrphidae and Stratiomyidae did not react against nematodes, and only a slight deposition of capsular material was provoked by the presence of pieces of agar in *E. tenax* haemolymph.

Only *Drosophila* does not appear to follow the rule observed in other Diptera that low blood cell counts correlate with humoral encapsulation. However, Nappi & Streams (1969) have shown that parasitization of *Drosophila* larvae is immediately followed by an increase in the number of circulating haemocytes, which is caused by a stimulation of stationary haemocytes. The effective haemocyte concentration in *Drosophila* should, therefore, be thought of as being higher than 2,000 cells/mm^3, as is indicated in Table 1.

As is also obvious from Table 1, the total haemocyte counts of most other insects, except Diptera, were above 6,000 cells/mm^3 and averaged 23,000 cells/mm^3. Only Heteroptera have relatively few blood cells, but in spite of this, those heteropteran species tested reacted by cellular encapsulation. Nearly all dipteran species with humoral encapsulation have aquatic larvae—only the larvae of *Microchrysa* live in humid soil. A correlation between the number of blood cells and an aquatic habitat was therefore conceivable. This hypothesis could not be confirmed, however, as demonstrated by Lingg (1976), who tested specimens of several other dipteran groups with aquatic larval stages (Tanypodinae, Simuliidae, Limnobiinae and Tabanidae) without finding humoral encapsulation.

Investigation of humoral encapsulation in *Chironomus*

The mechanisms controlling humoral encapsulation were mainly investigated in *Chironomus* larvae, which are easily reared in the laboratory, and whose haemolymph displays fast and heavy reactions under *in vivo* and *in vitro* conditions (Table 2).

In *Chironomus* larvae, humoral encapsulation efficiently reacted against nematodes (Wülker 1961; Götz 1969), fungi (Götz 1973; Götz & Vey 1974), bacteria (Götz, Enderlein & Roettgen in preparation), and certain

Table 2. Encapsulation in haemolymph of dipteran larvae after incubation with different provoking agents.

Dipteran Species	Number of larvae tested	Sephadex G–50	SE-Sephadex C–50	Agar	*H. contorta*	*M. hiemalis*	*A. niger*	Start (min)	End (min)	Colour of capsule
C. riparius (Chironominae)	30	xxx	x	xxx	xxx	xxx	xx	3.5	15	Brown
C. melanotus (Chironominae)	20	xx	xx	xx	xx	xx	xx	2.5	15	Brown
Endochironomus tendens (Chironominae)	25	xxx	xx	xx	xxx	xxx	xx	1.5	15	Red brown
Glyptotendipes sp. (Chironominae)	10	xxx						0.5	15	Red brown
Orthocladiinae (Chironomidae)	15	x	xx	x	–					Light brown
Prodiamesa olivacea (Diamesinae)	30	x/o	o	x/o	o		o	30.0		
Psectrotanypus varius (Tanypodinae)	15	o	o	o	o	o	o			
Chaoborus sp. (Culicidae)	30	xx	x	x	–	x	x	3.5	20	Brown
Culex pipiens (Culicidae)	30	xx	x	x	–	x		1.5	20	Black
Anopheles sp. (Culicidae)	10	x	x		–	x		1.5	20	Brown
Psychoda sp. (Psychodidae)	30	x	x	x	–	x	xx	9.0	120	Black
Simulium sp. (Simuliidae)	30	o	o	o	o	o	o			
Liponeura sp. (Blepharoceridae)	20	o	o	o	o	o	o			

Limnobiinae (Limnobiidae)	4	o	o		o					
Exechia separata (Fungivoridae)	20	o	o	o	o	o	o			
Tipula sp. (Tipulidae)	6	o	o	o	o	o	o			
Eristalomya tenax (Syrphidae)	25	xx	–	(x)	–	xx	xx	3.5	30	Brown
Tubifera pendula (Syrphidae)	3	x	x	–	–	x	x	4.0		Brown black
Microchrysa sp. (Stratiomyidae)	15	xx	x	–	–	xx	x	4.0	30	Black
Tabanus sp. (Tabanidae)	4	o			o	o				
Phormia sp. (Tachinidae)	20	o	o	o	o	o	o			
D. melanogaster (Drosophilidae)	10	o	o	o	o	o	o			

The provoking agents used were a polydextran (Sephadex G–50), a cation exchanger (SE-Sephadex C–50), a polysaccharide (agar), infective stages of the entomophilic nematode *Hydromermis contorta*, and germinating spores of the fungi *Mucor hiemalis* and *Aspergillus niger*. The foreign objects were either injected into the dipteran larvae with finely drawn glass capillaries, or incubated *in vitro* with fresh larval haemolymph. (x) = humoral encapsulation, (o) = cellular encapsulation, (–) = no reaction. Differences in the speed and intensity of the encapsulation reaction are indicated by the number of symbols used. Because the number of larvae available were sometimes restricted, some test combinations were omitted. (From Lingg 1976.)

Table 3. Scheme of visible events during humoral encapsulation in Diptera (*Chironomus*).

Phase	Time	Events during humoral encapsulation
Phase 1	3 min:	Droplets of sticky material precipitate on foreign surface
(Deposition of capsule material)	10 min:	Precipitated material forms a complete cover of soft capsule material around the provoking agent
Phase 2 (Hardening	10–60 min:	Capsule material hardens; the capsule loses its elasticity and bursts if mechanically overstressed
and tanning of capsule	1–12 h:	Capsular wall continuously thickens until it reaches a diameter of several μm
material)	1–12 h:	Capsule material attains brown colouration

Time = approximate time after introduction or invasion of provoking agent.

foreign materials (Vey & Götz 1975; Wilke 1979; Götz 1986). Phagocytosis played only a minor role in *Chironomus* larvae, and other cellular defence reactions or anti-bacterial factors were not detected.

Process of humoral encapsulation in *Chironomus* larvae

Humoral encapsulation starts within minutes of an invading parasite entering the haemocoel of a *Chironomus* larva. The same reaction occurs *in vitro* when foreign organisms, tissues or certain organic compounds are incubated with the isolated haemolymph of *Chironomus* larvae (Fig. 1B). Droplets of capsule material appear on the foreign surface, quickly increase in number and size, and finally cohere to form a complete coating around the foreign object. At first the capsule material is soft and sticky, but then it hardens and takes on a brown pigmentation. This biphasic course of humoral encapsulation (Table 3) can best be observed during the incubation of isolated haemolymph with nematode larvae. The nematodes, although initially free-swimming at the onset of encapsulation, soon become sticky and increasingly adhere to glass or other surfaces with which they come into contact. After about 10–15 min, they are completely trapped within a capsule that now begins to harden. It is then no longer sticky and elastic, and breaks if mechanically overstressed, e.g. by the movements of an enclosed nematode. During the next few hours, the capsule wall continues to increase in thickness, and melanizes.

Since the capsule material closely attaches to the surface of a parasite, it forms an exact template of the encapsulated object. Electron microscopic studies on humoral encapsulation have revealed that the capsule has an irregular fibrillar nature (Fig. 2A). The two phases of humoral encapsulation, the attachment of sticky material and the following solidification and pigmentation of this material, find their ultrastructural expression in an increase of fibrillar elements leading to greater electron density (Fig. 2B).

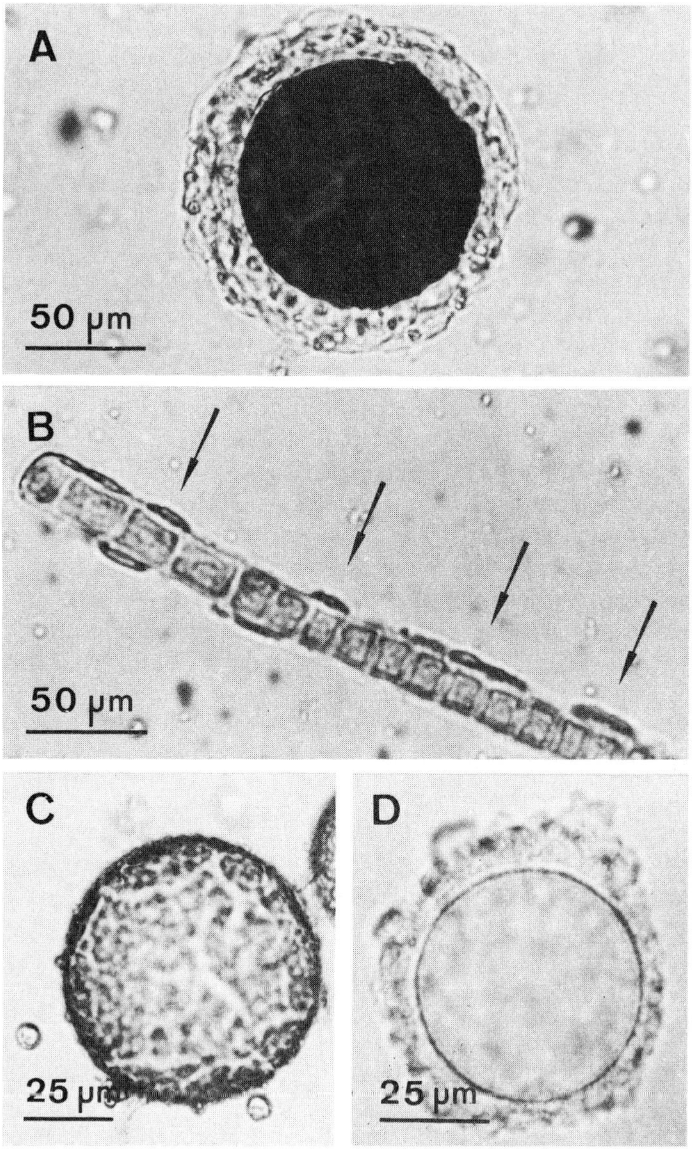

Fig. 1. Cellular and humoral encapsulation in haemolymph from different dipteran species. A: Cellular encapsulation of Dowex particle after injection into *Tipula* sp. larva. Note multicellular capsule around foreign particle. B: Beginning of humoral encapsulation of a blue-green alga immediately after incubation in haemolymph of *Endochironomus tendens*. Deposition of the capsule material (arrows) occurs without participation of haemocytes. C,D: Effect of phenylthiourea (PTU) on humoral encapsulation of Sephadex G beads in haemolymph of *Eristalomya tenax*; C: without, D: with PTU. Presence of PTU leads to the formation of a voluminous but soft and non-melanized capsule. (Photographs courtesy of W. Lingg.)

Characteristics of *Chironomus* haemolymph

Chironomus haemolymph contains only about 2,000 haemocytes per mm^3. Maier (1969) identified plasmatocytes (80%), granular cells (18%), adipohaemocytes (1.5%), oenocytoids and others (0.3%). Participation of these haemocytes in encapsulation was never observed under the light microscope. Electron microscopy revealed intact haemocytes next to encapsulated fungal spores, or latex particles, without any signs of active involvement (Götz & Vey 1974). However, some disintegrated haemocytes were occasionally found in the neighbourhood of encapsulated nematodes. This haemocyte disintegration may indicate haemocyte involvement in the encapsulation reaction, but could also be an artifact that is produced during preparation of the material for electron microscopy.

Under *in vivo* conditions, deposition of capsule material is restricted to the surface of the invading or injected foreign material, and occurs without further alteration of the haemolymph. Isolated haemolymph, however, begins to alter spontaneously upon contact with air, and this is evident by the appearance of fine granular material within the haemolymph. Approximately half an hour after bleeding, isolated haemolymph has lost its encapsulation capacity if kept at room temperature. Cooled or frozen haemolymph retains its encapsulation capacity. If provoking agents (living organisms, foreign tissues or certain organic compounds) are incubated with freshly bled haemolymph, such granule formation in the haemolymph is less extensive. Nevertheless, encapsulated objects are often surrounded by islands of granular material, which settle to the bottom of the preparation chamber. These observations indicate a biochemical relationship between encapsulation and coagulation in *Chironomus* haemolymph.

Chemical nature of the capsule material

The capsule material resulting from humoral encapsulation is an unusually resistant substance. It is insoluble in organic solvents and strong acids, and only treatment with strong alkali leads to its degradation. Bleaching of the deep-brown capsular substance occurs after incubation with strong oxidizing agents, such as peracetic acid or hydrogen peroxide (Poinar & Leutenegger 1971; Vey & Götz 1975). The capsule material is also completely resistant to a variety of enzymes capable of breaking down proteins, lipoproteins or polysaccharides (Götz & Vey in press). Consequently, encapsulated parasites remain in the haemocoel of *Chironomus* larvae without showing signs of disintegration, and without causing visible irritation to the host. On the basis of these observations and additional histochemical tests, Götz & Vey (1974) concluded that the capsule material consists of a protein-polyphenol complex, comparable to the sclerotized proteins which are responsible for the hardening and tanning of the insect cuticle.

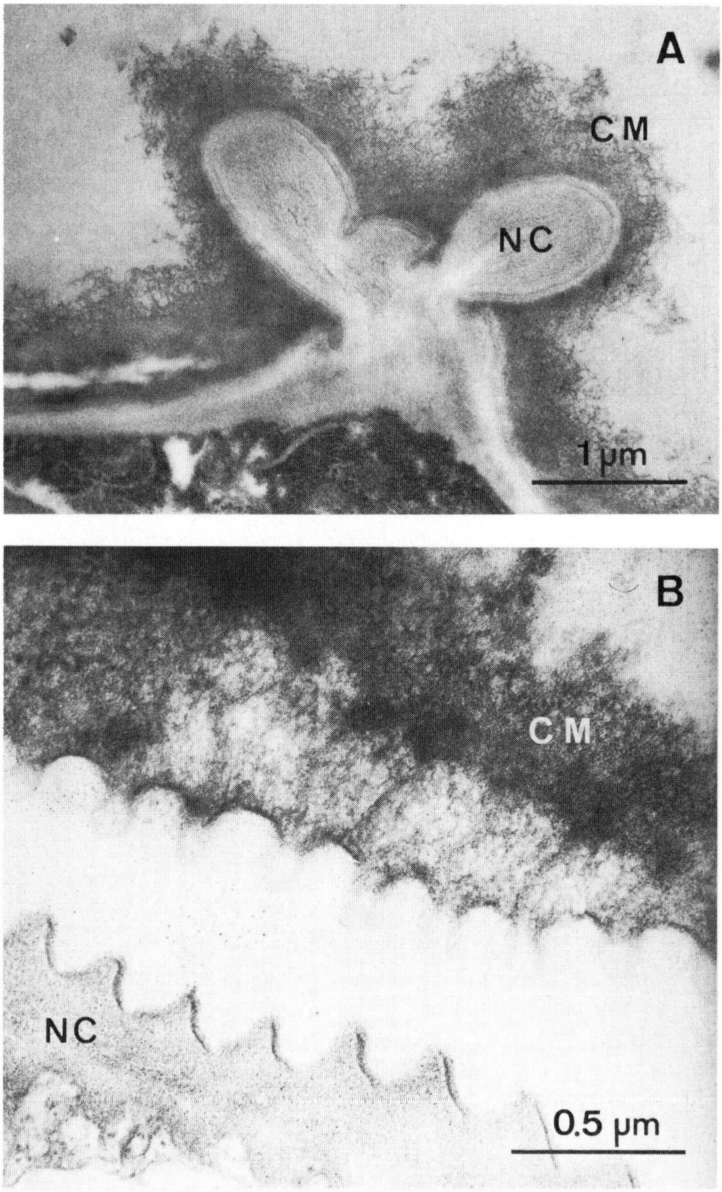

Fig. 2. Ultrastructure of capsule material in the course of humoral encapsulation of nematodes (infective larvae of *Hydromermis contorta*) in haemolymph of *Chironomus riparius*. A: Fresh capsule material (phase 1 of humoral encapsulation) is seen as an irregular fibrillar material attaching to the nematode cuticle. B: During hardening and melanization (phase 2 of humoral encapsulation) the fibrillar structure becomes less obvious, and the electron density of the capsule material significantly increases.
NC = nematode cuticle; CM = capsule material.

Such sclerotization (or melanization as it is sometimes known) is the result of phenoloxidase activity. Phenoloxidase transforms phenolic compounds into quinones, which then readily bind with the secondary and tertiary amino groups of proteins. This leads to a crosslinking of the protein chains and the formation of a resistant network of macromolecules (Andersen 1985). The phenolic quinones are also capable of autopolymerization. The resulting quinone-polymer is pure melanin, and this gives the sclerotized cuticle its brown to black appearance. It is known from the albino mutants of insects that melanin is not necessary for the hardening process. Nevertheless, under normal conditions melanin formation (possibly resulting from a surplus of quinones) always accompanies sclerotization (Richards 1978). Cuticle phenoloxidase has been investigated by Ashida & Dohke (1980), and was found to exist as a pro-enzyme that requires a second enzyme (which is also present in the cuticle) for its activation.

Pro-phenoloxidases were also found in the haemolymph and in certain haemocytes of insects (Ashida 1971; Maier 1973; Schmit, Rowley & Ratcliffe 1977; Rowley, Ratcliffe & Leonard 1984). Maier (1973) demonstrated phenoloxidase activity in *Chironomus riparius* larvae by overlaying haemolymph smears with agar gel containing dihydroxyphenylalanine (DOPA). Under conditions which left the blood cells intact, phenoloxidase activity was restricted to some of the granular haemocytes. Disintegration of haemocytes (e.g. in hypotonic medium) resulted in a distribution of phenoloxidase activity throughout the entire haemolymph. In this case, melanization started from many foci in the haemolymph, which were thought to contain particles of an activating compound that also originated from the haemocytes.

Ashida and Söderhäll, who investigated phenoloxidases in *Bombyx mori* and crustaceans, established some of the biochemical pathway for the activation of haemolymph pro-phenoloxidase (see chapter by Smith & Söderhäll in this volume). Unestam & Söderhäll (1977) and Smith & Söderhäll (1983 a,b) showed that the activation of crustacean haemolymph pro-phenoloxidase is initiated by β-1,3-glucans, which are common in fungal cell walls. Further components of the pro-phenoloxidase activating pathway are divalent cations and a serine protease which is specifically inhibited by p-nitrophenyl-p′-guanidinobenzoate (Ashida & Söderhäll 1984).

Evidence for a pro-phenoloxidase activating pathway triggering humoral encapsulation

Freshly deposited capsule material from *Chironomus* haemolymph exhibits phenoloxidase activity. This can easily be demonstrated by placing a drop of haemolymph on a piece of filter paper. Since cellulose provokes humoral encapsulation, capsule material is deposited on the paper, where it remains, even after thoroughly washing off the haemolymph. If tyrosine, DOPA, or

Table 4. Inhibition of different steps of the pro-phenoloxidase activating pathway initiating humoral encapsulation: filter paper test[a] and photometer readings[b] (Harmstorf in preparation).

Steps of pro-phenoloxidase activating pathway	Inhibited by	Inhibition reversed by	Melanization of DOPA solution (filter paper test)[a]	Maximal phenoloxidase activity (photometer reading)[b]
Activation of serine protease	EDTA		–	0.003
	EDTA	Ca^{2+}	+	0.039
Activity of serine protease	STI		–	0.003
	STI	trypsin	+	0.033
Activity of phenoloxidase	PTU		–	0.003

[a] Freshly collected, cooled haemolymph of *Chironomus tentans* last instar larvae was mixed (1:1, v/v) with potassium phosphate buffer (0.1 M, pH 6.0), which contained the inhibiting or activating substances to be tested. The mixture was incubated for 30 min (at room temperature) with a 2 mm diameter disc of filter paper, which was then thoroughly washed in buffer, and placed in a buffered saturated DOPA solution. (+) = without inhibition of the pro-phenoloxidase activating pathway, the filter paper turned slightly black within 30 min of incubation in DOPA; after 3 h the filter paper was totally black. (–) = no melanization. Trypsin: 5 mg/ml; further concentrations, abbreviations and photometric methods are given in the legend of Fig. 3.
[b] Inhibition of the pro-phenoloxidase activating pathway prevents the formation of dopachrome (low photometer readings).

related phenolic compounds are added to such adsorbed capsule material, black sediments of melanin are produced (Wilke in preparation).

The addition of known inhibitors of phenoloxidase activation (Ashida & Söderhäll 1984), such as ethylenediaminotetraacetic acid (EDTA), soya bean trypsin inhibitor, or p-nitrophenyl-p′-guanidinobenzoate (p-NPGB), suppressed the deposition of activated phenoloxidase on the filter paper. The deposition of capsule material was not inhibited by phenylthiourea (PTU), but subsequent melanization of added tyrosine or DOPA solution was significantly delayed (Table 4). This was also demonstrated by measuring the phenoloxidase activity of *Chironomus* haemolymph with a Beckman DU6 spectrophotometer under the influence of activating or inhibiting agents (Fig. 3 and Table 4). As in crustaceans (see chapter by Smith & Söderhäll in this volume), activation of *Chironomus* haemolymph pro-phenoloxidase depends on the preceding activation of a serine protease, which in turn is dependent on the presence of divalent cations. Activation of *Chironomus* pro-phenoloxidase was, therefore, prevented by incubating the haemolymph with EDTA. The subsequent addition of calcium ions reversed the effect of the EDTA. The activity of the serine protease was suppressed by a trypsin inhibitor such as soya bean trypsin inhibitor (STI), but phenoloxidase activity was restored after the addition of excess trypsin. Finally, the activity of the phenoloxidase itself was reduced by PTU, which acts as a competitive inhibitor (Harmstorf in preparation). It is obvious from these results that pro-phenoloxidase and other components

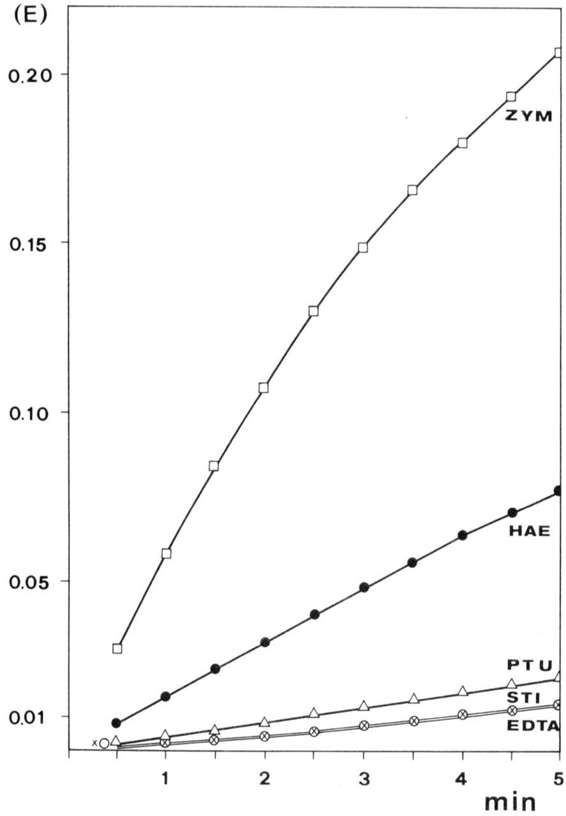

Fig. 3: Activity of *Chironomus* haemolymph phenoloxidase after incubation with activating and inhibiting agents.
5 µl of freshly collected, cooled haemolymph from *C. tentans* last instar larvae were added to 5 µl of potassium phosphate buffer (0.1 M, pH 6.0) which contained the activating or inhibiting substance to be tested. After 15 min incubation at 30°C, 480 µl of buffered saturated DOPA solution were added, and the absorbence measured at 490 nm over a 5 min period (readings every 0.5 min) with a Beckman 6 DU spectrophotometer. EDTA = ethylenediaminotetraacetic acid (20 mg/ml), STI = soya bean trypsin inhibitor (10 mg/ml), PTU = phenylthiourea (saturated aqueous solution), HAE = haemolymph + potassium phosphate buffer without activating or inhibiting agents, ZYM = zymosan, a cell wall preparation of yeast cells (supernatant of 2 mg/ml). Increased phenoloxidase activity was recorded after the addition of zymosan, whereas addition of three different inhibitors of the pro-phenoloxidase activating pathway led to a very low level of phenoloxidase activity.

of a pro-phenoloxidase activating system are present in the haemolymph of *Chironomus* larvae.

What triggers the beginning of the pro-phenoloxidase activation? It occurs when parasites invade the haemocoel or when foreign materials are injected into it. It has been assumed that the injury itself might lead to the release of factors from the cuticle or hypodermal cells, which then initiate the activation pathway. However, if this was the case, one should expect a total melanization of the haemolymph; but the deposition of activated pro-phenoloxidase is restricted to the foreign surfaces only. Furthermore, humoral encapsulation still occurs even days after the initial injury, for example, on the surface of growing hyphal filaments developing from introduced fungal spores. Finally, only living organisms and tissues and certain organic compounds provoke encapsulation and melanization, whilst exposure to other inanimate materials (e.g. glass powder or charcoal) does not lead to the deposition of phenoloxidase. Therefore, the occurrence and intensity of humoral encapsulation must depend on the type of foreign material used. Vey & Götz (1975) and Wilke (1979) tested a variety of materials for their ability to provoke humoral encapsulation. They found strong provokers, such as living organisms, foreign tissues, silk, agar, cellulose, polyacrylamide gel, Dowex, Latex, neutral and anionic polydextran (Sephadex C and G), whilst dead organisms, hair, feathers, methacrylate, epoxy-resin, cationic polydextran (Sephadex A), and celluloid proved to be weak provokers. Non-provokers include polyethylene, polyvinylchloride, paraffin oil, vaseline, glass powder, activated charcoal, iron, colloid gold, powdered calcium carbonate, and powdered calcium sulphate. We can conclude from these results that the essential stimulus for initiating the pro-phenoloxidase activating cascade comes from the provoking material itself, e.g. from certain molecules or special physicochemical qualities (see Lackie 1983) on its surface. The recognition mechanism that triggers encapsulation is unknown. A similar pro-phenoloxidase activating system seems to be involved in both types of defence reaction, in humoral as well as in cellular encapsulation.

Comparison of humoral and cellular encapsulation

The differences between humoral and cellular encapsulation are not necessarily fundamental; however, quantitative differences between both types of defence reactions are quite obvious. Cellular encapsulation (see the following chapter by Ratcliffe) is initiated by the attachment of granular haemocytes on the provoking material. The haemocytes partly disintegrate and deposit an electron-dense material, possibly containing phenoloxidase, on the foreign surface. During the following hours numerous haemocytes, mainly plasmatocytes, continue to accumulate and form the middle and outer regions of the multicellular capsule. In general, the plasmatocytes take on an extremely flattened appearance and form concentric layers around the enclosed object. The presence of microtubules and cell contacts (tight junctions) demonstrates the mechanical strength of such a cellular capsule. Melanization first occurs in the innermost region of the capsule, adjacent to the provoking surface, where the granular cells have released their vesicles; the melanization may finally spread throughout the entire capsule. The time for the formation of a cellular capsule varies from several hours to several days.

In contrast, humoral encapsulation produces a complete cover around a foreign object within 10 to 30 min. In Chironomidae, the capsule was always observed to be cell-free, whereas in Culicidae, disintegrated blood cells were sometimes found enclosed in the homogenous matrix of capsule material. Therefore, the main difference between humoral and cellular encapsulation is the degree of cell participation in capsule formation. In cellular encapsulation, the amount of pro-phenoloxidase released from the attaching granular cells is relatively small, and the bulk of the capsule is formed by cells. In the case of humoral encapsulation, the capsule is totally or predominantly cell-free and, instead, consists of a material that is not released from attaching cells, but is deposited directly out of the haemolymph. Plasmatocytes were not seen to be involved in the formation of humoral capsules. At the present time, we do not know if the precursors necessary for humoral encapsulation are always present in the haemolymph, or if they are only released from certain blood cells after provocation by foreign surfaces. In any case, freshly bled haemolymph of certain Diptera (e.g. *Chironomus*) represents a humoral system which contains the components required for the formation of melanotic capsules.

Efficiency of humoral encapsulation as compared to cellular encapsulation

A comparison of cellular encapsulation in *Galleria mellonella* and humoral encapsulation in *Chironomus* larvae revealed that the *Chironomus* defence system was clearly superior in counteracting pathogenic bacteria (Götz, Enderlein *et al.* in preparation). This advantage may be due to the rapidity with which the deposition of capsule material in humoral encapsulation occurs. This reduces the time available to pathogens for releasing toxins or enzymes, which would otherwise weaken the immune system of the insect host. However, actively moving parasitic nematodes may be strong enough to mechanically break the humoral capsules that surround their bodies, and therefore often escape this defence reaction. Furthermore, entomophilic nematodes have evolved mechanisms that reduce the intensity at which humoral encapsulation occurs (Götz & Vey in press). Consequently, nematode parasitism is more common in chironomid populations than infections due to haemocoelic bacteria or fungi.

Acknowledgements

I wish to thank Mrs Bettina Harmstorf for her results (quoted in Fig. 3) concerning the activation of *Chironomus* pro-phenoloxidase, Mr W. Lingg for results and micrographs of encapsulation in Diptera, Mrs Ingrid Thoss for technical assistance and Mr Mirko Whitfield for his help with the English version of the manuscript.

References

Andersen, S.O. (1985). Sclerotization and tanning of the cuticle. In *Comprehensive insect physiology, biochemistry and pharmacology* 3: 59–74. (Eds Kerkut, G.A. & Gilbert, L.J.). Pergamon Press, Oxford, New York.

Anderson, R.M. & May, R.M. (1982). Coevolution of hosts and parasites. *Parasitology* **85**: 411–26.

Ashida, M. (1971). Purification and characterization of pre-phenoloxidase from hemolymph of the silkworm *Bombyx mori. Archs Biochem. Biophys.* **144**: 749–62.

Ashida, M. & Dohke, K. (1980). Activation of pro-phenoloxidase by the activating enzyme of the silkworm, *Bombyx mori. Insect Biochem.* **10**: 37–47.

Ashida, M. & Söderhäll, K. (1984). The prophenoloxidase activating system in crayfish. *Comp. Biochem. Physiol.* **77B**: 21–6.

Bronskill, J.F. (1962). Encapsulation of rhabditoid nematodes in mosquitoes. *Can J. Zool.* **40**: 1269–75.

Chen, C.C. & Laurence, B.R. (1985). An *in vitro* study on the encapsulation of microfilariae. *Devl comp. Immunol.* **9**: 174.

Christensen, B.M., Sutherland, D.R. & Gleason, L. N. (1984). Defense reactions of mosquitoes to filarial worms: Comparative studies on the response of three different mosquitoes to inoculated *Brugia pahangi* and *Dirofilaria immitis* microfilariae. *J. Invert. Path.* **44**: 267–74.

Esslinger, J.H.Y. (1962). Behavior of microfilariae of *Brugia pahangi* in *Anopheles quadrimaculatus. Am. J. trop. Med. Hyg.* **11**: 749–58.

Götz, P. (1969). Die Einkapselung von Parasiten in der Hämolymphe von *Chironomus* Larven (Diptera). *Zool. Anz.* (Suppl.) **33**: 610–7.

Götz, P. (1973). Immunreaktionen bei Insekten. *Naturw. Rdsch. Stutt.* **26**: 367–75.

Götz, P. (1986). Encapsulation in arthropods. In *Immunity in invertebrates*: 153–70. (Ed. Brehélin, M.). Springer-Verlag, Berlin, Heidelberg.

Götz, P. & Boman, H.G. (1985). Insect immunity. In *Comprehensive insect physiology, biochemistry and pharmacology* 3: 453–85. (Eds Kerkut, G.A. & Gilbert, L.J.). Pergamon Press, Oxford, New York.

Götz, P., Enderlein, G. & Roettgen, I. (In preparation). *Immune reactions of* Chironomus *larvae (Insecta: Diptera) against bacteria.*

Götz, P., Roettgen, I. & Lingg, W. (1977). Encapsulement humoral en tant que réaction de défense chez les Diptères. *Annls Parasit. hum. comp.* **52**: 95–7.

Götz, P. & Vey, A. (1974). Humoral encapsulation in Diptera (Insecta): defence reactions of *Chironomus* larvae against fungi. *Parasitology* **68**: 193–205.

Götz, P. & Vey, A. (In press). Humoral encapsulation in insects. In *Hemocytic and humoral immunity in arthropods*. (Ed. Gupta, A.P.). John Wiley & Sons, New York.

Kaaya, G.P. & Ratcliffe, N.A. (1982). Comparative study of hemocytes and associated cells of some medically important dipterans. *J. Morph.* **173**: 351–65.

Lackie, A.M. (1983). Effect of substratum wettability and charge on adhesion *in vitro* and encapsulation *in vivo* by insect haemocytes. *J. Cell Sci.* **63**: 181–90.

Lingg, W. (1976). *Experimentelle Untersuchungen zur humoralen Einkapselung.* Thesis: University of Freiburg, West Germany.

Maier, W. (1969). Die Hämocyten der Larven von *Chironomus thummi* (Dipt.). *Z. Zellforsch. mikrosk. Anat.* **99**: 54–63.

Maier, W.A. (1973). Die Phenoloxidase von *Chironomus thummi* und ihre Beeinflussung durch parasitäre Mermithiden. *J. Insect Physiol.* **19**: 85–95.

Maier, W.A. (1976). Arthropoden als Wirte und Überträger menschlicher Parasiten: Pathologie und Abwehrreaktion der Wirte. *Z. Parasitenk.* **48**: 151–79.

Nappi, A.J. & Streams, F.A. (1969). Haemocytic reactions of *Drosophila melanogaster* to the parasites *Pseudeucoila mellipes* and *P. bochei*. *J. Insect Physiol.* **15**: 1551–66.

Poinar, G.O., Jr., Hess, R.T., Hansen, E. & Hansen, J.W. (1979). Laboratory infection of blackflies (Simuliidae) and midges (Chironomidae) by the mosquito mermithid, *Romanomermis culicivorax*. *J. Parasit.* **64**: 613–5.

Poinar, G.O., Jr., Hess, R.T. & Petersen, J.J. (1979). Immune responses of mosquitoes against *Romanomermis culicivorax* (Mermithidae: Nematoda). *J. Nematol.* **11**: 110–6.

Poinar, G.O., Jr., & Leutenegger, R. (1971). Ultrastructural investigation of the melanization process in *Culex pipiens* (Culicidae) in response to a nematode. *J. Ultrastruct. Res.* **36**: 149–58.

Ratcliffe, N.A. (1982). Cellular defence reactions of insects. *Fortschr. Zool.* **27**: 223–44.

Richards, A.G. (1978). The chemistry of insect cuticle. In *Biochemistry of insects*: 205–32. (Ed. Rockstein, M.). Academic Press, New York, San Francisco and London.

Rowley, A.F., Ratcliffe, N.A. & Leonard, C.M. (1984). Role of agglutinins and components of the phenoloxidase system in the recognition of foreignness by insect hemocytes. *Int. Congr. Ent.* **17**: 760. (Abstract.)

Schmit, A.R., Rowley, A.F. & Ratcliffe, N.A. (1977). The role of *Galleria mellonella* hemocytes in melanin formation. *J. Invert. Path.* **29**: 232–4.

Smith, V.J. & Söderhäll, K. (1983a). Induction of degranulation and lysis of haemocytes in the freshwater crayfish, *Astacus astacus*, by components of the prophenoloxidase activating system *in vitro*. *Cell Tissue Res.* **233**: 295–303.

Smith, V.J. & Söderhäll, K. (1983b). β-1,3-glucan activation of crustacean hemocytes *in vitro* and *in vivo*. *Biol. Bull. mar. biol. Lab.* Woods Hole **164**: 299–314.

Unestam, T. & Söderhäll, K. (1977). Soluble fragments from fungal cell walls elicit defence reactions in crayfish. *Nature, Lond.* **267**: 45–6.

Vey, A. & Götz, P. (1975). Humoral encapsulation in Diptera (Insecta): Comparative studies *in vitro*. *Parasitology* **70**: 77–86.

Wilke, U. (1979). Humorale Infektabwehr bei *Chironomus*-Larven. *Verh. dt. zool. Ges.* **72**: 315.

Wülker, W. (1961). Untersuchungen über die Intersexualitat der Chironomiden (Dipt.) nach *Paramermis* Infektion. *Arch. Hydrobiol.* (Suppl.) **25**: 127–81.

Symp. zool. Soc. Lond. (1986) No. 56 21–43

Insect cellular immunity and the recognition of foreignness

N. A. RATCLIFFE

Biomedicine and Physiology Research Group,
School of Biological Sciences,
University College, Swansea, SA2 8PP, Wales, U.K.

Synopsis

The cellular defence reactions of insects are described and problems associated with research and possible directions for future study in this field are identified. Emphasis is placed on vector species wherever possible. The need for more accurate ways of identifying the various categories of insect blood cells is discussed, as is the use of lectin-binding specificities and monoclonal antibodies for characterizing the blood cells. The possible role of the coagulation-type cells in the recognition of non-self materials is considered and a model for this process is presented. In this model, the coagulation cells degranulate in the presence of foreign surfaces and discharge their contents which include the prophenoloxidase cascade. It is proposed that the activation of this cascade produces opsonic materials to which the plasmatocytes respond. It is then concluded that non-self discrimination in insects is a two-phase process involving the co-operation of different cell types. Other models discussed include the possible role of surface charge and functional aspects of the agglutinins. Consideration is also given to the interaction of the cellular with the humoral defence reactions, and to the relevance of laboratory experiments to natural conditions in the field.

Introduction

There have been a number of reviews published recently on insect immunity and the recognition of foreignness (e.g. Gupta 1979; Lackie 1980, 1981a; Ratcliffe 1982; Götz & Boman 1985; Ratcliffe & Rowley in press). This present brief overview rather than representing a reiteration therefore seeks to identify some of the problems involved with research in this field and indicates possible directions for future study. Wherever possible, special attention is directed towards the Hemiptera and Diptera as members of these orders include many of the most important vectors of parasites plaguing mankind.

Cells and tissues involved

The blood cells or haemocytes usually mediate the insect cellular defences although other tissues such as the haemopoietic organs, nephrocytes/

ZOOLOGICAL SYMPOSIUM No. 56
ISBN 0-19-854004-3

pericardial cells, the gut and the fat body may also have important, as yet only partially defined, roles.

Haemocytes

These may be free, sessile or arranged into aggregates termed haemopoietic or phagocytic organs of different complexities (Jones 1970).

There have been, and still are, many problems associated with the classification of insect blood cells. This situation was alleviated by Jones (1962) whose scheme categorized nine haemocyte types and by Price & Ratcliffe (1974) who identified six basic types of cells. Despite these advances, many difficulties still remain and, although unfashionable, additional work in this field is urgently required. The literature describing haemocyte types is often confused both by the use of vertebrate terms for the blood cell types and by a number of ultrastructural studies which insist on classifying cells mainly on the basis of their granular inclusions (e.g. Brehélin & Zachary 1986). Electron microscopy is indeed useful in insect haematology but the findings should be interpreted in the light of functional studies and must surely take into account too the vast diversity of species involved. For example, it is not surprising that the number and structure of the granules in the phagocytic cells of the Lepidoptera and Orthoptera should vary but there seems little justification, on this basis alone, in recognizing these as separate categories of cells. Some variation is surely to be expected in a group of animals which may include as many as three million species!

It would now be most useful to identify the blood cells involved in the insect cellular defences by means of their surface determinants. Such studies have been successfully undertaken in the annelids (e.g. Roch & Valembois 1978), molluscs (e.g. Cheng, Huang, Karadogan, Renwrantz & Yoshino 1980; Schoenberg & Cheng 1980; Yoshino & Davis 1983; Yoshino & Granath 1983, 1985; Renwrantz, Daniels & Hansen 1985), and tunicates (Schlumpberger, Weissman & Scofield 1984) using lectin-binding specificities and monoclonal antibodies against haemocyte surface antigens. Not only have haemocyte sub-populations in these groups been identified by these techniques but Schlumpberger *et al.* (1984) have also isolated the cell types of *Botryllus* sp. by means of Percoll gradients and fluorescence-activated cell sorting, and are now studying the haemocytes responsible for allodiscrimination in this species. In insects, both Rizki & Rizki (1983) and Nappi & Silvers (1984) have utilized lectin-binding to investigate changes in the immune competence of haemocytes from *Drosophila* temperature-sensitive mutants. They showed that elevated temperature correlated both with an increased immune competence, in the form of enhanced encapsulation reactions, and with an increase in the percentage of blood cells binding wheat germ agglutinin (WGA). Nappi & Silvers (1984) also reported that

WGA-negative cells participated in the encapsulation reactions, so that further studies utilizing other lectins to characterize these cells may now be required since WGA-binding only reflects changes in N-acetyl-glucosamine residues on the haemocyte membrane.

Despite the aforementioned problems, six common types of haemocytes are found in insects (Price & Ratcliffe 1974): the prohaemocytes, plasmatocytes, granular cells, cystocytes (coagulocytes), spherule cells and oenocytoids.

The prohaemocytes are small cells with a high nuclear:cytoplasmic ratio, rarely constitute more than a few per cent of the total cell population and may represent the progenitor for other types of cells.

The plasmatocytes are larger haemocytes and highly variable in structure. The cytoplasm may enclose few (e.g. some Lepidoptera) or many (e.g. Orthoptera) granules and this variation has led to confusion in their identification (Ravindranath 1978). They are, however, usually amoeboid and their spreading and phagocytic capacities *in vitro* differentiate them from other cell types. Plasmatocytes comprise *c.* 30–60% of the total haemocyte population and are involved in phagocytic and encapsulation-type reactions.

Granular cells, as the name implies, contain numerous granules which they frequently discharge rapidly *in vitro* and *in vivo* (e.g. Lepidoptera). They may, however, be more stable in other groups such as the Phasmida. This degranulation process, which occurs in contact with foreign surfaces, is most important as it may form the basis of many immunological and haemostatic processes in insects (see pp. 26–27, 34). Granular cells, like plasmatocytes, form 30–60% of the haemogramme.

Cystocytes superficially resemble prohaemocytes except that the thin rim of cytoplasm is usually granular. They are highly unstable *in vitro* and in species containing both granular cells and cystocytes, it is the cystocytes that are more unstable and play a greater role in haemolymph coagulation. Granular cells and cystocytes could be classified together as they are both involved in coagulation and in modulation of the cellular defences (see pp. 26–27, 34). Cystocytes usually compose up to 40–60% of the haemogramme and are present in most insects except some Lepidoptera.

Spherule cells are characterized by 1–3μm inclusions which are not usually discharged *in vitro*. These cells are thus highly stable, usually represent less than 5% of the total haemocytes (although much higher numbers have been observed in a few species) and play an unknown, if any, role in the immune defences.

The final cell type, the oenocytoid, is easily identified as the cells are large, 15–30μm in diameter, have a homogeneous cytoplasm, a small eccentric nucleus and are highly stable *in vitro*. They rarely make up more than 1–2% of the total blood cells and contain components of the prophenoloxidase system so that they may be involved in melanization and the cellular defence reactions (see p. 33).

The haemocytes of the Diptera vary considerably from the basic cell categories outlined above. Furthermore, members of this Order are particularly difficult to experiment with since not only are they frequently very small and short-lived but the adults may contain very few blood cells (e.g. mosquitoes and sandflies) or many of the cells present often rapidly lyse, fragment or aggregate on exposure to air (e.g. tsetse flies). The haemolymph may also immediately clot and melanize (e.g. leather-jackets), but particularly confusing is the unusual structure of many dipteran haemocytes (Kaaya & Ratcliffe 1982). Cystocytes are usually absent and instead peculiar anucleate cellular fragments, termed thrombocytoids (Zachary & Hoffmann 1973), are often present. These unusual structures rapidly agglutinate *in vitro* and are responsible for wound healing and encapsulation reactions. Other unusual haemocyte types in the Diptera include the spindle cells of, for example, newly emerged tsetse fly adults (East, Molyneux & Hillen 1980; Kaaya & Ratcliffe 1982), which may represent muscle remnants from reorganization of the pupa, and the crystal cells of *Drosophila melanogaster* which rapidly release their crystalline inclusions to induce melanization (Rizki 1978) and may therefore be equivalent to the oenocytoids of other species.

Very limited studies on the Hemiptera indicate that typical prohaemocytes, plasmatocytes, granular cells/cystocytes and possibly oenocytoids are present (Wigglesworth 1973) but that the number of blood cells free in the circulation varies remarkably according to the nutritional state. Thus, attempts to bleed *Rhodnius* sp. or *Dipetalogaster* sp. which have not been fed for some weeks is impossible, while insects fed a few days previously readily yield large volumes of haemolymph with a high concentration of blood cells (L. Petry, L. Renwrantz & N. A. Ratcliffe unpublished) (and see also chapter by Molyneux, Takle, Ibrahim & Ingram, this volume). These observations may explain why some workers believe that hemipterans contain very few free haemocytes with the majority of the blood cells being attached to the internal organs (Wigglesworth 1965).

Sessile cells and other organs involved in the cellular defences

Apart from the transiently sessile haemocytes in, for example, the Hemiptera, other populations of permanently attached cells are involved in the insect cellular defences. Thus, haemopoietic or phagocytic organs have been described in many insect orders (Jones 1970). They may vary from a few aggregates of haemocytes in the abdomen of mosquitoes to highly organized structures originating from the heart wall in orthopteroid insects. In *Locusta migratoria*, they contain fully differentiated plasmatocytes, granular cells and oenocytoids, and have been shown to remove and encapsulate micro-organisms from the blood (Hoffmann, Brehélin & Hoffmann 1974). It was also reported that the haemopoietic organs of *L. migratoria* contain a special cell type, termed the reticular cell, which does not enter the circulation but produces proteinaceous antibacterial factors which are presumably released into the blood (Brehélin & Hoffmann 1980).

Other tissues in the insect body are capable of responding to invading parasites. The cells lining the gut of *Amphimallon majalis*, infected by *Bacillus popilliae*, endocytose and destroy the bacteria (Kawanishi, Splittstoesser & Tashiro 1978), while those of the hindgut of *Blattella germanica* and the ileum of *Acheta pennsylvanicus* encapsulate the nematode *Physaloptera maxillaris* (Cawthorn & Anderson 1977). The possible role of other organs in the insect body in the clearance of invading micro-organisms requires further investigation, especially in the light of reports of the removal of test particles from the circulation by miscellaneous organs in the annelids, molluscs, crustaceans and echinoderms (detailed in Ratcliffe, Rowley, Fitzgerald & Rhodes 1985). The fat body of insects, which is extremely extensive in some species, also plays a key role in the immune defences. Organ cultures of fat body removed from immunized cecropia pupae of *Hyalophora cecropia*, waxmoth larvae of *Galleria mellonella*, and from larvae of the tobacco hornworm, *Manduca sexta*, have been shown to synthesize a range of immune proteins (Faye & Wyatt 1980; DeVerno, Chadwick, Aston & Dunphy 1984; Dunn, Dai, Kanost & Geng 1985). This organ has also been reported to produce agglutinins in the dipteran *Sarcophaga peregrina* (Kubo, Komano, Okada & Natori 1984). Apart from these organs and the haemocytes themselves, other cells, which have been likened to the reticuloendothelial system of vertebrates (Wigglesworth 1970), have a defensive function in insects. These cells have been termed pericardial cells, nephrocytes, diaphragm cells and garland cells, and are often located around the dorsal vessel (Rowley & Ratcliffe 1981). They may phagocytose micro-organisms (Cameron 1934) but usually pinocytose colloids and waste products to detoxify the blood (Hollande 1922). More recently, however, in the blowfly, *Calliphora erythrocephala*, they have been reported to synthesize lysozyme (Crossley 1972).

Cellular defence reactions

The cellular defences of insects include haemolymph coagulation, phagocytosis and encapsulation-type responses which have been reviewed in detail elsewhere (e.g. Ratcliffe & Rowley in press). These responses are described briefly in isolation here, but it must be emphasized that, in many cases, not only are they dependent on each other but they also interact intimately with the humoral defences to generate the overall immune reactivity of the insect host.

Haemolymph coagulation

In many insects, haemolymph coagulation occurs immediately following wounding and prevents excess loss of body fluids as well as the entry of would-be pathogens and parasites. Usually coagulation is mediated by factors discharged from granular cells and/or cystocytes which probably interact with plasma proteins to induce gelation. Although it is known that the speed, pattern and extent of coagulation vary markedly from species to species (Grégoire 1970), our knowledge of the biochemical events involved is strictly limited. Brehélin (1979a,b) and Bohn, Barwig & Bohn (1981) have, however, used modern analytical techniques to investigate haemolymph coagulation in *Locusta migratoria* and *Leucophaea maderae*, respectively. In *L. migratoria*, Brehélin (1979b) isolated and characterized a plasma coagulogen (= a clotting protein) which is a glycolipoprotein complex of six polypeptide chains with a molecular weight of *c.* 1×10^6 Daltons. This plasma coagulogen has recently been identified as lipophorin (Gellissen 1983). Bohn *et al.* (1981), however, with *L. maderae*, showed by immunofluorescence that both haemocyte and plasma coagulogens exist and interact during haemolymph coagulation. The haemocyte coagulogen can induce gelation in the absence of plasma but the plasma coagulogen is dependent for activation upon factors derived from the haemocytes. Further clues as to the processes involved in haemolymph coagulation in insects are provided by the detection of the prophenoloxidase system in the cystocytes of *L. migratoria* (Hoffman, Porte & Joly 1970) and in the granular cells of *Galleria mellonella* (Schmit, Rowley & Ratcliffe 1977), which are the cells responsible for coagulation in these species. In decapod crustaceans, activation of a complex prophenoloxidase cascade from haemocyte lysates has been shown to accompany the clotting process (Söderhäll 1981, 1982). This activation occurs through a complex enzyme cascade, the prophenoloxidase cascade, located in the blood cells and probably involving one or more serine proteases (Söderhäll & Smith 1984). A number of such proteases have also been detected in the haemocyte lysates from the cockroach, *Blaberus craniifer* (Leonard, Söderhäll & Ratcliffe 1985). Thus, in both crustaceans and insects, a complex cascade involving several serine proteases, one or more coagulogens and prophenoloxidases, and probably a number of protease inhibitors (Eguchi, Haneda & Iwamoto 1982), are involved in the haemolymph coagulation process. The importance of this reaction cannot be overemphasized as its activation may catalyse many of the other cellular, and possibly also humoral, defences.

In the Hemiptera, haemolymph coagulation has been observed as a result of granule discharge from the cystocytes (L. Petry *et al.* unpublished). In many Diptera, however, such a gelation process is wanting and instead small cytoplasmic fragments, termed thrombocytoids, aggregate and interdigitate to prevent excessive haemolymph loss (Zachary & Hoffmann 1973).

Phagocytosis

Once the physicochemical barriers, such as the gut and cuticle, have been breached, and providing that entrapment in any haemolymph coagulum formed has been avoided, then invading micro-organisms are released into the circulation and exposed to the freely circulating blood cells. Recently, it has been shown that in the waxmoth, *G. mellonella*, phagocytosis is the primary cellular defence reaction in the circulation and can effectively deal with doses of bacteria below *c.* 10^3 μl^{-1} haemolymph (Ratcliffe & Walters 1983). Above this concentration, phagocytosis is augmented by nodule formation (see p. 29). This is a generalization since this latter study also showed that the extent to which the cellular processes are elicited depends very much upon the nature of the invading organism.

Insect blood cells have been shown to ingest viruses, bacteria, protozoans and fungi both *in vivo* and *in vitro* (Jones 1962; Salt 1970; Whitcomb, Shapiro & Granados 1974), but the rate of uptake depends not only on the nature of the invader but also on environmental factors such as temperature, season and pH (Whitcomb *et al.* 1974) as well as the immune status of the host.

Research on phagocytosis in insects has mainly concentrated upon bacterial uptake and there are comparatively few studies on haemocyte interactions with other micro-organisms such as protozoans. In both refractory and susceptible *Culex pipiens* mosquitoes infected with *Plasmodium cathemerium* or *Plasmodium relictum*, however, Huff (1934) has reported that the phagocytes are non-responsive to the oocysts in the stomach wall, although in one case they reacted strongly to an adjacent bacterial infection of this organ. The latter indicates that the haemocytes of vector species are capable of ingesting invading micro-organisms and this has been confirmed by reports of the phagocytosis of several flagellates in *Rhodnius prolixus* (Tobie 1968; and chapter by Molyneux *et al.* in this volume), by the uptake of *Plasmodium gallinaceum* in *C. pipiens* and *Aedes aegypti* (Weathersby & McCall 1968), and by the adherence of many stages of *Plasmodium berghei* to the haemocytes of *Anopheles stephensi in vitro* (Foley 1978). More recently, the phagocytosis of bacteria by the plasmatocytes of *Glossina morsitans morsitans* has also been reported *in vivo* (Kaaya, Ratcliffe & Alemu 1986).

Detailed studies of the phagocytosis of bacteria in insects have been mainly undertaken in simple monolayer or suspension cultures. Plasmatocytes are the major phagocytic cells although granular cells (Neuwirth, 1974) and cystocytes (Brehélin, Zachary & Hoffmann 1978) may be involved too. Phagocytosis by these cells in insects, as with vertebrates, is a dynamic process probably involving chemotaxis, attachment, ingestion and killing phases. Limited details of these stages have been described previously (Ratcliffe & Rowley 1979), although unequivocal proof for the presence of chemotaxis and recognition factors is wanting, and the killing of intracellular bacteria has only been investigated biochemically in one species (Anderson, Holmes & Good 1973a).

Chemotaxis has not been studied in detail in insects but has been inferred from observations of the premature movement of haemocytes from their posterior location to encapsulate the nematode *Heterotylenchus autumnalis* in *Musca* spp. (Nappi & Stoffolano 1972), and from the selective attachment in *G. mellonella* of plasmatocytes to form multicellular sheaths around nodules and capsules *in vivo* (Ratcliffe & Rowley 1979) (see p. 30).

The attachment phase of phagocytosis in vertebrates and invertebrates is mediated by both serum and cytophilic recognition factors. In insects, Mohrig, Schittek & Hanschke (1979a, b) have shown that the haemocytes of *G. mellonella* can be stimulated to ingest a normally non-phagocytosed *Bacillus thuringiensis* strain, provided that the larvae receive a prior injection of latex beads. They were also able to transfer the stimulatory factor in cell-free haemolymph to naive larvae. Ratcliffe & Rowley (1983) also reported the presence of a factor in the haeomolymph of *Periplaneta americana* which increases the ingestion of a *Bacillus cereus* isolate by the haemocytes of this species. Additional proof for the role of recognition factors in phagocytosis by insect haemocytes has been provided by Leonard, Ratcliffe & Rowley (1985) utilizing monolayer cultures from three insect species. Activation of the prophenoloxidase system in these monolayers by microbial products significantly enhanced the ingestion of test particles. These results are discussed in detail elsewhere (see pp. 33–4).

Following attachment, micro-organisms are ingested by the formation of pseudopods and are withdrawn within phagocytic vacuoles into the cytoplasm. Unlike the attachment phase, ingestion is energy-dependent and fuelled by glycolysis (Anderson, Holmes & Good 1973b). The final stage, during which the ingested bacteria are killed, has been studied in the cockroach, *Blaberus craniifer*, by Anderson *et al.* (1973a). Six out of ten bacterial species were effectively killed *in vitro* by the cockroach haemocytes and the bactericidal effect was shown not to be due to a myeloperoxidase-H_2O_2-halide system. More recently, C. Leonard,, A.F. Rowley & N. A. Ratcliffe (unpublished) have found that activation of the prophenoloxidase system in the haemocyte lysates of *G. mellonella* or *B. craniifer* generates a bactericidal effect. Walters & Ratcliffe (1981) have also detected acid phosphatase, β-glucuronidase and β-glucosaminidase activity in *G. mellonella* blood cells. The latter two enzymes are bacteriolytic and since lysozyme is also present in insect haemocytes (Anderson & Cook 1979; Zachary & Hoffmann 1984), then the killing of sequestered organisms in insect phagocytes probably involves a multiplicity of factors.

Encapsulation-type responses

If large numbers of micro-organisms or parasites that are too large to be ingested by single blood cells invade the haemocoel of insects then multicellular sheaths, called nodules and capsules, respectively, are often formed to confine the invaders.

Once in the circulation, micro-organisms make random and rapid contact with the haemocytes and if the pathogen/parasite has the appropriate stimulatory product (β1, 3-glucan, lipopolysaccharide etc.), the granular cells, cystocytes and/or crystal cells will react by discharging their contents on to the surface of the non-self material. This degranulation process corresponds to the activation of the prophenoloxidase system and the transgressor becomes covered in sticky proteinaceous substances which may represent recognition markers (Ratcliffe & Rowley in press; details on p. 34). The invaders, depending upon their size and numbers, will now be phagocytosed or encapsulated.

If the invading micro-organisms exceed *c.* $10^3 \mu l^{-1}$ haemolymph, phagocytosis is often rapidly augmented by nodule formation (Ratcliffe & Walters 1983). In *G. mellonella*, the first stage of nodule formation is initiated by random contact between the granular cells and the bacteria. Rapid degranulation of the cells then occurs which become sticky and form a localized clot to entrap masses of bacteria. Thus, non-self recognition is probably instantaneous and the prophenoloxidase complex is immediately activated. After *c.* 2h post-injection, the second stage begins during which the plasmatocytes begin to attach and flatten on to the central bacteria/granular cell core which is now melanizing. By 24h the plasmatocytes form a typical multicellular sheath containing a mass of melanized, degenerating cells and entrapped bacteria (Ratcliffe & Gagen 1977). The delay in attachment of the plasmatocytes to form the outer sheath may indicate that the cells are responding to some factor emanating from the melanizing core. Alternatively, the plasmatocytes may be the only haemocytes which, following random contact with the aggregates, are able to attach owing to the nature of their cell surface receptors and/or some charge difference between the haemocytes and the cell aggregates (see p. 36). The extent of nodule formation in *G. mellonella* can be gauged by the presence of numerous black clumps attached to the gut, fat body and other organs by 12h post-injection. Regarding the fate of bacteria within the nodules, it has recently been shown, using four strains of bacteria, that by 4h post-injection over 90% of micro-organisms sequestered are effectively killed. The survivors of the two pathogenic strains used subsequently multiplied to break out the nodules, while the viability of the two non-pathogenic strains remained at only 1–4% (Walters & Ratcliffe 1983). The mode of killing within the nodules may have involved the synthesis of toxic quinones during the melanization of the central core and/or lysosomal enzymes which have been detected in these structures (Walters & Ratcliffe 1981).

Nodules have been reported to occur in response to endotoxin, bacteria, fungi and protozoans (see references in Ratcliffe & Rowley in press), and are extremely effective at clearing large numbers of micro-organisms within a very short time (Gagen & Ratcliffe 1976). Furthermore, and as detailed above, nodule formation, like the other cellular defences, is a biphasic process involving the interaction of the coagulation-inducing cells (granular cells and/or cystocytes) with the phagocytic plasmatocytes.

When insects are invaded by parasites that are too large to be ingested by individual haemocytes then the parasites frequently become surrounded by multicellular sheaths, termed capsules, which closely resemble the nodules described above. Parasites encapsulated include acanthocephalans, cestodes, nematodes, insect parasitoids, large protozoans, fungi and, sometimes, bacteria (see references in Ratcliffe & Rowley in press). In many dipterans and some hemipterans, however, a supposedly non-cellular or humoral capsule is formed composed of layers of melanin (Götz & Boman 1985). Details of humoral encapsulation are discussed by Peter Götz in another chapter of this volume. What, however, seems to be the factor determining the type of capsule formed is the number of blood cells present in the host. Götz, Roettgen & Lingg (1977) clearly showed that humoral capsules were formed in eight out of 12 dipterans examined and that these same eight species had less than 4,000 haemocytes mm^{-3} of haemolymph. The remaining four species formed cellular capsules and had more than 8,000 blood cells mm^{-3} of haemolymph. The importance of humoral encapsulation should be emphasized as it is the means by which many vector species, including *Glossina* spp., deal with invading parasites and non-self materials (Kaaya *et al.* 1986). Humoral encapsulation has been reported mainly in mosquitoes and midges in response to nematodes, although it also occurs around fungi, bacteria and the eggs of parasitoids (see references in Ratcliffe & Rowley in press). The end result is the formation of a pigmented sheath of melanin around the parasite. There is some discussion as to whether the melanin is solely derived from precursors in the haemolymph or if a cellular input also occurs (Götz 1969; Götz & Vey 1974).

Regarding the more common cellular encapsulation process, this reaction, like nodule formation, also seems to involve the interaction of the granular cells and/or cystocytes with the plasmatocytes in a biphasic process (Ratcliffe & Rowley 1979). The end result is a multicellular sheath of plasmatocytes composed of layers of cells, flattened to a variable extent from the outside to the inside of the capsule. Adjacent to the parasite is often a necrotic layer of melanized cells, probably representing the remnants of the coagulation cells which initiated the encapsulation process.

It must be emphasized, however, that the above descriptions are generalizations made from studies on just a few insect species. Variations in encapsulation reactions do occur; for example, in *Anopheles quadrimaculatus* parasitized by the microfilariae of the nematode, *Brugia pahangi*, Chen & Laurence (1985) have shown that the host reaction involves cellular and humoral responses. Furthermore, the formation of the multicellular capsule wall may involve charge differences between the surfaces of the haemocytes and the non-self material (Lackie 1983; Ratner & Vinson 1983; Lackie, Takle & Tetley 1985; see p. 36).

Finally, another type of variation has recently been observed in the encapsulation reaction of *Drosophila melanogaster* against the parasitoid wasp, *Leptopilina boulardi*. Carton & Boulétreau (1985) showed that the encapsulation ability of the *D. melanogaster* hosts was a variable trait and that within a given population this variability was under partial genetic control. These observations are important as individuals with a more efficient encapsulation ability may be selected for, making the overall population more resistant to the parasite (but see also chapter by Anderson, this volume). This then may lead to a progressive failure of biological control programmes based on the release of insect parasitoids.

Cellular defences in relation to the overall immunity of the host

So far, the cellular defence reactions of insects have been considered in isolation from the overall immune potential of the host. This brief section indicates when the cellular defences function following pathogen or parasite invasion and describes the interactive nature of the cellular and humoral defence reactions.

Following Chadwick & Aston's (1978) original model for insect immunity, we recently divided the insect immune defences into three interacting phases, viz.: 1. an immediate and inductive, 2. a mainly cellular and synthetic, and 3. a mainly humoral and recovery stage (Ratcliffe & Rowley in press). The first phase rapidly follows penetration of the physicochemical barriers as a result of wounding or defective moulting and, as its name implies, occurs instantaneously, and subsequently probably leads to activation of the later events in insect immunity. Wounding effects the release of haemolymph and leads to the first cellular defence reaction during which the fragile coagulation-type cells release their contents to induce haemolymph gelation and sealing of the wound. As mentioned previously, haemolymph coagulation is probably instigated by a complex prophenoloxidase cascade which leads not only to coagulation, but also to other factors which may mediate non-self recognition and later events in the immune response (see

p. 33). The second stage of insect immunity follows penetration of the invader into the haemocoel. It occurs immediately contact is made between the foreign organism and the freely circulating haemocytes, and results in a rapid clearance of the haemolymph by phagocytosis and/or encapsulation-type responses. During this mainly cellular stage, the immune proteins are synthesized in the fat body. The final stage is characterized by the activity of these antimicrobial proteins manufactured during the second, mainly cellular, phase of insect immunity. Thus, the cellular defences function during the early events of insect immunity and, if they fail to contain the invader, it is subsequently exposed to powerful antimicrobial proteins released into the circulation.

Regarding the interactions of cellular and humoral components of insect immunity, recent work indicates that these are closely integrated. For example, the agglutinins (Amirante 1976) and the prophenoloxidase system (Leonard, Söderhäll & Ratcliffe 1985), which probably generate the recognition molecules of insect immunity, are partially or wholly synthesized in the blood cells. De Verno *et al.* (1984) and Dunn *et al.* (1985) have also recently produced evidence indicating that haemocytes interact with the fat body during the production of the immune proteins by this organ. Thus, bactericidal activity of *G. mellonella* fat body was stimulated *in vitro* by the addition of haemocytes (De Verno *et al.* 1984), while in *Manduca sexta*, experiments suggested that bacteria might be phagocytosed by the circulating haemocytes and partially degraded to generate a signal (fragments of peptidoglycan) which induces the synthesis of antibacterial proteins by the fat body (Dunn *et al.* 1985).

Relevance of laboratory investigations to the field situation

Most information on the insect defence reactions has been gleaned from laboratory experiments during which mutant strains of bacteria are introduced, often at artificially high doses, into the haemocoel by injection. Apart from the work of Bucher (1959), there is little evidence reported for the natural occurrence of the cellular defences in insects collected from the field. Recently, in an attempt to correct this anomaly, J. B. Walters & N. A. Ratcliffe (unpublished) examined, for evidence of natural infections and cellular defence reactivity, 15 species of insects from seven Orders collected from a range of natural terrestrial and aquatic environments. They found a remarkably high percentage of bacterial infections in lepidopterans such as *Pieris brassicae* (74% infected) and odonatans such as *Aeshna grandis* (78% infected), which contrasted with the low infectivity rates of the hemipterans, *Notonecta glauca* and *Corixa punctata* (19% and 11%, respectively). The natural occurrence of phagocytosis observed varied from 20–30% of the haemocytes in the lepidopterans, odonatans, and some orthopterans, to a complete lack of phagocytosis in the hemipterans and coleopterans examined. Not surprisingly, there thus appeared to be a correlation between percentage infection and the amount of phagocytosis observed. Nodules too were present in 7–20% of the lepidopterans and orthopterans but were usually necrotic and small, and rarely as well-developed as those seen in laboratory insects. Overall, the incidence of phagocytosis and nodule formation was not as high as would be expected considering the large number of infections. There also appeared to be a correlation between the blood volume and numbers of circulating haemocytes with the incidence of cellular defence reactivity. Thus, the lepidopterans and odonatans, both with large numbers of blood cells, frequently showed evidence of phagocytosis and nodule formation, while the converse was true of the hemipterans which usually contained very few free blood cells. Another significant observation was the fact that all insects with viral inclusion bodies also contained bacteria in the haemolymph. Possibly the bacteria stressed the immune system and allowed the manifestation of latent viral diseases (Steinhaus 1958). Together these infections would probably suppress the immune system and this may explain the reduced cellular responsiveness recorded. Other important determinants would be environmental in origin and would include temperature, humidity and pollution stressors.

Non-self recognition

There are a number of factors present in the haemolymph of insects which may act as recognition molecules. These substances include: 1. components of the prophenoloxidase cascade and 2. the agglutinins.

Components of the prophenoloxidase cascade

The possibility that the prophenoloxidase system plays a role in the host defences resulted from the observation that melanization often accompanies insect immune reactions (Salt 1970; Nappi 1975; Ratcliffe & Rowley 1979). Attempts to inhibit melanization and thus the defence reactions have produced conflicting results (Ratcliffe & Rowley in press). This confusion probably originates from a lack of understanding of the events involved in the prophenoloxidase cascade in which it has been shown in crustaceans, at least, by Söderhäll and his co-workers that components of the cascade *prior to* phenoloxidase and melanin production are probably the factors mediating the immune response (Söderhäll 1982; Söderhäll & Smith 1984). Proof, however, for the involvement of the prophenoloxidase system in insect immunity is mainly confined to two studies in which the effects of laminarin (a β1, 3-glucan component of fungal cell walls), dextran (an α D-glucan) and lipopolysaccharide were examined on the *in vitro* phagocytosis of heat-killed *Bacillus cereus* by the haemocytes of three insect species (Ratcliffe,

Leonard & Rowley 1984; Leonard, Ratcliffe & Rowley 1985). These experiments are very similar to those originally undertaken by Smith & Söderhäll (1983a, b) in their studies on crustacean immune recognition, and produced comparable results for insects. They showed that in *G. mellonella* the addition of laminarin or lipopolysaccharide significantly enhanced the phagocytic activity of the haemocytes compared with the saline and dextran controls (Table 1). In an additional control, an inhibitor of serine proteases, p-nitrophenyl-p′-guanidobenzoate, was incubated with the laminarin, since proteases are involved in activation of prophenoloxidase (Söderhäll 1982; Söderhäll & Smith 1984; Leonard, Söderhäll *et al.* 1985), and was found to abrogate the stimulatory effect of the laminarin. Similar results were obtained with all three species tested (Table 1). Spectrophotometric assays for phenoloxidase generation by *G. mellonella* haemocyte lysates gave complementary results since laminarin but not dextran activated the prophenoloxidase cascade (Table 2). These experiments also showed that the plasmatocytes were the main phagocytic cells in the monolayers (Table 1), while subsequent work, utilizing purified cell populations from *G. mellonella* (G. P. M. Mead, N. A. Ratcliffe & L. Renwrantz unpublished), has shown that the prophenoloxidase system may be present in the coagulation-type cells (granular cells in *G. mellonella*) and oenocytoids but not in the plasmatocytes (Schmit *et al.* 1977).

From the above results, a model has been derived for non-self recognition and activation of the cellular defences of insects. Contact of the granular cells/cystocytes with appropriate foreign surfaces will result in activation and discharge of the prophenoloxidase system. During this activation, Smith & Söderhäll (1983a, b) have suggested in crustaceans that 'sticky' proteins, which act as opsonins, are produced to cover the non-self invaders. Indeed, the increased stickiness of granular cells following activation has recently been described (Leonard, Ratcliffe *et al.* 1985). Furthermore, the discharge of granular cell contents onto foreign surfaces during the initial stages of encapsulation and nodule formation is also well documented (Schmit & Ratcliffe 1977; Ratcliffe & Gagen 1977). Thus, the release of the prophenoloxidase system from the coagulation-type cells probably initiates the cellular defence reactions. The second stage then occurs during which the plasmatocytes respond to the coated foreign surfaces either by ingesting small particles or by attaching to larger parasites to form multicellular sheaths. It is therefore likely that cellular immunity in insects represents a co-operative venture between different types of blood cells.

Much of the above account of non-self recognition and cellular reactivity in insects is supposition and is open to criticism and modification. For example, an alternative explanation for the enhanced phagocytosis in the presence of laminarin could be that this substance is binding to agglutinins on the outside of both the test bacteria and the blood cells. Laminarin would

Table 1. Effect of laminarin on phagocytosis of *Bacillus cereus* by haemocytes of three different insect species[h].

Treatment	% Haemocytes phagocytic	% Plasmatocytes phagocytic	% Cystocytes/ granular cells phagocytic	Number of bacteria/ 100 haemocytes
G. mellonella				
Laminarin (1 mg ml^{-1})	8.0 ± 1.8[a,b]	14.1 ± 3.7[b]	4.3 ± 1.4[b]	17.4 ± 3.1[b]
GIM control[e]	1.2 ± 0.7	2.0 ± 0.6	0.7 ± 0.5	1.8 ± 0.9
0.01 mM pNPGB[f]	1.3 ± 0.5	2.4 ± 1.0	0.7 ± 0.4	2.2 ± 1.1
Laminarin ± pNPGB	1.5 ± 0.6	2.8 ± 1.6	0.5 ± 0.4	2.2 ± 0.8
Dextran (1 mg ml^{-1})	1.9 ± 0.6	2.7 ± 1.1	1.0 ± 0.5	2.6 ± 0.8
L. maderae				
Laminarin (1 mg ml^{-1})	9.8 ± 1.6[b,c]	15.9 ± 4.3[b]	4.7 ± 1.6[b]	26.8 ± 6.8[b]
L-J control[g]	1.5 ± 0.7	2.2 ± 1.1	0.9 ± 0.8	2.2 ± 1.9
0.01 mM pNPGB	0.9 ± 0.5	1.2 ± 0.6	0.6 ± 0.4	1.2 ± 0.9
Laminarin + pNPGB	1.4 ± 0.7	0.7 ± 0.6	1.2 ± 0.8	2.3 ± 1.7
B. craniifer				
Laminarin (1 mg ml^{-1})	4.6 ± 0.6[b,d]	9.6 ± 1.4[b]	1.2 ± 0.4[b]	10.7 ± 1.8[b]
Carlson's control	0.7 ± 0.3	1.2 ± 0.4	0.6 ± 0.4	1.5 ± 0.6
0.01 mM pNPGB	0.9 ± 0.2	1.2 ± 0.4	0.8 ± 0.4	1.3 ± 0.5
Laminarin + pNPGB	1.1 ± 0.2	1.5 ± 0.6	0.8 ± 0.3	1.6 ± 0.5

[a] Mean value of 20 monolayers (from a total of 28 insects in four experiments) ± SD.
[b] $P \leq 0.05$ compared with controls.
[c] Mean value of 10 monolayers (from a total of 14 insects in two experiments) ± SD.
[d] Mean value of 10 monolayers (from a total of 8 insects in two experiments) ± SD.
[e] GIM = Grace's insect medium.
[f] pNPGB = p-nitrophenyl-p′-guanidobenzoate.
[g] L-J = Landureau and Jollè's medium.
[h] From Leonard, Ratcliffe *et al.* (1985).

Table 2. Effect of β 1,3-glucans on the phenoloxidase activity of *G. mellonella* haemocyte lysate supernatant[d].

	Phenoloxidase activity (units enzyme activity/mg protein)
Laminarin (0.5 mg ml^{-1})[a]	69 ± 21[b,c]
Laminarin (0.05 mg ml^{-1})	53 ± 8[c]
Laminarin G (0.05 mg ml^{-1})	46 ± 18[c]
Dextran (0.5 mg ml^{-1})	3 ± 2
Trypsin (0.5 mg ml^{-1})	1275 ± 544[c]
Cacodylate control	6 ± 5

[a] Final concentration in reaction mixture.
[b] Mean value ± SD, $n = 5$.
[c] $P \leq 0.05$ compared with controls.
[d] From Leonard, Ratcliffe *et al.* (1985).

thus be forming a bridging molecule and increasing bacterial adherence to the cells. Likewise, insect haemocytes are known to contain agglutinins (Amirante 1976) so that degranulation and activation of the prophenoloxidase system by laminarin would also probably release agglutinin which could then bind to sugar determinants on the outsides of the bacteria and the blood cells. The abrogation of laminarin-mediated enhanced phagocytosis by serine protease inhibitors (see above) does, however, seem to favour the involvement of the prophenoloxidase cascade but additional experiments are required to eliminate these alternative explanations for the results obtained.

Finally, attention should be drawn to the recent work of Ratner & Vinson (1983) and Lackie and her colleagues (Lackie 1983; Lackie *et al.* 1985; Takle & Lackie 1985) who consider that initial recognition of foreignness may involve the physicochemical properties of the foreign surface. Electrostatic charges on the haemocytes and the non-self materials are seen as being particularly important. It may be significant that the haemocytes of *Schistocerca gregaria* are more negatively charged than their *Periplaneta americana* counterparts and are also generally less responsive to a variety of non-self materials *in vivo* (Lackie 1981a). Furthermore, a model for non-self discrimination could be derived in which the unstable coagulation cells respond to charged groups on the foreign surface by degranulating and thus releasing recruitment signals for other cells (Lackie 1983). One problem that needs to be addressed in this model is whether, once in the haemocoel, foreign surfaces coated with haemolymph proteins do in actual fact retain their surface charge differences (see, however, chapter by Lackie in this volume).

Agglutinins

Agglutinins (often referred to as lectins) are widely distributed in the body fluids of invertebrates, including insects (Ey & Jenkin 1982) in which they have been reported to be synthesized in the haemocytes (Amirante 1976) and fat body cells (Komano, Nozawa, Mizuno & Natori 1983). Only recently has definitive proof for the role of agglutinins as recognition molecules in invertebrates been provided. Renwrantz & Stahmer (1983) used the purified agglutinin of the bivalve mollusc, *Mytilus edulis*, to test for its opsonic activity *in vitro*. The purified molecule enhanced the phagocytic uptake of yeast cells by the haemocytes of *M. edulis* to 55% in comparison with the 5% of saline-incubated controls. This report clearly illustrates the value of using purified molecules for tests on putative recognition molecules in invertebrates.

Such studies are still wanting in insects although circumstantial evidence indicates a role for agglutinins in insect recognition processes. Agglutinins, for example, have been detected on the surfaces of plasmatocytes and granular cells (Yeaton 1980; Komano *et al.* 1983) but not on other non-phagocytic cells. Lackie (1981b) has also shown a positive correlation between the level of discrimination and the range of specific agglutinins present in different insects. Komano & Natori (1985) too have reported the involvement of the agglutinin of *Sarcophaga peregrina* larvae in the clearance and lysis of ^{51}Cr-labelled sheep erythrocytes from the haemocoel. Even if the agglutinins lack opsonic activity, they may agglutinate foreign organisms and facilitate their removal by the blood cells (see, however, chapter by Renwrantz in this volume). In addition, lectins present in the different regions of the gut of *Rhodnius prolixus* and *Glossina austeni*, with agglutinating activity for *Trypanosoma cruzi*, may control the development of such parasites outside the haemocoel (Pereira, Andrade & Ribeiro 1981; Ibrahim, Ingram & Molyneux 1984). Lectins present on the peritrophic membrane of larvae of the blowfly, *Calliphora erythrocephala*, may also regulate symbiotic and pathogenic micro-organisms (Peters, Kolb & Kolb-Bachofen 1983).

Finally, it is not inconceivable that the agglutinins may bind microbial and other non-self materials to the coagulation cells and trigger the prophenoloxidase cascade. This binding may thus represent the recognition phase in the insect immune defences. The recent reports of the binding of endotoxin by *Limulus polyphemus* lectin (Rostam-Abadi & Pistole 1982) and the presence of an endotoxin-binding protein on the blood cells of this species (Liang, Sakmar & Liu 1980) are relevant in this context. These observations are particularly interesting as they could indicate a possible interactive mechanism between the agglutinins and the prophenoloxidase cascade.

Acknowledgements

I am grateful to the Science and Engineering Research Council (Grant Nos. GR/B/60958, GR/D/2168.4 and GR/D/54644), The Leverhulme Trust and The British Council for financial support. I also wish to thank Professor L. Renwrantz, University of Hamburg, for helpful discussion of aspects of this work.

References

Amirante, G.A. (1976). Production of heteroagglutinins in haemocytes of *Leucophaea maderae* L. *Experientia* **32**: 526–8.

Anderson, R.S. & Cook, M.L. (1979). Induction of lysozyme-like activity in the hemolymph and hemocytes of an insect, *Spodoptera eridania*. *J. Invert. Path.* **33**: 197–203.

Anderson, R.S., Holmes, B. & Good, R.A. (1973a). *In vitro* bactericidal capacity of *Blaberus craniifer* hemocytes. *J. Invert. Path.* **22**: 127–35.

Anderson, R.S., Holmes, B. & Good, R.A. (1973b). Comparative biochemistry of phagocytizing insect hemocytes. *Comp. Biochem. Physiol.* **46B**: 595–602.

Bohn, H., Barwig, B. & Bohn, B. (1981). Immunochemical analysis of hemolymph clotting in the insect *Leucophaea maderae* (Blattaria). *J. comp. Physiol.* **143**: 169–84.

Brehélin, M. (1979a). Mise en évidence de l'induction de la coagulation plasmatique par les hémocytes chez *Locusta migratoria*. *Experientia* **35**: 270–1.

Brehélin, M. (1979b). Hemolymph coagulation in *Locusta migratoria*: evidence for a functional equivalent of fibrinogen. *Comp. Biochem. Physiol.* **62B**: 329–34.

Brehélin, M & Hoffmann, J.A. (1980). Phagocytosis of inert particles in *Locusta migratoria* and *Galleria mellonella*: study of ultrastructure and clearance. *J. Insect Physiol.* **26**: 103–11.

Brehélin, M. & Zachary, D. (1986). Insect haemocytes; a new classification to rule out the controversy. In *Immunity in invertebrates*: 36–48. (Ed. Brehélin, M.). Springer-Verlag, Berlin & Heidelberg.

Brehélin, M., Zachary, D. & Hoffmann, J.A. (1978). A comparative ultrastructural study of blood cells from nine insect orders. *Cell Tissue Res.* **195**: 45–57.

Bucher, G.E. (1959). Bacteria of grasshoppers of western Canada. III. Frequency of occurrence, pathogenicity. *J. Insect Path.* **1**: 391–405.

Cameron, G.R. (1934). Inflammation in the caterpillars of Lepidoptera. *J. Path. Bact.* **38**: 441–66.

Carton, Y. & Boulétreau, M. (1985). Encapsulation ability of *Drosophila melanogaster*: a genetic analysis. *Devl comp. Immunol.* **9**: 211–9.

Cawthorn, R.J. & Anderson, R.C. (1977). Cellular reactions of field crickets (*Acheta pennsylvanicus* Burmeister) and German cockroaches (*Blatella germanica* L.) to *Physaloptera maxillaris* Mollin (Nematoda: Physalopteroidea). *Can J. Zool.* **55**: 368–75.

Chadwick, J.M. & Aston, W.P. (1978). An overview of insect immunity. In *Animal models of comparative and developmental aspects of immunity and disease*: 1–14. (Eds Gershwin, M.E. & Cooper, E.L.). Pergamon Press, New York.

Chen, C.C. & Laurence, B.R. (1985). An ultrastructural study on the encapsulation of microfiliariae of *Brugia pahangi* in the haemocoel of *Anopheles quadrimaculatus*. *Int. J. Parasit.* **15**: 421–8.

Cheng, T.C., Huang, J.W., Karadogan, H., Renwrantz, L.R. & Yoshino, T.P. (1980). Separation of oyster hemocytes by density gradient centrifugation and identification of their surface receptors. *J. Invert. Path.* **36**: 35–40.

Crossley, A.C.S. (1972). The ultrastructure and function of pericardial cells and other nephrocytes in an insect: *Calliphora erythrocephala*. *Tissue Cell* **4**: 529–60.

De Verno, P.J., Chadwick, J.S., Aston, W.P. & Dunphy, G.B. (1984). The *in vitro* generation of an antibacterial activity from the fat body and hemolymph of non-immunized larvae of *Galleria mellonella*. *Devl comp. Immunol.* **8**: 537–46.

Dunn, P.E., Dai, W., Kanost, M.R. & Geng, C. (1985). Soluble peptidoglycan fragments stimulate antibacterial protein synthesis by fat body from larvae of *Manduca sexta*. *Devl comp. Immunol.* **9**: 559–68.

East, J., Molyneux, D.H. & Hillen, N. (1980). Haemocytes of *Glossina*. *Ann. trop. Med. Parasit.* **74**: 471–4.

Eguchi, M., Haneda, I. & Iwamoto, A. (1982). Properties of protease inhibitors from the haemolymph of silkworms. *Bombyx mori*, *Antheraea pernyi* and *Philosamia cynthia ricini*. *Comp. Biochem. Physiol.* **71B**: 569–76.

Ey, P.L. & Jenkin, C.R. (1982). Molecular basis of self/non-self discrimination in the Invertebrata. In *The reticuloendothelial system: a comprehensive treatise* 3: 321–91. (Eds Cohen, N. & Sigel, M.M.). Plenum Press, New York.

Faye, I. & Wyatt, G.R. (1980). The synthesis of antibacterial proteins in isolated fat body from cecropia silkmoth pupae. *Experientia* **36**: 1325–6.

Foley, D.A. (1978). Innate cellular defense by mosquito hemocytes. *Comp. pathobiol.* **4**: 113–44.

Gagen, S.J. & Ratcliffe, N.A. (1976). Studies on the *in vivo* cellular reactions and fate of injected bacteria in *Galleria mellonella* and *Pieris brassicae* larvae. *J. Invert. Path.* **28**: 17–24.

Gellissen, G. (1983). Lipophorin as the plasma coagulogen in *Locusta migratoria. Naturwissenschaften* **70**: 45–6.

Götz, P. (1969). Die Einkapselung von Parasiten in der Haemolymphe von *Chironomus* Larven (Diptera). *Zool. Anz.* (Suppl.) **33**: 610–7.

Götz, P. & Boman, H.G. (1985). Insect immunity. In *Comprehensive insect physiology, biochemistry and pharmacology* 3: 453–85. (Eds Kerkut, G.A. & Gilbert, L.J.). Pergamon Press, Oxford.

Götz, P., Roettgen, I. & Lingg, W. (1977). Encapsulement humoral en tant que réaction de défense chez les Diptères. *Annls Parasit. hum. comp.* **52**: 95–7.

Götz, P. & Vey, A. (1974). Humoral encapsulation in Diptera (Insecta): defence reactions of *Chironomus* larvae against fungi. *Parasitology* **68**: 193–205.

Grégoire, C. (1970). Haemolymph coagulation in arthropods. *Symp zool. Soc. Lond.* No. 27: 45–74.

Gupta, A.P. (Ed.) (1979). *Insect hemocytes: Development, forms, functions, and techniques.* Cambridge University Press, Cambridge, London etc.

Hoffmann, D., Brehélin, M. & Hoffmann, J.A. (1974). Modifications of the hemogram and of the hemocytopoietic tissue of male adults of *Locusta migratoria* (Orthoptera) after injection of *Bacillus thuringiensis. J. Invert. Path.* **24**: 238–47.

Hoffmann, J.A., Porte, A. & Joly, P. (1970). Sur la localisation d'une activité phénoloxydasique dans les coagulocytes de *Locusta migratoria* L. (Orthoptère). *C.r. hebd. Séanc. Acad. Sci., Paris*: **270D**: 629–31.

Hollande, C. (1922). La cellule péricardiale des insectes (cytologie, histochimie, rôle physiologique). *Archs Anat. microsc.* **18**: 85–307.

Huff, C.G. (1934). Comparative studies on susceptible and insusceptible *Culex pipiens* in relation to infections with *Plasmodium cathemerium* and *P. relictum. Am. J. Hyg.* **19**: 123–47.

Ibrahim, E.A.R., Ingram, G.A. & Molyneux, D.H. (1984). Haemagglutinins and parasite agglutinins in haemolymph and gut of *Glossina. Tropenmed. Parasit.* **35**: 151–6.

Jones, J.C. (1962). Current concepts concerning insect hemocytes. *Am. Zool.* **2**: 209–46.

Jones, J.C. (1970). Hemocytopoiesis in insects. In *Regulation of hematopoiesis*: 7–65. (Ed. Gordon, A.S.). Appleton, New York.

Kaaya, G.P. & Ratcliffe, N.A. (1982). Comparative study of hemocytes and associated cells of some medically important dipterans. *J. Morph.* **173**: 351–65.

Kaaya, G.P., Ratcliffe, N.A. & Alemu, P. (1986). Cellular and humoral defenses of

Glossina: reactions against bacteria, trypanosomes and experimental implants. *J. med. Ent. Honolulu* **23**: 30–43.

Kawanishi, C.Y., Splittstoesser, C.M. & Tashiro, H. (1978). Infection of the European chafer, *Amphimallon majalis*, by *Bacillus popilliae*: II. Ultrastructure. *J. Invert. Path.* **31**: 91–102.

Komano, H. & Natori, S. (1985). Participation of *Sarcophaga peregrina* humoral lectin in the lysis of sheep red blood cells injected into the abdominal cavity of larvae. *Devl comp. Immunol.* **9**: 31–40.

Komano, H., Nozawa, R., Mizuno, D. & Natori, S. (1983). Measurement of *Sarcophaga peregrina* lectin under various physiological conditions by radioimmunoassay. *J. biol. Chem.* **258**: 2143–7.

Kubo, T., Komano. H., Okada, M. & Natori, S. (1984). Identification of haemagglutinating protein and bactericidal activity in the haemolymph of adult *Sarcophaga peregrina* on injury of the body wall. *Devl comp. Immunol.* **8**: 283–91.

Lackie, A.M. (1980). Invertebrate immunity. *Parasitology* **80**: 393–412.

Lackie, A.M. (1981a). Immune recognition in insects. *Devl comp. Immunol.* **5**: 191–204.

Lackie, A.M. (1981b). The specificity of the serum agglutinins of *Periplaneta americana* and *Schistocerca gregaria* and its relationship to the insects' immune response. *J. Insect Physiol.* **27**: 139–43.

Lackie, A.M. (1983). Effect of substratum wettability and charge on adhesion *in vitro* and encapsulation *in vivo* by insect haemocytes. *J. Cell Sci.* **63**: 181–90.

Lackie, A.M., Takle, G.B. & Tetley, L. (1985). Haemocytic encapsulation in the locust *Schistocerca gregaria* (Orthoptera) and in the cockroach *Periplaneta americana* (Dictyoptera). *Cell Tissue Res.* **240**: 343–51.

Leonard, C., Ratcliffe, N.A. & Rowley, A.F. (1985). The role of prophenoloxidase activation in non-self recognition and phagocytosis by insect blood cells. *J. Insect Physiol.* **31**: 789–99.

Leonard, C., Söderhäll, K. & Ratcliffe, N.A. (1985). Studies on prophenoloxidase and protease activity of *Blaberus craniifer* haemocytes. *Insect Biochem.* **15**: 803–10.

Liang, S.-M., Sakmar, T.P. & Liu, T.-Y. (1980). Studies on *Limulus* amoebocyte lysate. III. Purification of an endotoxin-binding protein from *Limulus* amoebocyte membranes. *J. biol. Chem.* **255**: 5586–90.

Mohrig, W., Schittek, D. & Hanschke, R. (1979a). Immunological activation of phagocytic cells in *Galleria mellonella*. *J. Invert. Path.* **34**: 84–7.

Mohrig, W., Schittek, D. & Hanschke, R. (1979b). Investigations on cellular defense reactions with *Galleria mellonella* against *Bacillus thuringiensis*. *J. Invert. Path.* **34**: 207–12.

Nappi, A.J. (1975). Parasite encapsulation in insects. In *Invertebrate immunity*: 293–326. (Eds Maramorosch, K. & Shope, R.E.). Academic Press, New York.

Nappi, A.J. & Silvers, M. (1984). Cell surface changes associated with cellular immune reactions in *Drosophila*. *Science, N.Y.* **225**: 1166–8.

Nappi, A.J. & Stoffolano, J.G., Jr. (1972). Distribution of haemocytes in larvae of *Musca domestica* and *Musca autumnalis* and possible chemotaxis during parasitization. *J. Insect Physiol.* **18**: 169–79.

Neuwirth, M. (1974). Granular hemocytes, the main phagocytic blood cells in *Calpodes ethlius* (Lepidoptera, Hesperiidae). *Can. J. Zool.* **52**: 783–4.

Pereira, M.E.A., Andrade, A.F.B. & Ribeiro, J.M.C. (1981). Lectins of distinct specificity in *Rhodnius prolixus* interact selectively with *Trypanosoma cruzi*. *Science, N.Y.* **211**: 597–600.

Peters, W., Kolb, H. & Kolb-Bachofen, V. (1983). Evidence for a sugar receptor (lectin) in the peritrophic membrane of the blowfly larva, *Calliphora erythrocephala* Mg. (Diptera). *J. Insect Physiol.* **29**: 275–80.

Price, C.D. & Ratcliffe, N.A. (1974). A reappraisal of insect haemocyte classification by the examination of blood from fifteen insect orders. *Z. Zellforsch. mikrosk. Anat.* **147**: 537–49.

Ratcliffe, N.A. (1982). Cellular defence reactions of insects. *Fortschr. Zool.* **27**: 223–44.

Ratcliffe, N.A. & Gagen, S.J. (1977). Studies on the *in vivo* cellular reactions of insects: an ultrastructural analysis of nodule formation in *Galleria mellonella*. *Tissue Cell* **9**: 73–85.

Ratcliffe, N.A., Leonard, C. & Rowley, A.F. (1984). Phenoloxidase activation, non-self recognition and cell co-operation in insect immunity. *Science, N.Y.* **226**: 557–9.

Ratcliffe, N.A. & Rowley, A.F. (1979). Role of hemocytes in defense against biological agents. In *Insect hemocytes: Development, forms, functions and techniques*: 331–414. (Ed. Gupta, A.P.). Cambridge University Press, Cambridge, London etc.

Ratcliffe, N.A. & Rowley, A.F. (1983). Recognition factors in insect hemolymph. *Devl comp. Immunol.* **7**: 653–6.

Ratcliffe, N.A. & Rowley, A.F. (In press). Insect responses to parasites and other pathogens. In *Immunology, immunoprophylaxis and immunotherapy of parasitic infections*. (Ed. Soulsby, E.J.L.). C.R.C. Press, U.S.A.

Ratcliffe, N.A., Rowley, A.F. Fitzgerald, S.W. & Rhodes, C.P. (1985). Invertebrate immunity—basic concepts and recent advances. *Int. Rev. Cytol.* **97**: 183–350.

Ratcliffe, N.A. & Walters, J.B. (1983). Studies on the *in vivo* cellular reactions of insects: clearance of pathogenic and non-pathogenic bacteria in *Galleria mellonella* larvae. *J. Insect Physiol.* **29**: 407–15.

Ratner, S. & Vinson, S.B. (1983). Phagocytosis and encapsulation: cellular immune responses in arthropods. *Am. Zool.* **23**: 185–94.

Ravindranath, M.H. (1978). The individuality of plasmatocytes and granular hemocytes of arthropods—a review. *Devl comp. Immunol.* **2**: 581–94.

Renwrantz, L., Daniels, J. & Hansen, P.-D. (1985). Lectin-binding to hemocytes of *Mytilus edulis*. *Devl comp. Immunol.* **9**: 203–10.

Renwrantz, L. & Stahmer, A. (1983). Opsonizing properties of an isolated hemolymph agglutinin and demonstration of lectin-like recognition molecules at the surface of hemocytes from *Mytilus edulis*. *J. comp. Physiol.* **149B**: 535–46.

Rizki, T.M. (1978). The circulatory system and associated cells and tissues. In *The genetics and biology of Drosophila* **2b**: 397–452. (Eds Ashburner, M. & Wright, T.R.F.). Academic Press, London.

Rizki, T.M. & Rizki, R.M. (1983). Blood cell surface changes in *Drosophila* mutants with melanotic tumours. *Science, N.Y.* **220**: 73–5.

Roch, P. & Valembois, P. (1978). Evidence for concanavalin A-receptors and their redistribution on lumbricid leukocytes. *Devl comp. Immunol.* **2**: 51–64.

Rostam-Abadi, H. & Pistole, T.G. (1982). Lipopolysaccharide-binding lectin from the horseshoe crab, *Limulus polyphemus*, with specificity for 2-keto-3-deoxyoctonate (KDO). *Devl comp. Immunol.* **6**: 209–18.

Rowley, A.F. & Ratcliffe, N.A. (1981). Insects. In *Invertebrate blood cells* **2**: 421–88. (Eds Ratcliffe, N.A. & Rowley, A.F.). Academic Press, London.

Salt, G. (1970). *The cellular defence reactions of insects.* Cambridge University Press, Cambridge. (*Cambridge monographs in experimental biology* No. 16).

Schlumpberger, J.M., Weissman, I.L. & Scofield, V.L. (1984). Separation and labeling of specific subpopulations of *Botryllus* blood cells. *J. exp. Zool.* **229**: 401–11.

Schmit, A.R. & Ratcliffe, N.A. (1977). The encapsulation of foreign tissue implants in *Galleria mellonella* larvae. *J. Insect Physiol.* **23**: 175–84.

Schmit, A.R., Rowley, A.F. & Ratcliffe, N.A. (1977). The role of *Galleria mellonella* hemocytes in melanin formation. *J. Invert. Path.* **29**: 232–4.

Schoenberg, D.A. & Cheng, T.C. (1980). Lectin-binding specificities of hemocytes from two strains of *Biomphalaria glabrata* as determined by microhemadsorption assays. *Devl comp. Immunol.* **4**: 617–28.

Smith, V.J. & Söderhäll, K. (1983a). *β*-1, 3-glucan activation of crustacean hemocytes *in vitro* and *in vivo*. *Biol. Bull. mar. biol. Lab. Woods Hole* **164**: 299–314.

Smith, V.J. & Söderhäll, K. (1983b). Induction of degranulation and lysis of haemocytes in the freshwater crayfish, *Astacus astacus*, by components of the prophennoloxidase activating system *in vitro*. *Cell Tissue Res.* **233**: 295–303.

Söderhäll, K. (1981). Fungal cell wall *β*-1, 3-glucans induce clotting and phenoloxidase attachment to foreign surfaces of crayfish hemocyte lysate. *Devl comp. Immunol.* **5**: 565–73.

Söderhäll, K. (1982). Prophenoloxidase activating system and melanization—a recognition mechanism of arthropods? A review. *Devl comp. Immunol.* **6**: 601–11.

Söderhäll, K. & Smith, V.J. (1984). The prophenoloxidase activating system—a complement-like pathway in arthropods. In *Infection processes of fungi*: 1–6. (Eds Aist, J. & Roberts, D.W.). Rockefeller Foundation Press, New York.

Steinhaus, E.A. (1958). Stress as a factor in insect disease. *Int. Congr. Ent.* **10**(4): 725–30.

Takle, G. & Lackie, A.M. (1985). Surface charge of insect haemocytes, examined using cell electrophoresis and cationized ferritin-binding. *J. Cell Sci.* **75**: 207–14.

Tobie, E.J. (1968). Fate of some culture flagellates in the hemocoel of *Rhodnius prolixus*. *J. Parasit.* **54**: 1040–6.

Walters, J.B. & Ratcliffe, N.A. (1981). A comparison of the immune response of the waxmoth, *Galleria mellonella*, to pathogenic and non-pathogenic bacteria. In *Aspects of developmental and comparative immunology* **1**: 147–52. (Ed. Solomon, J.B.). Pergamon Press, Oxford.

Walters, J.B. & Ratcliffe, N.A. (1983). Studies on the *in vivo* cellular reactions of insects: fate of pathogenic and non-pathogenic bacteria in *Galleria mellonella* nodules. *J. Insect Physiol.* **29**: 417–24.

Weathersby, A.B. & McCall, J.W. (1968). The development of *Plasmodium gallina-*

ceum Brumpt in the hemocoels of refractory *Culex pipiens pipiens* Linn. and susceptible *Aedes aegypti* (Linn.). *J. Parasit.* **54**: 1017–22.

Whitcomb, R.F., Shapiro, M. & Granados, R.R. (1974). Insect defense mechanisms against microorganisms and parasitoids. In *The physiology of Insecta* **5**: 447–536. (Ed. Rockstein, M.). Academic Press, New York & London.

Wigglesworth, V.B. (1965). *The principles of insect physiology.* 6th edn. Methuen, London.

Wigglesworth, V.B. (1970). The pericardial cells of insects: analogue of the reticuloendothelial system. *J. Reticuloendothel. Soc.* **7**: 208–16.

Wigglesworth, V.B. (1973). Haemocytes and basement membrane formation in *Rhodnius. J. Insect Physiol.* **19**: 831–44.

Yeaton, R.W. (1980). *Lectins of a North American silkmoth* (Hyalophora cecropia): *their molecular characterization and developmental biology.* Ph.D. Diss.: University of Pennsylvania.

Yoshino, T.P. & Davis, C.D. (1983). Surface antigens of *Biomphalaria glabrata.* (Gastropoda) hemocytes: evidence for linkage-independence of some hemolymph-like surface antigens and con A receptor-bearing molecules. *J. Invert. Path.* **42**: 8–16.

Yoshino, T.P., & Granath, W.O. Jr., (1983). Identification of antigenically distinct hemocyte subpopulations in *Biomphalaria glabrata* (Gastropoda) using monoclonal antibodies to surface membrane markers. *Cell Tissue Res.* **232**: 553–64.

Yoshino, T.P. & Granath, W.O., Jr. (1985). Surface antigens of *Biomphalaria glabrata* (Gastropoda) hemocytes: functional heterogeneity in cell subpopulations recognized by a monoclonal antibody. *J. Invert. Path.* **45**: 174–86.

Zachary, D. & Hoffmann, J.A. (1973). The haemocytes of *Callipora erythrocephala* (Meig.) (Diptera). *Z. Zellforsch mikrosk. Anat.* **141**: 55–73.

Zachary, D. & Hoffmann, D. (1984). Lysozyme is stored in the granules of certain haemocyte types in *Locusta. J. Insect Physiol.* **30**: 405–11.

Symp. zool. Soc. Lond. (1986) No. 56 45–58

Antibacterial immune proteins in insects

H. G. BOMAN

Department of Microbiology,
University of Stockholm,
S–106 91 Stockholm, Sweden

Synopsis

Many insects respond to a bacterial infection by the production of a potent antibacterial activity. This phenomenon is easy to study in diapausing pupae of the cecropia moth (*Hyalophora cecropia*) which respond to injected bacteria by a selective synthesis of immune RNA and some 15–20 immune proteins. From cecropia haemolymph we have purified two new classes of antibacterial proteins called cecropins and attacins and also a lysozyme. The three main cecropins have molecular weights around 4,000 and they have all been sequenced. Cecropin A and B have been synthesized by the solid-phase method. The attacins have molecular weights around 20,000 and the F-form has been sequenced. The lysozyme has also been sequenced. Immune RNA was used for the preparation of a cDNA bank and clones have been isolated corresponding to the precursor of cecropin B, the lysozyme and the two main forms of attacin. DNA sequencing revealed corresponding primary structures. Acidic electrophoresis in combination with an antibacterial assay has been used to demonstrate cecropin- and attacin-like proteins in some other insects such as tsetse flies.

Introduction

Humoral immunity can be induced in insects by an injection of either live, non-pathogenic bacteria or heat-killed pathogens (for a review see Götz & Boman 1985). This phenomenon can be analysed at the molecular level using pupae of the cecropia moth, *Hyalophora cecropia*. When a diapausing pupa is immunized it turns on predominantly the genes for immunity while the rest of the animal remains in a dormant state. Immunized pupae of cecropia are therefore a system for biological enrichment of the RNA and the proteins which are synthesized from the genes for immunity. We have taken advantage of this fact both in the purification of 15 different immune proteins and in the isolation of immune RNA, later to be used for the preparation of a cDNA bank. After a short period of RNA synthesis the insects respond to live bacteria by the production of a potent antibacterial activity which is due to the synthesis of 15–20 immune proteins. To this group of proteins belong a lysozyme and two novel classes of antibacterial proteins, the cecropins and the attacins. The three principal cecropins, A, B, and D,

ZOOLOGICAL SYMPOSIUM No. 56
ISBN 0-19-854004-3

are small basic proteins (molecular weight around 4,000 Daltons) with a comparatively long hydrophobic region. The attacins are larger (molecular weight around 20,000 Daltons) and there are two main forms, basic and acidic or neutral.

During the last five years parallel sequence work was carried out on the protein and on the DNA levels. This programme has so far produced the complete amino acid sequences for five cecropins, one lysozyme and one attacin. In addition we have obtained cDNA sequences corresponding to one lysozyme, one cecropin and the two major forms of attacin. This chapter is a summary of our structural work supplemented with data concerning the antibacterial activities.

Lysozyme

The first antibacterial factor to be identified in insect haemolymph was lysozyme and it has been claimed that this enzyme is the main antibacterial factor responsible for immunity of vaccinated insects (Mohrig & Messner 1968). We now know that this is not correct because insects can eliminate many lysozyme-resistant bacteria (Boman 1982). The lysozymes from *Galleria mellonella* and *Bombyx mori* were purified by Powning & Davidson (1973) and the *Galleria* enzyme has continued to interest other investigators (Jarosz 1979; Jollès, Schoentgen, Croizier, Croizier & Jollès 1979). The cecropia lysozyme was isolated in connection with our purification of cecropin A and B (Hultmark, Steiner, Rasmuson & Boman 1980). The complete amino acid sequence of the enzyme was recently worked out and a cDNA clone containing the lysozyme information was isolated and sequenced (Fig. 1; Å. Engström, Xanthopoulos, Boman & Bennich 1985). Cecropia lysozyme is composed of 120 amino acids, has a molecular weight of 13.8 Kilodaltons and shows great similarity with vertebrate lysozymes of the chicken type. The amino acid residues responsible for the catalytic activity and for the binding of substrate are essentially conserved. When the sequence translated from the cDNA clone was compared to the amino acid sequence there was an almost complete agreement. However, at two positions, 15 and 66, there were differences which were resolved as follows.

In the beginning of our purification work on the cecropia lysozyme we obtained only one peak. As time went on and our separation methods improved we usually obtained two peaks. Sequence studies on these two lysozymes revealed one component with Arg-15 and one with Leu-15. Furthermore, in the sequence work lysozyme with Leu-15 always gave two signals at position 66, one for Ser and one for Thr. It is therefore likely that there are in the cecropia population three allelic variants with the following alternatives: (i) Arg-15, Ser-66; (ii) Leu-15, Ser-66; and (iii) Leu-15, Thr-66. These three variants can all be derived from each other by point mutations

```
                                25
TGC CGT TCG TGG CAG TTC GCT TTG CAT TGC GAT GCG AAA CGT TTC
Cys Arg Ser Trp Gln Phe Ala Leu His Cys Asp Ala|Lys Arg Phe
--------- -10 ------- Leader peptide --------- -1|  1

                                    75
ACG AGA TGC GGG TTA GTG CAG GAG CTT AGG AGA CGA GGC TTC GAT
Thr Arg Cys Gly Leu Val Gln Glu Leu Arg Arg Arg Gly Phe Asp
                        10

        100
GAA ACT TTG ATG AGT AAC TGG GTC TGC CTT GTC GAG AAC GAA AGC
Glu Thr Leu Met Ser Asn Trp Val Cys Leu Val Glu Asn Glu Ser
    20                                      30      ---

                150
GGA CGG TTT ACC GAT AAA ATC GGT AAA GTT AAC AAG AAC GGA TCT
Gly Arg Phe Thr Asp Lys Ile Gly Lys Val Asn Lys Asn Gly Ser
                        40

                                    |------ Probe ------|
                    200
CGA GAC TAC GGC CTC TTC CAG ATC AAT GAC AAA TAC TGG TGC AGT
Arg Asp Tyr Gly Leu Phe Gln Ile Asn Asp Lys Tyr Trp Cys Ser
    ---
    50                                      60

                            250
AAG GGA TCC ACT CCT GGA AAG GAT TGC AAC GTG ACT TGT AAT CAG
Lys Gly Ser Thr Pro Gly Lys Asp Cys Asn Val Thr Cys Asn Gln
                        70

                                    300
CTA CTG ACT GAC GAC ATT AGC GTG GCA GCT ACG TGC GCG AAG AAG
Leu Leu Thr Asp Asp Ile Ser Val Ala Ala Thr Cys Ala Lys Lys
    80                                      90

                                        350
ATT TAC AAA CGC CAC AAG TTT GAC GCT TGG TAC GGA TGG AAA AAT
Ile Tyr Lys Arg His Lys Phe Asp Ala Trp Tyr Gly Trp Lys Asn
                        100

                                                    400
CAC TGT CAA CAT GGA CTG CCA GAT ATT AGC GAC TGT TAG  AGACGA
His Cys Gln His Gly Leu Pro Asp Ile Ser Asp Cys Stop
    110                                     120

CTTATTATAGCCTTC
```

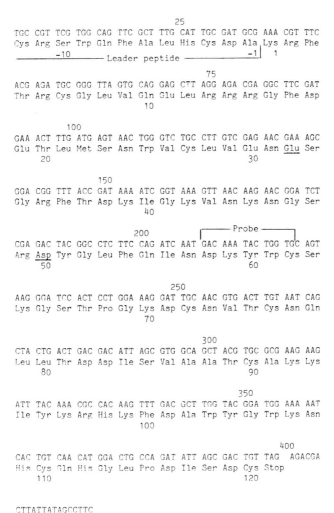

Fig. 1. Nucleotide and deduced amino acid sequences of the cecropia lysozyme. Part of the leader peptide is indicated by a line under the respective sequence. The principal structure of the oligonucleotide probe (without the alternative wobble bases) is indicated by a line above the respective sequence. The active sites of the enzyme, residues Glu-32 and Asp-50, are underlined.

because Arg-Leu and Ser-Thr replacements can be obtained by transversion of a single base (Å. Engström, Xanthopoulos *et al.* 1985).

Lysozyme is bactericidal to only some Gram-positive bacteria like *Bacillus megaterium* and *Micrococcus luteus*. However, cecropin A and B also destroy these two bacteria and no bacterium has yet been found to be lysozyme-sensitive and cecropin-resistant. Thus, the main function of the lysozyme may not be to kill sensitive bacteria but to remove the murein sacculus which is left after the action of cecropins and attacins.

Three main cecropins and one cDNA clone

Cecropins were discovered in 1979 when we succeeded in separating them from the cecropia lysozyme. Once separated, cecropin A and B were isolated simultaneously with the lysozyme (Hultmark, Steiner *et al.* 1980). Two years later we found cecropin D as well as three minor forms believed to be precursors (Hultmark, Engström, Bennich, Kapur & Boman 1982). In collaboration with the Shanghai Institute of Biochemistry, we have also isolated cecropin D and B from the Chinese oak silk moth, *Antheraea pernyi* (Qu, Steiner, Engström, Bennich & Boman 1982). Figure 2 shows the structure of these five cecropins with the three sequences from *Hyalophora cecropia* (H.c.) in the middle flanked by the corresponding two forms of the Chinese oak silk moth, *Antheraea pernyi* (A.p.). It is clear that all cecropins are in principle similar with a strongly basic N-terminal region and a long hydrophobic stretch in the C-terminal half. The high degree of homology shown by the five cecropins (boxed-in residues in Fig. 2) also suggests that they have evolved through gene duplications.

The antibacterial activity of the three main cecropins A, B and D, was assayed against nine different bacterial species (Hultmark, Engström, Bennich *et al.* 1982). Table 1 shows that cecropin A and B were highly active against several Gram-positive and Gram-negative bacteria while the D-form only showed a high activity against *E. coli.* Cecropin B was the best agent against *P. aeruginosa*, *X. nematophilus*, *B. subtilis* and *M. luteus*. Several of the bacteria tested, like *S. marcescens*, *P. aeruginosa*, and *S. faecalis*, are known to be insect pathogens and it is clear that cecropins A and B provide a good protection against these organisms. *X. nematophilus* and *B. thuringiensis* are both obligate insect pathogens. The former lives in symbiosis with a nematode and helps the nematode to kill insects. The nematode in turn helps the bacteria by making a proteolytic enzyme that selectively degrades the cecropins and the attacins (Götz, Boman & Boman 1981).

It turned out to be difficult to find lysozyme and cecropin clones using messenger selection and immunoprecipitation. With a synthetic oligonucleotide probe three cecropin clones were found of which the two larger were cecropin B clones (v. Hofsten, Faye, Kockum, Lee, Xanthopoulos, Boman, Boman, Engström, Andreu & Merrifield 1985). The sequence data in Fig. 3 brought some new pieces of information. First, clone pCP902 contains a precursor sequence of 26 amino acid residues in the N-terminus which is not present in the mature cecropin B. The signal peptide probably makes up the first 22 of these 26 amino acids, leaving a Pro-containing tetrapeptide before

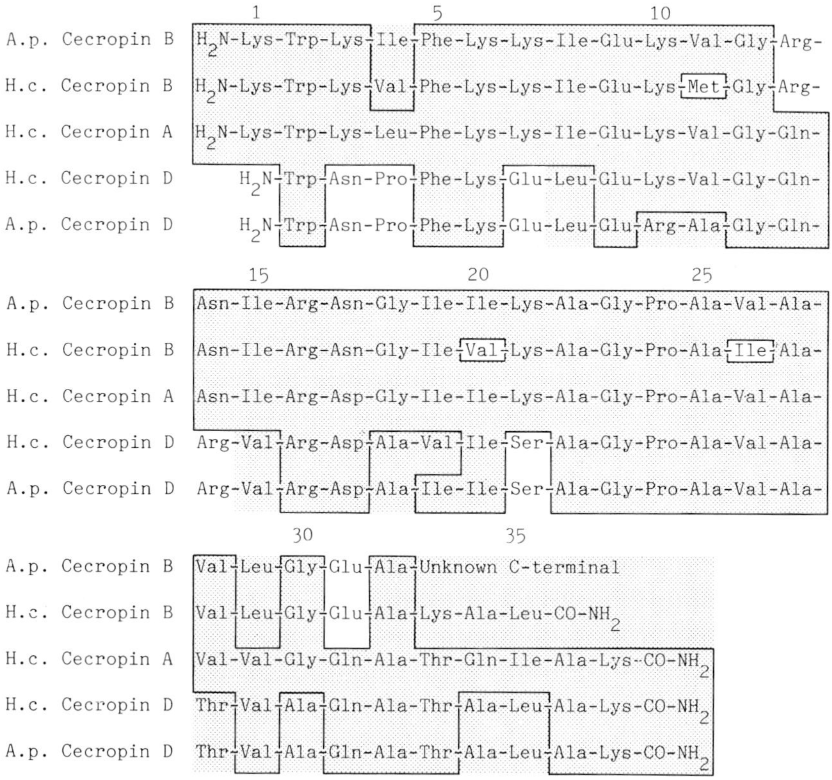

Fig. 2. Sequences for cecropin A, B and D from *Hyalophora cecropia* (H.c) and B and D from *Antheraea pernyi* (A.p.). Identical amino acid residues are boxed in and residues with essentially similar polarities are shaded.

the start of the mature cecropin B. This tetrapeptide, Ala-Pro-Glu-Pro, is identical to the first four amino acid residues of the prosequence of melittin (Kreil, Haiml & Suchanek 1980).

Secondly, the amide group in the C-terminus of the mature cecropin B must be derived from the Gly residue which terminates the coding part of the DNA-sequence. The mechanism of amidation is thus analogous to the one found for melittin and several brain hormones. We can therefore conclude that in order to obtain the mature cecropin B, the precursor molecule will have to be processed in two to three steps at both ends.

Solid-phase synthesis of cecropin A and B

A synthetic programme was initiated in order to confirm the structure of the cecropins and then continued with the aim of investigating a possible corre-

Table 1. Antibacterial spectra of the three major cecropins, A, B and D.

Bacterial species	Strain	Lethal conc. (μM) for cecropin		
		A	B	D
Escherichia coli	D21	0.2	0.3	0.4
	D31	0.2	0.3	0.4
Pseudomonas aeruginosa	OT97	4.8	1.9	100
Xenorhabdus nematophilus	Xn21	4.8	1.9	19
Serratia marcescens	Db11	4.2	4.5	14
Micrococcus luteus	Ml11	4.6	1.3	21
Bacillus megaterium	Bm11	0.7	0.4	41
Bacillus subtilis	Bs11	61	18	95
Bacillus thuringiensis	Bt11	80	133	95
Streptococcus faecalis	AD–4	15	7.3	95
	DS16	74	24	88

Thin agar plates were seeded with the respective test bacteria. Small wells were punched in the plates and loaded with a dilution series of each sample. After overnight incubation at 30°C the inhibition zones were recorded and lethal concentrations calculated as described by Hultmark, Engström, Andersson *et al.* (1983).

```
                          25                         50
TGTTTACCTATATATCTAAATTTAATATTTCGTTTATTAAAATTTAATATATTTCGACG ATG AAT TTC
                                                            Met Asn Phe
                                                            |___-25____

      75                              100
TCA AGG ATA TTT TTC TTC GTG TTC GCT TTG GTT CTG GCT TTG TCA ACA GTT TCG
Ser Arg Ile Phe Phe Phe Val Phe Ala Leu Val Leu Ala Leu Ser Thr Val Ser
____________-20_____________ Leader peptide __________-10______________

                                                 Probe
125                             150           |¯¯¯¯¯¯¯¯¯¯¯¯¯¯¯¯| 175
GCT GCA CCG GAG CCG AAA TGG AAA GTC TTC AAG AAA ATT GAA AAA ATG GGT CGC
Ala Ala Pro Glu Pro Lys Trp Lys Val Phe Lys Lys Ile Glu Lys Met Gly Arg
-5_______________-1||_1__________5___________________10________________

                                200                              225
AAC ATT CGA AAC GGT ATT GTC AAG GCT GGA CCA GCG ATC GCG GTT TTA GGC GAA
Asn Ile Arg Asn Gly Ile Val Lys Ala Gly Pro Ala Ile Ala Val Leu Gly Glu
_____15__ Cecropin B ______20__________25___________________30_________

                         250
GCC AAA GCG CTA GGA TAA  ATTAATTTTAATTT
Ala Lys Ala Leu Gly Stop
__________35_ NH2
```

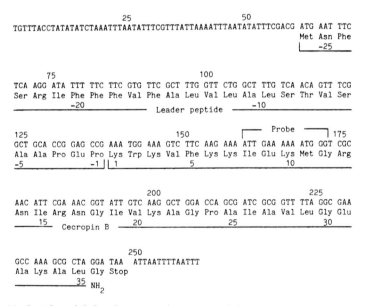

Fig. 3. Nucleotide and deduced amino acid sequences of clones pCP901 and pCP902 coding for the precursor of cecropin B. The structure of the leader peptide and the mature cecropin B molecule is indicated by a line under the respective parts of the amino acid sequence.

lation between the secondary structure and the antibacterial activity. A beginning was made with the synthesis of cecropin A(1–33), at that time thought to be the full cecropin A (Merrifield, Vizioli & Boman 1982). This was followed by the synthesis of the complete cecropin A and some truncated analogues (Andreu, Merrifield, Steiner & Boman 1983). In addition, it was shown that cecropin A has a C-terminus blocked by an amide group. Also cecropin B was synthesized with the structure given in Fig. 3 (v. Hofsten *et al.* 1985). As in the case of cecropin A the natural and the synthetic samples were found to be indistinguishable.

From Fig. 2 it can be seen that in the N-terminal region of the cecropins, polar and hydrophobic side chains are interspaced in a regular pattern. Steiner (1982) and Merrifield *et al.* (1982) observed that this amino acid distribution is likely to produce an amphipathic alpha helix. Such structures have often been implicated in membrane activities and the cecropins are indeed strongly lytic against a variety of bacteria. Cecropins will also lyse artificial liposomes (H. Steiner, unpublished results).

As a continuation of the synthetic programme a series of analogues of cecropin A were synthesized in which residues 2, 6 and 8 were altered in such a way that the nature of the side chain was changed from hydrophobic to hydrophilic or vice versa (Andreu, Merrifield, Steiner & Boman 1985). We also replaced residues 4 or 8 by Pro in order to break the alpha helix. From this work we could conclude that Trp-2 is an essential residue and that the antibacterial mechanism as such does not require an extended alpha helix. However, the two Pro analogues did not show any activity against *M. luteus*, so the broad-spectrum activity was lost when the alpha helix was disrupted.

Two main forms of attacin

The attacins were first isolated by molecular sieving as an antibacterial fraction with molecular weight considerably larger than the cecropins (Hultmark, Engström, Andersson, Steiner, Bennich & Boman 1983). Subsequent studies revealed as many as six different components (A–F) which could be fractionated according to isoelectric point. To our surprise they all turned out to react with antisera prepared against our immune protein P5 isolated several years earlier (Pye & Boman 1977). At that time we had not been able to find any antibacterial activity of P5, a discrepancy which later could be traced to the type of assay used. An Ouchterlony immunodiffusion plate revealed that all six attacins shared one antigenic determinant, while another one was common only to attacins E and F, the two acidic forms. The N-terminal sequences for five of the attacins indicated that the three basic forms all have similar sequences while the two acidic forms are identical but slightly different from the basic (Hultmark, Engström, Andersson *et al.*1983). These data strongly suggested the existence of only two different genes, one for the basic and one for the neutral or acidic form. This was born nout by the isolation and sequencing of clones pCP517 and pCP521 (Kockum, Faye, v. Hofsten, Lee, Xanthopoulos & Boman 1984). Figure 4 shows that the two main attacins are very similar with as much as 79% homology at the amino acid level. On the DNA level the homology is

76% for the coding region in contrast to only 36% in the region beyond the stop signal. Thus, as in the case of the cecropins, it seems likely that the attacins have arisen through gene duplications.

Å. Engström, Engström, Tao, Carlsson & Bennich (1984) worked out the full amino acid sequence for attacin F. There is a complete agreement between these protein data and amino acid residues 1–184 deduced from pCP521. However, pCP521 codes for an extended protein of 188 amino acid residues (attacin E). The difference corresponds to a tetrapeptide, Ser-Lys-Tyr-Phe, which is also coded for in the clone for the basic attacin. Since this peptide contains one positive charge, an incomplete proteolytic removal could account for four attacins with different charges. It cannot be decided at present if such a processing is the result of an artificial proteolytic cleavage or if it has a natural function.

In the case of pCP517 we obtained 36 amino acid residues of the leader sequence. Since neither the length nor the composition of this region is typical for a signal sequence, we believe attacins to be made from a pre-proform. Similar conclusions had earlier been reached from the size of the proteins produced by *in vitro* translation of immune mRNA (Lee, Edlund, Ny, Faye & Boman 1983). In addition there may be a further trimming also at the N-terminus of the basic attacin because evidence for a pyroglutamate group was found at the N-terminus of the basic attacin (Å. Engström, Engström *et al.* 1984). We therefore believe that the mature protein starts with the Gln residue indicated in Fig. 4. A partial cyclization of this residue would involve a partial loss of one positive charge. Thus, together with the removal of the tetrapeptide at the C-terminus it is possible by different steps of processing to account for all six forms of attacins observed.

The antibacterial spectra of the attacins at first seemed rather narrow with good activity only against *E. coli* and two other types of bacteria originating from the gut of an *Antheraea* larva (Hultmark, Engström, Andersson *et al.* 1983). A study of the mechanism of action on *E. coli* demonstrated that the two main attacins both act on the outer membrane (P. Engström, Carlsson, Engström, Tao & Bennich 1984). In particular it was shown that attacin facilitates the action of cecropin and lysozyme, thereby enabling these three immune proteins to work in consonance. For *E. coli* the lethal concentration of the attacins is around 1 μM while the physiological concentration in an immunized pupa is as high as 50-60 μM (P. Engström *et al.* 1984). It was found that, using this physiological concentration, attacins did act on a variety of bacteria.

How common are cecropins and attacins?

The cecropins and the attacins were first defined by their separation properties, and in particular by the acidic electrophoresis in combination with an

```
GTC TTG GTT GGC GTC AAC AGC CGG TAC GTG CTC GTT GAA GAG CCT GGT TAC TAT GAC AAA CAG TAC GAG GAA
VAL LEU VAL GLY VAL ASN SER ARG TYR VAL LEU VAL GLU GLU PRO GLY TYR TYR ASP LYS GLN TYR GLU GLU
pCP517

                                                 → Basic Attacin
CAG CCG CAG CAG TGG GTC AAC TCC AGA GTA CGT CGG CAA GCG GGT GCT CTC ACT ATC AAC TCT GAC GGT ACC
GLN PRO GLN GLN TRP VAL ASN SER ARG VAL ARG ARG GLN ALA GLY ALA LEU THR ILE ASN SER ASP GLY THR
                                            ASP ALA HIS GLY ALA LEU THR LEU ASN SER ASP GLY THR
                                     pCP521 GAC GCG CAC GGA GCC CTT ACG CTC AAC TCC GAT GGT ACC
                                            → Acidic Attacin                   10

TCA GGT GCT GTG GTC AAG GTA CCT ATA ACT GGG AAT GAA AAC CAC AAG TTC AGT GCT CTT GGC TCC GTT GAT
SER GLY ALA VAL VAL LYS VAL PRO ILE THR GLY ASN GLU ASN HIS LYS PHE SER ALA LEU GLY SER VAL ASP
SER GLY ALA VAL VAL LYS VAL PRO PHE ALA GLY ASN ASP LYS ASN ILE VAL SER ALA ILE GLY SER VAL ASP
TCT GGT GCT GTG GTT AAA GTA CCC TTT GCT GGT AAC GAC AAG AAT ATA GTA AGC GCT ATC GGT TCC GTA GAC
                        20                                  30

CTT ACT AAC CAA ATG AAA TTG GGA GCT GCT ACA GCT GGA TTG GCT TAT GAC AAC GTC AAC GGA CAC GGA GCG
LEU THR ASN GLN MET LYS LEU GLY ALA ALA THR ALA GLY LEU ALA TYR ASP ASN VAL ASN GLY HIS GLY ALA
LEU THR ASP ARG GLN LYS LEU GLY ALA ALA THR ALA GLY VAL ALA LEU ASP ASN ILE ASN GLY HIS GLY LEU
TTA ACT GAT AGG CAG AAA CTA GGC GCT GCA ACC GCT GGA GTG GCA CTG GAT AAT ATA AAC GGT CAC GGA CTA
        40                                  50                                          60

ACC CTA ACA AAA ACT CAT ATC CCT GGG TTC GGT GAC AAG ATG ACG GCT GCC GGC AAA GTA AAT CTC TTC CAT
THR LEU THR LYS THR HIS ILE PRO GLY PHE GLY ASP LYS MET THR ALA ALA GLY LYS VAL ASN LEU PHE HIS
SER LEU THR ASP THR HIS ILE PRO GLY PHE GLY ASP LYS MET THR ALA ALA GLY LYS VAL ASN VAL PHE HIS
AGT CTC ACG GAT ACA CAC ATC CCC GGG TTC GGA GAC AAG ATG ACA GCA GCC GGC AAA GTG AAT GTC TTC CAC
                                70                                  80

AAC GAC AAC CAC GAT TTC AGT GCC AAA GCA TTC GCC ACT AAA AAC ATG CCA AAT ATT CCT CAA GTT CCG AAC
ASN ASP ASN HIS ASP PHE SER ALA LYS ALA PHE ALA THR LYS ASN MET PRO ASN ILE PRO GLN VAL PRO ASN
ASN ASP ASN HIS ASP ILE THR ALA LYS ALA PHE ALA THR ARG ASN MET PRO ASP ILE ALA ASN VAL PRO ASN
AAT GAT AAC CAC GAC ATC ACA GCG AAG GCT TTC GCC ACC AGA AAC ATG CCG GAT ATT GCT AAT GTA CCT AAT
                90                                  100

TTC AAC ACT GTC GGT GCC GGA GTG GAC TAT ATG TTC AAA GAT AAG ATT GGT GCA TCT GCG AAT GCC GCT CAC
PHE ASN THR VAL GLY ALA GLY VAL ASP TYR MET PHE LYS ASP LYS ILE GLY ALA SER ALA ASN ALA ALA HIS
PHE ASN THR VAL GLY GLY GLY ILE ASP TYR MET PHE LYS ASP LYS ILE GLY ALA SER ALA SER ALA ALA HIS
TTC AAC ACT GTC GGT GGC GGA ATA GAC TAT ATG TTC AAA GAT AAG ATT GGT GCA TCT GCG AGC GCC GCT CAC
110                                     120                                     130

ACC GAT TTC ATC AAC CGC AAC GAC TAC TCT CTG GGC GGG AAA CTG AAT CTC TTC AAG ACT CCG ACC ACA TCG
THR ASP PHE ILE ASN ARG ASN ASP TYR SER LEU GLY GLY LYS LEU ASN LEU PHE LYS THR PRO THR THR SER
THR ASP PHE ILE ASN ARG ASN ASP TYR SER LEU ASP GLY LYS LEU ASN LEU PHE LYS THR PRO ASP THR SER
ACG GAC TTT ATC AAT CGC AAC GAC TAC TCT CTT GAC GGG AAA CTG AAC CTC TTC AAG ACT CCT GAT ACC TCG
                        140                                 150

CTG GAC TTC AAC GCC GGT TGG AAG AAG TTC GAT ACG CCC TTC TTT AAG TCC TCG TGG GAA CCC AGC ACT AGT
LEU ASP PHE ASN ALA GLY TRP LYS LYS PHE ASP THR PRO PHE PHE LYS SER SER TRP GLU PRO SER THR SER
ILE ASP PHE ASN ALA GLY PHE LYS LYS PHE ASP THR PRO PHE MET LYS SER SER TRP GLU PRO ASN PHE GLY
ATT GAT TTC AAC GCC GGT TTC AAG AAG TTC GAT ACA CCT TTC ATG AAG TCC TCT TGG GAG CCT AAC TTC GGA
        160                                 170                                         180

TTC TCG TTT TCT AAA TAT TTC
PHE SER PHE SER LYS TYR PHE TAA CTT AAA ACA GTC ATT AAC TTA AAA CGT AAA CTA ATT ATT ACA AAA CTT
PHE SER LEU SER LYS TYR PHE TGA TTA GTA TTT TAA TTT TAA TTC TAT ATA TAT AAA TTT AGA TGT ATA TGT
TTC TCA CTT TCT AAA TAT TTC
                        188

AAT GTT AAG GGA ACC ATT TTT CTT AAC ATA GTT AAT TGT TTG TAA ATA AAA TAT AAA GTT TTA ATA AAT TGT
ATA TAT ATA TAT TTT TTT TTT ATT AAT ATG ATA TCA CTA AAT GTA TTT ACT CCT TCG ATT ATT ATT ACT TTT

AGT TAC AAA TTG
TTT GTT TAA AGA AGT CCG CCT AAT AAA GAT AAT TTG
```

Fig. 4. Nucleotide sequences of the inserts in two attacin clones, pCP517 (uppermost line) and pCP521 (bottom line). The amino acid sequence corresponding to pCP517 is given below the nucleotide sequence, for pCP521 above. The line between the amino acid sequences is broken when the nucleotide sequences differ. Amino acid substitutions and the signals for polyadenylation are boxed in. The numbers indicate amino acid positions for both attacins.

antibacterial assay on top of the gel (Hultmark, Steiner *et al.* 1980). With this technique cecropin-like substances were demonstrated in seven other lepidopteran species (Hoffmann, Hultmark & Boman 1981). However, it was emphasized that ultimately it will be necessary to establish the identity of a cecropin by sequence analysis. This was first done in case of the Chinese oak silk moth (Fig. 2) in which the D-form was the major cecropin (Qu *et al.* 1982). The D-form is also the dominating cecropin in *Manduca sexta* and an incomplete sequence for this protein was reported at a recent meeting (P.E. Dunn, personal communication). Complete sequences for cecropins from Chinese and Japanese strains of *Bombyx mori* have been obtained (X.-M. Qu pers. comm.; Shiba, Ueki, Kubota, Teshima, Sugiyama, Oba & Kikushi 1984). Moreover, Okada & Natori (1985) recently reported the sequence of sarcotoxin I from the flesh fly *Sarcophaga peregrina*. These data clearly show that sarcotoxin I is a cecropin with about 40% homology to cecropin A from cecropia. More preliminary results indicate that antibacterial compounds are found also in *Drosophila* (C. Flyg & G. Dalhammar pers. comm.; M. Robertson & J.H. Postlethwait pers. comm.) as well as in *Locusta* and *Calliphora* (Hoffmann, Zachary, Lambert, Keppi & Hoffmann 1985).

Medically important vectors like mosquitoes and tsetse are comparatively small insects and therefore unsuitable for biochemical studies. Despite this difficulty we recently performed a series of experiments on the response of tsetse to a bacterial infection. So far we have found that an injection of bacteria in tsetse (10^4 viable *E.coli*/fly) produced an antibacterial activity in the haemolymph which lasted from one to four days (G.P. Kaaya, C. Flyg & H. Boman unpubl.). Characterization of this antibacterial activity with acidic electrophoresis showed two antibacterial spots with mobilities corresponding to cecropin A and attacin from cecropia.

Discussion

The cecropia moth has three different cecropins and at least two different attacins. So far we have not been able to document any real differences in function between these multiple antibacterial factors. In general, cecropin B is slightly more potent than the A-form, while cecropin D has a narrower antibacterial spectrum. However, in no case do we have a bacterium on which only a single factor acts. This raises the question of the survival value of multiple forms of very similar molecules. One alternative is that they simply represent proteins 'in the middle' of an evolution towards separate functions. A second possibility is of course that both cecropins and attacins have separate target organisms which we just have not found. A third alternative, which perhaps is the most likely one, is that each of the cecropins and attacins has separate targets on most of the organisms on which they act. If so, this would in itself be of survival value to the insect because it would make it virtually impossible for a susceptible bacterium to produce mutants which are resistant to the humoral immunity of an insect. One line of evidence supporting this alternative was obtained when we isolated mutants of *E. coli* which showed an increased susceptibility to cecropin D but an unaltered response to the A and B forms (Sidén & Boman 1983). In addition it is clear that, during the evolution of cecropia and the Chinese oak silk moth, the only mutations which have been tolerated in cecropin D are strictly conservative replacements (see Fig. 2). This fact also speaks for a high survival value of the intact structure of cecropin D.

The exoskeleton of insects is an effective barrier against invasions of different microbes. Still, infections occur when insects are wounded and when parasites attack. However, the most common route of infection is through the gut. A special type of infection may occur during metamorphosis when the gut wall is dissolved by histolytic enzymes and the intestinal flora are released into the haemolymph. This would require an immune system with the capacity to eliminate a sizeable number of bacteria within a very short time regardless of the type of bacterium that is released. The only demand for specificity would be to avoid self-destruction and this seems to be fulfilled by the combined effects of the cecropins and attacins. The role of the lysozyme would be to dissolve the murein sacculus which is left over as an empty bag after the membranes have been ruined by the detergent-like action of the cecropins.

The insects have avoided the complicated method of self-recognition which is used by the vertebrates. The ideal immune substances for an insect would be the ones which have the widest action against foreign intruders without causing any self-destruction. It is therefore interesting to compare immune substances like cecropins with a toxin such as melittin which in an ideal situation should meet rather similar demands. Both cecropins and melittin are short polypeptides containing a strongly basic part and a long stretch of hydrophobic amino acid residues. However, the polarity is reversed because in the cecropins the N-terminal region is basic while in melittin the C-terminal is positively charged. Both kinds of molecules have alpha amidated C-terminal ends formed from glycine residues. Both proteins are also made as precursor molecules, presumably in order to minimize self-destruction that could be caused by very high concentrations at the site of synthesis. Even so, the mature molecules differ in their specificity: both will lyse bacteria but only melittin will lyse eucaryotic cells (Steiner, Hultmark, Engström, Bennich & Boman 1981). Moreover, the antibacterial spectrum of melittin is narrow (Boman 1982) so it could hardly qualify as an immune substance even if it were not toxic to insect cells. It is therefore possible that a synthetic programme exploring the properties of cecropin–

melittin hybrids would yield information concerning the specificity of these molecules.

Acknowledgements

The amino acid sequence work reported here has been possible only through the co-operation of Hans Bennich and Åke Engström in Uppsala and the synthetic work has to the same extent been due to Bruce Merrifield and David Andreu in New York. All work in Stockholm was supported by grants from the Swedish Natural Science Research Council.

References

Andreu, D., Merrifield, R.B., Steiner, H. & Boman, H.G. (1983). Solid-phase synthesis of cecropin A and related peptides. *Proc. natn. Acad. Sci. U.S.A.* **80**: 6475–9.

Andreu, D., Merrifield, R.B., Steiner, H. & Boman, H.G. (1985). N-Terminal analogs of cecropin A: Synthesis, antibacterial activity and conformational properties. *Biochemistry* **24**: 1683–8.

Boman, H.G. (1982). Humoral immunity in insects and the counter defence of some pathogens. *Zentbl. Bakt. ParasitKde Suppl.* **12**: 211–22.

Engström, Å., Engström, P., Tao, Z.-J., Carlsson, A. & Bennich, H. (1984). Insect immunity. The primary structure of the antibacterial protein attacin F and its relation to two native attacins from *Hyalophora cecropia*. *EMBO J.* **3**: 2065–70.

Engström, Å., Xanthopoulos, K.G., Boman, H.G. & Bennich, H. (1985). Amino acid and cDNA sequences of lysozyme from *Hyalophora cecropia*. *EMBO J.* **4**: 2119–22.

Engström, P., Carlsson, A., Engström, Å., Tao, Z.-J. & Bennich, H. (1984). The antibacterial effect of attacins from the silk moth *Hyalophora cecropia* is directed against the outer membrane of *Escherichia coli*. *EMBO J.* **3**: 3347–51.

Götz, P., Boman, A. & Boman, H.G. (1981). Interactions between insect immunity and an insect pathogenic nematode with symbiotic bacteria. *Proc. R. Soc. Lond.* (B) **212**: 333–50.

Götz, P. & Boman, H.G. (1985). Insect immunity. In *Comprehensive insect physiology, biochemistry and pharmacology* **3**: 453–85. (Eds Kerkut, G.A. & Gilbert, L.J). Pergamon Press, Oxford.

Hoffmann, D., Hultmark,D. & Boman, H.G (1981). Insect immunity: *Galleria mellonella* and other Lepidoptera have Cecropia-P9-like factors active against gram negative bacteria. *Insect Biochem.* **11**: 537–48.

Hoffmann, D., Zachary, D., Lambert, J., Keppi, E. & Hoffmann, J. (1985). Humoral antibacterial reactions in insects: a comparison between two models, *Locusta migratoria* (Orthoptera) and *Phormia terranovae* (Diptera). *Devl comp. Immunol.* **9**: 169.

v. Hofsten, P., Faye, I., Kockum, K., Lee, J.-Y., Xanthopoulos, K.G., Boman, I.A., Boman, H,G., Engström, Å., Andreu, D. & Merrifield, R.B. (1985). Molecular cloning, cDNA sequencing and chemical synthesis of cecropin B from *Hyalophora cecropia*. *Proc. natn. Acad. Sci. U.S.A.* **82**: 2240–3.

Hultmark, D., Engström, Å., Andersson, K., Steiner, H., Bennich, H. & Boman, H.G. (1983). Insect immunity. Attacins, a family of antibacterial proteins from *Hyalophora cecropia*. *EMBO J.* **2**: 571–6.

Hultmark, D., Engström, Å., Bennich, H., Kapur, R. & Boman, H.G. (1982). Insect immunity. Isolation and structure of cecropin D and four minor antibacterial components from cecropia (*Hyalophora cecropia*) pupae. *Eur. J. Biochem.* **127**: 207–17.

Hultmark, D., Steiner, H., Rasmuson, T. & Boman, H.G. (1980). Insect immunity. Purification and properties of three inducible bactericidal proteins from hemolymph of immunized pupae of *Hyalophora cecropia*. *Eur. J. Biochem.* **106**: 7–16.

Jarosz, J. (1979). Simultaneous induction of protective immunity and selective synthesis of hemolymph lysozyme protein in larvae of *Galleria mellonella*. *Biol. Zbl.* **98**: 459–71.

Jollès, J., Schoentgen, F., Croizier, G., Croizier, L. & Jollès, P. (1979). Insect lysozymes from three species of Lepidoptera: Their structural relatedness to the C (chicken) type lysozyme. *J. molec. Evol.* **14**: 267–71.

Kockum, K., Faye, I., v. Hofsten, P., Lee, J.-Y., Xanthopoulos, K.G., & Boman, H.G. (1984). Insect immunity. Isolation and sequence of two cDNA clones corresponding to acidic and basic attacins from *Hyalophora cecropia*. *EMBO J.* **3**: 2071–5.

Kreil, G., Haiml, L. & Suchanek, G. (1980). Stepwise cleavage of the pro part of promelittin by dipeptidylpeptidase IV. Evidence for a new type of precursor–product conversion. *Eur. J. Biochem.* **111**: 49–58.

Lee, J.-Y., Edlund, T., Ny, T., Faye, I. & Boman, H.G. (1983). Insect immunity. Isolation of cDNA clones corresponding to attacins and immune protein P4 from *Hyalophora cecropia*. *EMBO J.* **2**: 577–81.

Merrifield, R.B., Vizioli, L.D. & Boman, H.G. (1982). Synthesis of the antibacterial peptide cecropin A (1–33). *Biochemistry* **21**: 5020–31.

Mohrig, W. & Messner, B. (1968). Immunreaktionen bei Insekten. I. Lysozym als grundlegender antibakterieller Faktor im humoralen Abwehrmechanismus der Insekten. *Biol. Zbl.* **87**: 439–70.

Okada, M. & Natori, S. (1985). Primary structure of sarcotoxin I, an antibacterial protein induced in the hemolymph of *Sarcophaga peregrina* (flesh-fly) larvae. *J. biol. Chem.* **260**: 7174–7.

Powning, R.F. & Davidson, W.J. (1973). Studies on insect bacteriolytic enzymes. I. Lysozyme in haemolymph of *Galleria mellonella* and *Bombyx mori*. *Comp. Biochem. Physiol.* **45B**: 669–81.

Pye, A.E. & Boman, H.G. (1977). Insect immunity. III. Purification and partial characterization of immune protein P5 from haemolymph of *Hyalophora cecropia* pupae. *Infect. Immun.* **17**: 408–14.

Qu, X.-M., Steiner, H., Engström, Å., Bennich, H. & Boman, H.G. (1982). Insect immunity. Isolation and structure of cecropin B and cecropin D from pupae of the Chinese oak silk moth, *Antheraea pernyi*. *Eur. J. Biochem.* **127**: 219–24.

Shiba, T., Ueki, Y., Kubota, I., Teshima, T., Sugiyama, Y., Oba, Y. & Kikushi, M.

(1984). Structure of Lepidoperan, a self-defence substance produced by silkworm. In *Peptide chemistry 1983*: 209–14. (Ed. Munekada, E.). Protein Research Foundation, Osaka.

Sidén, I. & Boman, H.G. (1983). *Escherichia coli* mutants with an altered sensitivity to cecropin D. *J. Bacteriol.* **154**: 170–6.

Steiner, H.(1982). Secondary structure of the cecropins: Antibacterial peptides from the moth *Hyalophora cecropia. FEBS Lett.* **137**: 283–7.

Steiner, H., Hultmark, D., Engström, Å., Bennich, H. & Boman, H.G. (1981). Sequence and specificity of two antibacterial proteins involved in insect immunity. *Nature, Lond.* **292**: 246–8.

Symp. zool. Soc. Lond. (1986) No. 56 59–79

Cellular immune mechanisms in the Crustacea

Valerie J. SMITH
University Marine Biological Station, Millport, Isle of Cumbrae, Scotland, KA28 0EG

and Kenneth SÖDERHÄLL
Institute of Physiological Botany, University of Uppsala, Box 540, S751–21, Uppsala, Sweden.

Synopsis

Immunity in crustaceans rests largely with the phagocytic, encapsulating, agglutinating and lytic activities of the circulating blood cells (haemocytes). The mechanism(s) controlling these responses and the way(s) in which crustaceans discriminate self from non-self have been the subject of considerable research. Now evidence is accumulating in favour of the prophenoloxidase activating system as a complement-like pathway in arthropods. This not only mediates non-self recognition but also stimulates haemocyte activation and the subsequent cellular defences. The prophenoloxidase (proPO) system is a complex cascade of serine proteases and other factors that are specifically activated by endotoxin (LPS) or β 1,3-glucans. The cascade proteins reside principally in the granular haemocytes but small amounts are also present in the semigranular cells and release of the enzymes from the cells is achieved by exocytosis. In host defence, the proPO activating system provides the opsonins necessary for efficient phagocytosis and/or encapsulation of foreign entities, and participates in agglutination and clotting. Parasites or pathogens which successfully exploit crustaceans (or other arthropods) as hosts may inhibit or interfere with the activation or functioning of the proPO system within the haemolymph, and the ways this might be achieved are discussed.

Introduction

In their marine or freshwater habitats, crustaceans live in an environment rich in potentially harmful pathogens or parasites. Whilst some protection against microbial attack is afforded by the possession of a hard impenetrable carapace, there is still a need for an efficient internal immune network to deal with any opportunistic or pathogenic micro-organisms which may gain access to the haemocoel through wounds or during the moult. In common with other arthropods, host defence in crustaceans is achieved principally through the phagocytic, encapsulating, agglutinating,

ZOOLOGICAL SYMPOSIUM No. 56
ISBN 0-19-854004-3

microbicidal and clotting activities of the circulating blood cells. In some cases, however, micro-organisms or metazoans are able to evade or overcome the host's defence reactions, with the inevitable consequences of disease or parasitic exploitation. Not only does this present problems in terms of economic loss in aquaculture but it may also facilitate transmission of potentially harmful parasites to secondary hosts through crustacean vectors. Several instances where crustaceans serve as vectors have been described in the literature (see Gordon 1966 or Denny 1969), and examples are provided by the crayfish *Austropotamobius pallipes* hosting the fungus *Fusarium solani* [a root pathogen of soybeans which may also infect man and other vertebrates (O'Day, Akrabawi, Richmond, Jones & Clayton 1979; Vey & Vago 1973)]; the signal crayfish *Pacifastacus leniusculus* harbouring the plague fungus *Aphanomyces astaci* (Unestam 1981) and amphipods (e.g. *Gammarus* spp.) or isopods (e.g. *Armadillidium* spp.) carrying the developmental stages of various acanthocephalan, trematode or cestode helminths before transmission to the final, vertebrate host (Gordon 1966; Denny 1969; Sindermann 1970; Gemmell 1971; Seidenberg 1973; Nickol & Dappen 1982). Precisely how these and other parasites avoid cellular attack is still largely unknown, but it is reasonable to predict that some, at least, interfere with the initial recognition process or inhibit the immunological machinery of the host. As a prerequisite to the understanding of pathogenesis, it is necessary to know the biochemical basis for non-self recognition and haemocyte activation in the host. In this review the cellular and humoral defence strategies of crustaceans are described with special reference to the prophenoloxidase-activating system, and the ways in which this might be disturbed by parasites or pathogens during systemic infection are discussed.

Non-self recognition and prophenoloxidase activation

Central to any active cellular or humoral response to microbial or parasitic invasion is the initial recognition of foreignness by the host. In mammals, this is known to be accomplished through specific antibodies, T lymphocytes and/or the complement pathway. Invertebrates, by contrast, do not express immunoglobulin and do not have cells analogous to mammalian lymphocytes. However, a number of workers, notably Day, Gewurz, Johanssen, Finstad & Good (1970), and Bertheussen (1984), have suggested that invertebrates might possess molecules akin to complement, and evidence is now accumulating that, for arthropods, at least, the prophenoloxidase activating system constitutes such a complement-like pathway in host immunity (Söderhäll 1982; Söderhäll & Smith 1984, in press).

In crustaceans, the prophenoloxidase activating system comprises a complex cascade of serine proteases and other factors in the haemocytes that are specifically triggered by foreign molecules (Fig. 1) (Unestam & Söderhäll 1977; Söderhäll & Unestam 1979; Söderhäll & Häll 1984; Smith & Söderhäll 1983a; Ashida & Söderhäll 1984). The cascade reactions terminate in the conversion of the proenzyme to active phenoloxidase—the key enzyme in melanin synthesis (see review by Söderhäll & Smith in press), and phenoloxidase activity in haemocyte lysate supernatants (HLS) may be measured spectrophotometrically using L-dopa as substrate (Söderhäll & Unestam 1979). *In vitro*, prophenoloxidase may also be activated non-physiologically by a variety of agents or treatments including heat, detergents, or organic solvents (see review by Söderhäll & Smith in press), but these are more likely to be significant in laboratory manipulation of the cascade than in host defence.

As far as recognition is concerned, the proPO activating cascade serves as the 'receptor' for the non-self signals borne on, or released from, the surface of micro-organisms or parasites (see reviews by Söderhäll 1982; and Söderhäll & Smith in press), and the biochemical changes that ensue from this interaction then initiate cellular activity and/or exert antimicrobial effects of their own. For Gram negative bacteria and fungi the non-self signals on the cell walls are the lipopolysaccharides (LPS) and β, 1–3 glucans respectively (Unestam & Söderhäll 1977; Söderhäll & Unestam 1979; Söderhäll & Häll 1984), and proPO is specifically activated by these molecules even in concentrations as low as $10^{-9}-10^{-10}$ g ml^{-1} (Söderhäll & Häll 1984). Other carbohydrates, such as trehalose, glucose, mannose, mannitol, chitin, hyaluronic acid or dextran, fail to provoke proPO activation (Söderhäll & Unestam 1979), and pretreatment of LPS with polymyxin B abolishes the stimulatory influence of LPS on proPO (Söderhäll & Häll 1984). Curiously, concentrations of LPS above 10^{-4} g ml^{-1} actually prevent enzyme activation *in vitro* (Söderhäll & Häll 1984), but the significance of this to host-parasite interactions remains obscure. Following induction of the proPO cascade with glucans or LPS, a number of chemical transformations take place (Fig. 1). In particular, it appears that the serine proteases are activated by limited proteolysis, to produce active enzymes and a number of small, but as yet undefined, peptides (Fig. 1) (see review by Söderhäll & Smith in press). Five or more 'sticky' proteins, one of which is phenoloxidase, are also generated during activation (Söderhäll, Vey & Ramstedt 1984). These proteins bind strongly to the nearest adjacent surface (Söderhäll, Vey *et al.* 1984) and are now known to play important roles in host defence (see below). Calcium is required for the reactions (Söderhäll 1981) and, at concentrations of 5mM or below, calcium will trigger prophenoloxidase activity independent of glucans or LPS (Ashida & Söderhäll 1984). Calcium-mediated activation of proPO, as with heat or other non-physiological elicitors, is a useful way of selectively triggering the cascade *in vitro* for functional studies.

In addition to the serine proteases and other factors, the proPO activating

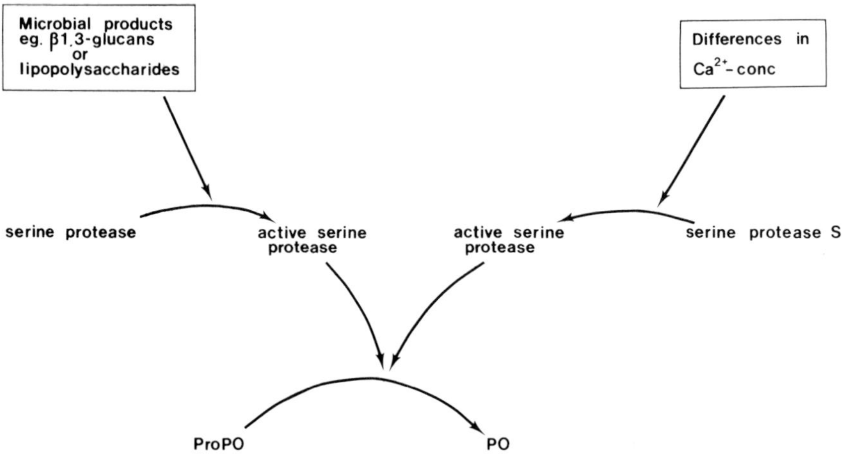

Fig. 1. The prophenoloxidase activating system of crustaceans. The proPO-system is confined to the semigranular and granular cells and can become activated by minute amounts (10^{-10} g/ml) of β1,3-glucans or lipopolysaccharides (LPS) or spontaneously at low Ca^{2+}-concentrations.

cascade must encompass certain regulatory or modulating molecules which serve to limit the enzymic reactions where they would be unnecessary or detrimental to the host. Little is known about proPO regulation in crustaceans or other arthropods, but it is reasonable to assume that inhibiting substances co-exist in the haemolymph with the proPO enzymes. Häll & Söderhäll (1982) have found a subtilisin inhibitor in the haemocytes of crayfish, with an identical inhibitor also present in the cuticle (Häll & Söderhäll 1983). The inhibitor is probably deposited in the cuticle by the haemocytes and is likely to be important in cuticular host defence, blocking or delaying the enzymic breakdown of the host cuticle by invading parasites (Häll & Söderhäll 1983). Whether the subtilisin inhibitor also influences proPO activation in the haemocoel is unknown, but recently, α_2 macroglobulin-like molecules have been found in the plasma of crayfish (Hergenhahn & Söderhäll 1985) and the crabs *Cancer borealis* and *Libinia emarginata* (Armstrong, Rossner & Quigley 1985). The molecules probably function in a manner similar to α_2 macroglobulin in vertebrates; i.e. to sequester proteases, such as those of complement, from the circulation (Starkey & Barrett 1977). Because of the potency of the proPO activating enzymes in host homeostasis (see below), it is essential for the host to possess a variety of regulatory mechanisms in the blood, but until more is known about the nature of these, caution should be exercised in the use of crude proPO fractions from arthropod haemolymph.

Regarding recognition and its evasion by pathogens and parasites, one strategy that could be adopted by a parasite is to conceal or disguise the non-self molecules carried on its surface from the proPO activating proteins in the host haemolymph. This may well occur with *Aphanomyces astaci* in European crayfish where the parasite ramifies extensively through the cuticle before entering the haemocoel (Unestam 1981) and it is possible that the lobster pathogen *Aerococcus viridens* var. *homari* (formerly *Gaffkya homari*) fails to elicit a strong defence reaction in the host (Cornick & Stewart 1968) because the presence of an alcianophilic acidic polysaccharide capsule around the bacterium (Kenne, Lindberg, Lindqvist, Lönngren, Arie, Brown & Stewart 1976) effectively disguises the foreign molecules on its surface. It remains to be confirmed whether proPO activation is impaired in gaffkaemic lobsters, but Johnson, Stewart & Arie (1981) have shown, in a detailed histological study, that the disease is a non-toxic, non-invasive bacteraemia against which the host is able to mount only a feeble and largely inadequate defence response. *A. viridens* is known to be mildly pathogenic for other decapod species (Cornick & Stewart 1968; Newman & Feng 1982), so virulence in lobsters may also be related to the rapid growth of the bacterium in lobster haemolymph (Cornick & Stewart 1968). Whatever the basis for susceptibility, the relatively more resistant species of decapod may act as reservoirs of the disease for lobsters. It has not been established if other micro-organisms evade recognition by 'masking', but with the crayfish pathogen *Psorosperium haeckali*, Vranckx & Durliat (1981) have suggested that changes in the biochemical structure of the parasite 'shell', particularly in its polysaccharide composition, may influence the host immunological responsiveness. Clearly, further research in this area is required.

An alternate strategy for evasion might be the production of inhibiting substances by the parasite against the host defence network. In insects, Stoltz & Cook (1983) have shown that parasitic wasps inject an inhibitor to phenoloxidase when depositing eggs inside the arthropod host and, possibly, parasites of crustaceans may similarly employ neutralizing or inhibitory substances against the host recognition system. Indirect evidence in favour of this is provided by the work of Söderhäll (1985) who has found that melittin, a component of bee sting venom, blocks β1,3-glucan activation of prophenoloxidase in crayfish haemocyte lysate supernatant (HLS) (Söderhäll 1985). However, for most parasites of crustaceans, the nature of any proPO inhibitors remains obscure.

Cellular and humoral defences in crustaceans

As with most invertebrates, host defence in crustaceans is brought about by the activities of the circulating haemocytes, together with various factors present, or released into the plasma from the cells. Traditionally, immunity has been considered to be either cellular or humoral, but in arthropods, because of the intrinsic lability of the haemocytes, the distinction between cell-mediated and humoral defence is far from clear. Undoubtedly, some so-called plasma factors are derived from the haemocytes, and cellular activity itself may be influenced by certain features or substances within the plasma. Therefore, in this review, humoral and cellular immunity will not be discussed separately, but for clarity a brief description of the different types of haemocyte found in crustaceans will be given, with an evaluation of their individual contribution to overall host defence.

Haemocyte types

Typically, crustaceans possess three distinct populations of circulating haemocytes; the hyaline cells, the semigranular cells and the granular cells (Bauchau 1981). These may be separated from whole haemolymph by density gradient centrifugation (Söderhäll & Smith 1983; Smith & Söderhäll 1983b), and analysis of the various cell fractions reveals that the cells differ not only in morphology but also in biochemical composition and behaviour *in vitro* (Söderhäll, Smith & Johansson in press).

A detailed account of the blood cells of crustaceans has already been given by Bauchau (1981), so only those features examined more recently with isolated haemocyte populations will be discussed here.

First, with the hyaline cells, we now know that they are not only devoid of large intracellular granules but that they also lack phenoloxidase activity (Söderhäll & Smith 1983; Smith & Söderhäll 1983b). Phagocytosis is a basic property of these cells and perhaps related to it is their ability to attach strongly and spread extensively on glass surfaces (Söderhäll, Smith *et al.* in press). They appear to be unaffected by exposure to dissolved foreign molecules (β, 1–3 glucans or LPS) but phagocytic activity is stimulated by proPO activating components *in vitro* (Söderhäll, Smith *et al.* in press) (see below). Morphologically and functionally, the hyaline cells are probably equivalent to the plasmatocytes of insects, and seem to represent a fundamental type of amoebocyte cell present in nearly all invertebrates (V. J. Smith & K. Söderhäll unpublished).

The semigranular cells, by contrast, may well be unique to arthropods and characteristically contain a variable number of cytoplasmic granules with only small amounts of prophenoloxidase (Söderhäll & Smith 1983; Smith & Söderhäll 1983b). They are extremely fragile and readily degranulate or lyse during handling or under *in vitro* conditions (Söderhäll & Smith 1983; Söderhäll, Smith *et al.* in press). Phagocytosis is exhibited by these cells in some species, but this does not appear to be true for all, and phagocytic activity, when present, is abolished upon degranulation (Söderhäll, Smith *et al.* in press). Foreign molecules trigger degranulation of the semigranular cells, and with strong non-self signals such as with proPO activat-

Table 1. Summary of the biological characteristics of crustacean haemocytes.

	Hyaline cells	Semigranular cells	Granular cells
Phenoloxidase activity	Absent	Limited	Strong
Phagocytosis *in vitro*	Yes	Yes/No	No
Attachment	Strong	Strong	Weak
Spreading	Strong	Strong	Weak
Nodule/capsule formation	No	Yes	No[a]
Microbial killing	Absent	Weak	Strong[b]
Cytotoxicity	Not known	Yes	Yes
Response to β 1,3-glucans/LPS	None	Degranulation	None
Response to proPO proteins	Enhanced phagocytosis	Lysis	Degranulation

For references see text.
[a] Persson, Vey & Söderhäll (in prep.)
[b] Unpublished findings.

ing components, lysis commonly occurs (Smith & Söderhäll 1983b; Johansson & Söderhäll in press; Söderhäll, Smith *et al.* in press). Whether the semigranular cells are the 'explosive corpuscles' described by Hardy (1892) is a matter of opinion, but equivalent cells are seldom found in other invertebrate phyla (V. J. Smith & K. Söderhäll unpublished).

Lastly, the granular cells in crustaceans are distinguished by the presence of many large intracellular granules and the location of large amounts of the proPO activating components within these organelles (Johansson & Söderhäll in press). The cells are always non-phagocytic and they form only weak attachment, with limited spreading, on glass surfaces (Smith & Söderhäll 1983b; Söderhäll, Smith *et al.* in press). Degranulation is not usually observed in the granular cells following treatment with dissolved glucans or LPS, but in the presence of active proPO proteins they soon undergo exocytosis (Smith & Söderhäll 1983b; Johansson & Söderhäll in press; Söderhäll, Smith *et al.* in press). Granular type cells are present in many invertebrates though in some groups, e.g. the Limulidae, the cells are capable of ingesting non-self particles (Armstrong & Levin 1979), display active motility on glass substrates (Armstrong 1985), and are susceptible to degranulation by LPS (Armstrong & Rickles 1982). The biological characteristics of the different haemocytes in crustaceans are summarized in Table 1.

The relative proportions of each cell type vary according to the host. In crabs, for instance, the hyaline cells are predominant with the semigranular cells constituting only a small percentage of the total haemogram (Söderhäll & Smith 1983). In crayfish, by contrast, the semigranular cells are the most abundant with the hyaline cells by far the most infrequent cell type (Smith & Söderhäll 1983b; Söderhäll, Smith *et al.* in press). Because of the sensitivity of the various cell types to handling-stress or non-self molecules, it is often impossible to establish accurately the true identity of each cell type in mixed culture on *in vitro* monolayers (Smith & Ratcliffe 1978; Smith &

Söderhäll 1983a,b). Whenever possible, therefore, *in vitro* analyses of the cells should be made with isolated populations purified by density gradient centrifugation or other means. A convenient marker for cell purity, at least with the hyaline cells, is the presence/absence of phenoloxidase activity, and a method for rapidly checking this has been given by Söderhäll & Smith (1983).

Exocytosis, whilst posing problems in the manipulation of the haemocytes *in vitro*, is an extremely important process *in vivo*. Johansson & Söderhäll (in press) have made a careful study of the phenomenon in crayfish and found that in addition to foreign molecules (LPS, β 1,3- glucans), exocytosis is induced in the semigranular and granular cells by the calcium ionophore A21387, and the proPO activating enzymes thus released are in an inactive state. Degranulation of the granular cells, whether triggered by non-self molecules or A21387, is prevented by the anion channel blocker, SITS (4-acet-amido-4^1-isothiocyanatostilbene- $2,2^1$- disulfonic acid disodium salt), the calcium channel blocker, cobalt chloride, or calmidazolium (Johansson & Söderhäll in press), indicating that the process in crustaceans is essentially the same as that described for other organisms (Pollard, Pazoles, Creutz & Zinder 1979). With the granular cells of crayfish, melittin, from bee sting venom, has been found to induce cell lysis *in vitro*, but following the response melittin fails to trigger crayfish proPO activation, and actually prevents activation of the cascade by β 1,3- glucans (Söderhäll 1985).

In the light of recent findings on exocytosis with isolated haemocyte populations from crustaceans, we have recently proposed a scheme whereby cellular co-operation may operate during host defence in arthropods (Söderhäll & Smith in press; Söderhäll, Smith *et al.* in press). Because of their sensitivity to foreign agents, we identify the semigranular cells as the key haemocytes in the process and suggest that they release communicating signals (presumably proPO proteins) by exocytosis upon non-self stimulation (Söderhäll & Smith in press). These signals then act directly on the hyaline cells and/or cause the granular cells to discharge more proPO-activating components which in turn influence cellular responsiveness to homeostatic disturbance (Söderhäll, Smith *et al.* in press) (see below). Exocytosis thus serves to release previously cellbound factors into the haemolymph where they are biochemically activated and are free to exert their physiological function (Smith & Söderhäll 1983b).

Phagocytosis

Following the discovery of phagocytosis in the water flea, *Daphnia*, by Metchnikoff (1884), many analyses of the process have been made *in vitro* and *in vivo* (see reviews by Chorney & Cheng 1980; Ratcliffe, White, Rowley & Walters 1982). The earliest investigations were concerned with evaluating the types of materials ingested by crustacean blood cells but later studies focused on the search for recognition factors (opsonins) in the haemolymph (Chorney & Cheng 1980; Ratcliffe *et al.* 1982). The results of this work revealed that while crustacean haemocytes are capable of recognizing and engulfing a wide variety of biotic and abiotic particles, the presence/absence of opsonic factors in the blood is less clear. In a detailed study of phagocytosis in the crayfish, *Cherax destructor*, McKay & Jenkin (1970) and Tyson & Jenkin (1974) found that efficient uptake of bacteria or erythrocytes by the haemocytes was dependent upon factors present in the serum (i.e. the supernatant remaining after centrifugation of clotted whole, undiluted haemolymph). Equivalent findings were obtained by Paterson & Stewart (1974) with lobster, *Homarus americanus*, haemocytes, but Smith & Ratcliffe (1978) were unable to substantiate this for the shore crab, *Carcinus maenas*. The situation has now been clarified somewhat by the discovery of an opsonic role for the proPO activating cascade in crayfish and crabs (Smith & Söderhäll 1983a; Söderhäll, Smith *et al.* in press). These opsonins may be generated *in situ* in the culture system by inclusion of β 1,3- glucans (or other proPO elicitors) in the haemocyte:bacteria mixtures (Smith & Söderhäll 1983a), but confirmation that the sticky proteins liberated by glucan activation of the proPO cascade serve as the opsonins is provided by the enhanced uptake of the test particles observed after pretreatment of the foreign agents with the attaching proteins in glucan-activated supernatant from a haemocyte lysate (HLS) (Söderhäll, Smith *et al.* in press). It is highly likely that the enhanced degranulation, spreading and phagocytic behaviour seen in lobster haemocytes *in vitro* by Goldenberg, Huebner & Greenberg (1984), following incubation of the cells in LPS, was due to stimulation of the proPO activating cascade within the target cells in a manner similar to that observed in crayfish and crabs with β 1,3- glucans (Smith & Söderhäll 1983a). Precisely which of the attaching proteins in crustacean HLS have opsonic properties has not been established. Phenoloxidase itself does not appear to participate in the phenomenon as pretreatment of bacteria with heat-activated HLS (in which only prophenoloxidase is activated (Ashida & Söderhäll 1984)) does not promote uptake in the same way as glucan-activated HLS (where the entire cascade is switched on (Söderhäll, Smith *et al.* in press)). More importantly, the opsonins contained within the cascade must be liberated at the appropriate time and in the correct place if they are to be effective, and we have found that premature production of the sticky proteins on monolayer cultures of mixed crayfish or crab haemocytes fails to stimulate phagocytosis of bacteria, probably because the opsonins have attached to the glass substrate or other available surfaces and are no longer available to mediate uptake (Smith & Söderhäll 1983a). This finding has also been made for insects (Leonard, Ratcliffe & Rowley 1985) but in crustaceans, since the opsonins are donated by the granular cells, which are themselves non-phagocytic (see above), the response must involve some degree of cell communication/co-operation. How this is achieved at the molecular level is still enigmatic and much needs to be learned about phagocytosis in crustaceans, particularly about the nature of the communicating signals and the 'receptors' for the opsonins on the phagocyte surface.

Without the stimulatory effect of the proPO activating system on the haemocytes, phagocytosis in this group of arthropods would be very inefficient indeed. Measurements of the percentage uptake of foreign particles *in vitro* have yielded values of circa 5% in crabs (Smith & Ratcliffe 1978; Söderhäll, Smith *et al.* in press), circa 3% in crayfish (Söderhäll & Smith 1983; Söderhäll, Smith *et al.* in press) and circa 1% in lobsters (Paterson & Stewart 1974). Opsonization raises the percentage uptake three- or four-fold and significantly increases the number of particles taken up per cell (Smith & Söderhäll 1983a) but, even making allowances for the non-physiological conditions of the assay system and the large numbers of circulating phagocytes in crustaceans, phagocytosis must be augmented by other defence strategies *in vivo*. Nonetheless, the process is probably important in surveillance and the removal of opportunistic microbial invaders from the haemocoel; but whether any micro-organisms produce proteases or other inhibitory substances against the opsonins of the host is unknown.

Nodule/capsule formation

The main adjunct to phagocytosis in crustaceans (and other arthropods) is nodule or capsule formation. The response is extremely effective in containing the spread of infective agents within the haemocoel and brings the foreign material into close contact with the haemocytes (thereby facilitating cell-mediated killing or destruction of the invaders). Whatever the nature of the non-self material, the cellular reaction is always similar, i.e. aggregation and flattening of haemocytes around the foreign intruder, accompanied by the deposition of melanin either on the parasite surface or within the haemocyte matrix (for reviews see Bauchau 1981; Ratcliffe *et al.* 1982). The response seems to be the prerogative of the semigranular cells although some reports suggest that all the cells act in concert to bring about the phenomenon. Attachment and spreading of the haemocytes on the foreign material (or cellular complex) are key processes during the development of the nodules/capsules and, as the structures increase in size, they come to lodge in various vascular sinuses in the body. With the attachment of the haemocytes to the nodules or capsules, there is a corresponding decline in the number of circulating cells in the 'blood'. This may be temporarily detrimental to the host but mobilization or haemopoiesis of new haemocytes soon (i.e. within 48 h) returns the circulating cell count to normal (Ratcliffe *et al.* 1982). The mechanisms controlling mobilization and haemopoiesis have not been properly investigated, although in some species the haemopoietic tissue has been identified (e.g. Johnson 1980). The nature of the molecular signals which stimulate cell division have still to be described.

That the prophenoloxidase activating system participates in nodule/capsule formation is indicated by the presence of melanin in the completed structures. In addition, in crabs and crayfish, dissolved non-self molecules (β 1,3- glucans, but not other dissolved glycans or carbohydrates) specifically induce cell aggregation and haemocytopenia in exactly the same way as bacteria or other non-self particles (Smith & Söderhäll 1983a; Smith, Söderhäll & Hamilton 1984). Furthermore, enhanced encapsulation has been induced in crayfish by pretreatment of injected fungal spores with crayfish HLS and, in this case, larger capsules are formed more rapidly than around fungal spores coated in plasma or buffer (Söderhäll, Vey *et al.* 1984).

From studies with isolated crustacean haemocytes *in vitro*, it is likely that the sequence of events involved in nodule/capsule formation is as follows. The non-self molecules released from the surface of the foreign material trigger exocytotic release of proPO proteins from the semigranular cells. At this stage the proPO cascade is stable (i.e. inactive), but activation is rapidly induced by the non-self molecules, and the active proPO proteins thus generated stimulate further degranulation of proPO components from the granular cells. Haemocyte adherence to the foreign material then commences and haemocytes continue to attach to the complex as long as proPO proteins are generated by exocytosis (M. Persson, A. Vey & K. Söderhäll unpublished). The structure finally reaches equilibrium when sufficient cells surround the capsule to effect 'sealing'. We have no evidence that the proPO cascade liberates chemotactic factors for the haemocytes (K. Söderhäll, V. J. Smith & L. Cerenius unpublished), but our inability to obtain a positive cellular response to proPO components in crayfish HLS does not necessarily imply that chemotaxis plays no part in encapsulation as the techniques available for assaying chemotaxis are, by and large, unsuitable for invertebrate systems.

Only a few parasites fail to elicit the reaction in crustacean hosts, and others are able to survive and grow despite the presence of the haemocytic capsule. There are probably several reasons for this, but the situation in most parasite-host associations is poorly understood. With the acanthocephalan, *Polymorphus minutus*, for instance, following consumption of the parasite eggs by the intermediate host, *Gammarus pulex*, the acanthors hatch and migrate through the intestine wall to the haemocoel (Hynes & Nicholas 1957). Here the presence of the parasites elicits a cellular reaction by the amphipod and the larvae soon become surrounded by a capsule of haemocytes (Crompton 1964). *P. minutus* does not seem to be adversely affected by this host response, however, and proliferation of the acanthor surface membrane rapidly produces a coat of microvilli that helps dislodge the haemocytes and contributes to the development of an acellular envelope around each parasite (Crompton 1964, 1967; Butterworth 1969; Rotheram & Crompton 1972). Precisely how *P. minutus* resists cellular attack by *G. pulex* is unclear, but the formation of the acellular envelope appears to favour parasite survival in the amphipod, as *P. minutus* experimentally stripped of these structures usually die upon implantation into the host haemocoel (Crompton 1967; and see chapter by Lackie, this volume). In general, parasites survive better in the 'correct' (usual) host than an incorrect (unusual) host (Denny 1969; Gemmell 1971), but even with the usual host, internal factors, such as age, may strongly influence the ability of the host to resist infestation (Nickol & Dappen 1982). Possibly, some parasites might overcome the encapsulation response of the host either by rapid growth or by production of specific inhibitory or toxic substances against the haemocytes. As yet, virtually nothing is known about such tactics in crustaceans, so, again, further research in this area is urgently required.

Agglutinins, lysins and other antimicrobial factors

There has always been considerable interest in the presence/absence of agglutinating or other factors in crustacean haemolymph, partly because these molecules were once thought to represent 'recognition factors' (Acton & Weinheimer 1974) and partly because agglutinins are relatively easy to detect *in vitro*. Agglutinins have, consequently, been demonstrated in various crustaceans against a wide range of bacteria, erythrocytes or other particles (for reviews see Acton & Weinheimer 1974; and Tyson, McKay & Jenkin 1974), but their relationship to non-self recognition has remained obscure. Agglutinins are considered to aggregate foreign particles in the haemolymph (Acton & Weinheimer 1974; Ratcliffe *et al.* 1982), but the origin, mode of action and regulation of these molecules have not been adequately explained (see also chapter by Renwrantz, this volume).

With the prophenoloxidase activating system, induction of the enzymic cascade is known to liberate a number of 'sticky' or attaching proteins (Söderhäll, Vey *et al.* 1984) and it is reasonable to predict that these might account for some agglutination activity in crustacean haemolymph. These agglutinating proteins would be produced by the haemocytes, released into the haemolymph by exocytosis upon appropriate stimulation of the cells (see above), and following specific activation, would be directed against any type of particulate material. Possibly some regulation of their potency is afforded by α_2 macroglobulin present in the plasma (see above). Such a mechanism is attractive because it requires less genetic diversity than that demanded for the production of a wide spectrum of agglutinins to deal with all the different types of foreign material likely to be encountered, and is thus more 'economical' to the host. Separate specific carbohydrate-binding molecules (lectins) might well co-exist in crustacean haemolymph (Yeaton 1981) serving a variety of functions including, for example, transport or host defence, but until more is known about the lectins, their role in immunity remains speculative.

As far as lysins and other antimicrobial factors are concerned, several crustaceans have been reported to possess such molecules (Evans, Painter, Evans, Weinheimer & Acton 1968; Acton, Weinheimer & Evans 1969). Again, little is known about the nature, source or function of these factors, and detailed investigations such as those of Boman and his co-workers on insect cecropins (Boman, Faye, Pye & Rasmuson 1978; see also chapter by Boman, this volume) are wanting for crustaceans. With crayfish, Söderhäll, Wingren, Johansson & Bertheussen (1985) have shown that the semigranular cell and granular cells are cytotoxic towards a range of vertebrate tumour and non-tumour cell lines *in vitro*, and the response may be important in the destruction of metazoan parasites by the host. Söderhäll & Ajaxon (1982) have also found that the reactions leading to melanization are fungitoxic, and we have obtained some preliminary data indicating that the proPO cascade has a bactericidal effect (Söderhäll & Smith in press, and unpublished). Phenoloxidase itself is known to be toxic (Zlotkin, Gurevitz & Shulov 1973), and melanin and the intermediate compounds generated by phenoloxidase activity are stable free radicals that possess antimicrobial or cytotoxic properties (see review by Söderhäll & Smith in press), so the proPO activating system must help to maintain sterility of the haemolymph. Whether parasites or pathogens are able to protect themselves against the lytic principles in crustacean blood is unknown.

Clotting

Clotting is a fundamental process exhibited by all animals possessing a circulatory system. Not only does it serve to prevent blood loss and plug wounds upon injury, but it also entraps any opportunistic micro-organisms that may gain access to the haemocoel through the wound (Levin 1967). The mechanism governing clotting has been the subject of considerable investigation, not least because of the need to minimize haemolymph coagulation for *in vitro* experimentation, but also because it may have been conserved throughout evolution (Hardy 1892). In crustaceans, the phenomenon is a two-phase process involving cell aggregation and plasma gelation (see review by Grégoire 1970; Bauchau 1981; Ratcliffe *et al.* 1982). Material released from the granular haemocytes is thought to be responsible for both phases (Hardy 1892; Bang 1970) and, in lobsters, transglutamidase

Table 2. Summary of the immunological functions of the proPO system in crustaceans.

Opsonization	Yes
Cell attachment	Yes
Nodule/capsule formation	Yes
Agglutination	Needs confirmation
Fungicidal	Yes
Bactericidal	Needs confirmation
Lysis/cytotoxicity	Yes
Clotting	Yes
Wound repair	Needs confirmation
Haemopoiesis	Not known
Chemotaxis	Not demonstrated

For references see text.

has been identified as the enzyme catalysing the formation of peptide cross-links with the plasma coagulogen (Fuller & Doolittle 1971). Prophenoloxidase activation seems to participate in clotting since haemocyte lysate supernatants reacted with β 1,3- glucans can be seen to undergo a gelation reaction *in vitro* (Söderhäll 1981) and the coagulation reaction induced in this way utilizes a protease with specificity similar to that which effects conversion of prophenoloxidase to phenoloxidase (Söderhäll 1981) (Fig. 1). Calcium is always necessary for clotting to take place and the most efficient anticoagulants are those that chelate calcium and block prophenoloxidase activation (Söderhäll & Smith 1983, in press). We have previously compared and contrasted the clotting pathway of crustaceans with those in horseshoe crabs, insects or mammals (Söderhäll & Smith in press) so further discussion of the process will not be given here.

Conclusions

Crustaceans possess a variety of internal defence strategies which effectively protect the host against disease or parasitization. In responding specifically to foreign molecules, liberating opsonins, mediating nodule/capsule formation, generating agglutinins, lysins or cytotoxic factors, and in participating in coagulation, the prophenoloxidase activating system plays a number of crucial roles in these defences (Table 2). However, much still remains to be learnt about the proPO cascade in arthropod immunity particularly with respect to its interaction with invasive pathogens or parasites. Certainly, the ability of micro-organisms or parasites to survive and grow in crustaceans can have very profound economic or ecological consequences, as demonstrated, for example, by the fungus *Aphanomyces astaci* and the various helminth worms which infest crustaceans during their development.

With *A. astaci*, the parasite infects a variety of crayfish species, but in some strains, such as the American crayfish, *Pacifastacus leniusculus*, growth of the hyphae is kept in check by the defence network of the host (see review by Unestam 1981). *A. astaci* is thus harboured by *P. leniusculus* but in European crayfish, the fungus evades recognition by the host haemolymph, and a lethal infection is soon established in the cuticle (Unestam 1981). When implanted into European lakes, ponds or rivers for commercial purposes, *P. leniusculus* serves as the vector for the parasite, with the obvious devastating effect on endemic crayfish populations (Unestam 1981).

Regarding the helminths, many crustaceans (amphipods, copepods, isopods, crabs and lobsters) are known to harbour these parasites before transmission to the final vertebrate host. Some examples are listed in Table 3 along with their usual intermediate and final hosts. Unfortunately, while several reports documenting the life cycles of these and other worms have been published (see, for instance Denny 1969), rather less is known about the host response towards them or the strategies adopted by the parasites to avoid immunological attack. Clearly, where either the intermediate or the final host is of commercial importance, as with *Ascariophus* spp., *Rhabdochona uca*, *Microphallus nicolli*, *Strichocotyle nephropsis*, *Acanthocephalus dirus* or *Echinorhynchus truttae*, parasitization may result in undesirable economic loss. In such cases, there is a distinct need to control infestation through an understanding of the epidemiology, pathology and immunology of the parasites and hosts concerned. Likewise, it is essential to comprehend the effects of parasitization in unusual hosts, particularly if this is directly or indirectly detrimental to man. The condition known as anisakiasis serves to illustrate this point. Anisakiasis is a complaint common in Japan and northern European countries where fish is eaten raw, and it is characterized by serious damage to the gastric or intestinal submucosa (see Sindermann 1970). It is caused by *Anisakis* nematodes, which pass from the usual fish host (pollock, cod, salmon, herring or mackerel) to the human intestine if contaminated fish is eaten raw. The *Anisakis* larvae develop in a crustacean intermediary and, despite encapsulation by this host, retain infectivity in both the fish and the mammalian host (see Sindermann 1970).

In view of the significance of crustaceans as vectors of disease agents, and as marketable commodities in their own right, much can be gained from an insight into the complex molecular interactions which take place between the host and its parasites. Some progress has already been made with respect to the proPO activating system and as more is learned about cell communication, blood cell production and immunological modulation, it might be possible to comprehend the ways successful parasites neutralize or evade cellular and humoral attack in these arthropods.

Table 3. Some helminth parasites using crustaceans as intermediate hosts.

Parasite	Intermediate host	Final host(s)	Reference
Cestoda			
Laterioporus clerci	*Gammarus* spp.	Gulls	Denny (1969)
L. skajabini	*Gammarus* spp.	Ducks	Denny (1969)
L. mathevossicenae	*Gammarus* spp.	Ducks	Denny (1969)
Hymenolepis spiralibursata	*Gammarus* spp.	Ducks	Denny (1969)
Fimbriaria fasciolaris	*Gammarus* spp.	Ducks	Denny (1969)
Proteocephalus filicollis	*Eucyclops serratulus*	Three-spined sticklebacks	Gemmell (1971)
Schistocephalus solidus	Copepods	Powan	Gemmell (1971)
Diphyllobothrium latum	*Cyclops* spp.	Fish	Guttowa (1961)
Nematoda			
Streptocara crassicauda	*Gammarus* spp.	Ducks	Denny (1969)
Rhabdochona uca	*Uca mani*	Elasmobranch fish	Gordon (1966)[a]
Ascarophis spp.	Lobsters	Gadoid fish	Sindermann (1970)[a]
Trematoda			
Microphallus nicolli	*Callinectes sapidus*	Herring gulls	Sindermann (1970)[a]
Stichocotyle nephropis	*Nephrops norvegicus*/ *Homarus americanus*	Skates/rays	Sindermann (1970)[a]
Gorgodera amplicava	*Cambarus americanus*	Frogs	Gordon (1966)[a]
Paragonimus ringeri	Crabs/crayfish	Carnivorous vertebrates (incl. man)	Gordon (1966)[a]
Acanthocephala			
Polymorphus minutus	*Gammarus pulex*	Domestic ducks	Hynes & Nicholas (1957)
P. contortus	*G. lacustris*	Ducks	Denny (1969)
P. botulus	*Carcinus maenas*	Eider duck	Rayski & Garden (1961)
Plagiorhynchus cylindraceus	*Armadillidium vulgare*	Passerine birds	Nickol & Dappen (1982)
Acanthocephalus dirus	*Asellus intermedius*	Freshwater drum/blue-gill/ cat-fish	Seidenberg (1973)
Echinorhynchus truttae	*Gammarus pulex*	Salmon/trout	Awachie (1966)

[a] Review article.

References

Acton, T. & Weinheimer, P.F. (1974). Haemagglutinins: Primitive receptor molecules operative in invertebrate defence mechanisms. *Contemp. Top. Immunobiol.* **4**: 271–82.

Acton, R.T., Weinheimer, P.F. & Evans, E.E. (1969). A bactericidal system in the lobster *Homarus americanus. J. Invert. Path.* **13**: 463–64.

Armstrong, P.B. (1985). Adhesion and motility of the blood cells of *Limulus*. In *Blood cells of marine invertebrates*: 77–124. (Ed. Cohen, W.D.). A.R.Liss, New York.

Armstrong, P.B. & Levin, J. (1979). *In vitro* phagocytosis by *Limulus* blood cells. *J. Invert. Path.* **34**: 145–51.

Armstrong, P.B. & Rickles, F.R. (1982). Endotoxin induced degranulation of the *Limulus* amoebocyte. *Exp. Cell Res.* **140**: 15–24.

Armstrong, P.B., Rossner, M.T. & Quigley, J.P. (1985). α_2 Macroglobulin-like activity in the blood of chelicerate and mandibulate arthropods. *J. exp. Zool.* **236**: 1–9.

Ashida, M. & Söderhäll, K. (1984). The prophenoloxidase activating system in crayfish. *Comp. Biochem. Physiol.* **77B**: 21–6.

Awachie, J.B.E. (1966). The development and life history of *Echinorhynchus truttae* Schrank, 1788 (Acanthocephala). *J. Helminth.* **40**: 11–32.

Bang, F.B. (1970). Cellular aspects of blood clotting in the seastar and hermit crab. *J. Reticuloendoth. Soc.* **7**: 167–72.

Bauchau, A.G. (1981). Crustaceans. In *Invertebrate blood cells* **2**: 385–420. (Eds Ratcliffe, N.A. & Rowley, A.F.). Academic Press, London & New York.

Bertheussen, K. (1984). Complement and lysins in invertebrates. *Devl comp. Immunol. Suppl.* **3**: 173–81.

Boman, H.G. Faye, I., Pye, A. & Rasmuson, T. (1978). The inducible immunity system of giant silk moths. *Comp. Pathobiol.* **4**: 145–63.

Butterworth, P.E. (1969). The development of the body wall of *Polymorphus minutus* (Acanthocephala) in its intermediate host *Gammarus pulex. Parasitology.* **59**: 373–88.

Chorney, M.J. & Cheng, T.C. (1980). Discrimination of self and non-self in invertebrates. *Contemp. Top. Immunobiol.*: 37–54.

Cornick, J.W. & Stewart, J.E. (1968). Interaction of the pathogen *Gaffkya homari* with the natural defence mechanisms of *Homarus americanus. J. Fish. Res. Bd Can.* **25**: 695–709.

Crompton, D.W.T. (1964). The envelope surrounding *Polymorphus minutus* (Goeze, 1782) (Acanthocephala) during its development in the intermediate host, *Gammarus pulex. Parasitology* **54**: 721–35.

Crompton, D.W.T. (1967). Studies on the haemocytic reactions of *Gammarus* spp., and its relationship to *Polymorphus minutus* (Acanthocephala). *Parasitology* **57**: 389–401.

Day, N.K.B., Gewurz, H., Johanssen, R., Finstad, J. & Good, R.A. (1970). Complement and complement-like activity in lower vertebrates and invertebrates. *J. exp. Med.* **132**: 941–50.

Denny, M. (1969). Life-cycles of helminth parasites using *Gammarus lacustris* as an intermediate host in a Canadian lake. *Parasitology* **59**: 795–827.

Evans, E.E., Painter, B., Evans, M.L., Weinheimer, P. & Acton, R.T. (1968). An induced bactericidin in the spiny lobster, *Panulirus argus*. *Proc. Soc. exp. Biol. Med.* **128**: 394–8.

Fuller, G.M. & Doolittle, R.F. (1971). Studies of invertebrate fibrinogen. II. Transformation of lobster fibrinogen into fibrin. *Biochemistry* **10**: 1311–5.

Gemmell, D.K. (1971). *Some aspects of the biology of the tapeworms* Proteocephalus *spp. and* Schistocephalus solidus *(Muller)*. Ph.D. Thesis: University of Glasgow.

Goldenberg, P.Z., Huebner, E. & Greenberg, A.H. (1984). Activation of lobster hemocytes for phagocytosis. *J. Invert. Path.* **43**: 77–88.

Gordon, I. (1966). Crustacea—general considerations. *Mém. Inst. franç. Afr. noire* **77**: 27–86.

Grégoire, C. (1970). Haemolymph coagulation in arthropods. *Symp. zool. Soc. Lond.* No. 27: 45–74.

Guttowa, A. (1961). Experimental investigations on the systems "procercoids of *Diphyllobothrium latum* (L.)–Copepoda". *Acta Parasit. pol.* **9**: 371–408.

Häll, L. & Söderhäll, K. (1982). Purification and properties of a protease inhibitor from crayfish hemolymph. *J. Invert. Path.* **39**: 29–37.

Häll, L. & Söderhäll, K. (1983). Isolation and properties of a protease inhibitor in crayfish (*Astacus astacus*) cuticle. *Comp. Biochem. Physiol.* **76B**: 699–702.

Hardy, W.B. (1892). Blood corpuscles of the Crustacea together with a suggestion as to the origin of crustacean fibrin ferment. *J. Physiol., Lond.* **13**: 165–90.

Hergenhahn, G. & Söderhäll, K. (1985). α_2 Macroglobulin-like activity in plasma of the crayfish, *Pacifastacus leniusculus*. *Comp. Biochem. Physiol.* **81B**: 833–5.

Hynes, H.B.N. & Nicholas, W.L. (1957). The development of *Polymorphus minutus*. (Goeze 1782) (Acanthocephala) in the intermediate host. *Ann. trop. Med. Parasit.* **51**: 380–91.

Johansson, M. & Söderhäll, K. (In press). Exocytosis of the prophenoloxidase activating system from crayfish haemocytes. *J. comp. Physiol.*

Johnson, P.T. (1980). *Histology of the blue crab.* Callinectes sapidus. *A model for the Decapoda.* Praeger Press, New York.

Johnson, P.T., Stewart, J.E. & Arie, B. (1981). Histopathology of *Aerococcus viridans* var. *homari* infection (Gaffkemia) in the lobster, *Homarus americanus*, and a comparison with histological reactions to a Gram-negative species, *Pseudomonas perolens*. *J. Invert. Path.* **38**: 127–48.

Kenne, L., Lindberg, B., Lindqvist, B., Lönngren, J., Arie, B., Brown R.G. & Stewart, J.E. (1976). 4–0 [(S)-1- carboxyethyl] - D-glucose: A component of the extracellular polysaccharide materials from *Aerococcus viridans* var. *homari*. *Carbohydr. Res.* **51**: 287–90.

Leonard, C., Ratcliffe, N.A. & Rowley, A.F. (1985). The role of prophenoloxidase activation in non-self recognition and phagocytosis by insect blood cells. *J. Insect Physiol.* **31**: 789–99.

Levin, J. (1967). Blood coagulation and endotoxin in invertebrates. *Fed. Proc.* **26**: 1707–12.

McKay, D. & Jenkin, C.R. (1970). Immunity in the invertebrates. The role of serum factors in phagocytosis of erythrocytes by haemocytes of the freshwater crayfish (*Parachaeraps bicarinatus*). *Aust. J. exp. Biol. med. Sci.* **48**: 139–150.

Metchnikoff, E. (1884). Ueber eine Sprosskrankheit der Daphnien. *Virchows Arch.* **96**: 177–95.

Newman, M.C. & Feng, S.Y. (1982). Susceptibility and resistance of the rock crab, *Cancer irroratus*, to natural and experimental bacterial infection. *J. Invert. Path.* **40**: 75–88.

Nickol, B.B. & Dappen, G.E. (1982). *Armadillidium vulgare* (Isopoda) as an intermediate host of *Plagiorhynchus cylindraceus* (Acanthocephala) and isopod response to infection. *J. Parasit.* **68**: 570–5.

O'Day, D.M., Akrabawi, P.L., Richmond, L.D., Jones, B.R. & Clayton, Y. (1979). An animal model for *Fusarium solani* endophthalamitis. *Br. J. Ophthal.* **63**: 277–80.

Paterson, W.D. & Stewart, J.E. (1974). *In vitro* phagocytosis by haemocytes of the American lobster, (*Homarus americanus*). *J. Fish. Res. Bd Can.* **31**: 1051–6.

Pollard, H.B., Pazoles, C.J., Creutz, C.E. & Zinder, O. (1979). The chromaffin granule and possible mechanisms of exocytosis. *Int. Rev. Cytol.* **58**: 159–97.

Ratcliffe, N.A., White, K.N., Rowley, A.F. & Walters, J.B. (1982). Cellular defense systems of the Arthropoda. In *The reticuloendothelial system: A comprehensive treatise*: 167–255. (Eds Cohen, N. & Sigel, M.). Plenum Press, New York & London.

Rayski, C. & Garden, E.A. (1961). Life cycle of an acanthocephalan parasite of the eider duck. *Nature, Lond.* **192**: 185–6.

Rotheram, S. & Crompton, D.W.T. (1972). Observations on the early relationship between *Moniliformis dubius* (Acanthocephala) and the haemocytes of the intermediate host, *Periplaneta americana. Parasitology* **64**: 15–21.

Seidenberg, A.J. (1973). Ecology of the acanthocephalan, *Acanthocephalus dirus* (Van Cleave, 1931), in its intermediate host *Asellus intermedius* Forbes (Crustacea: Isopoda). *J. Parasit.* **59**: 957–62.

Sindermann, C.J. (1970). *Principal diseases of marine fish and shellfish.* Academic Press, London & New York.

Smith, V.J. & Ratcliffe, N.A. (1978). Host defence reactions of the shore crab, *Carcinus maenas* (L.). *in vitro. J. mar. biol. Ass. U.K.* **58**: 367–79.

Smith, V.J. & Söderhäll, K. (1983a). β 1,3–glucan activation of crustacean hemocytes *in vitro* and *in vivo. Biol. Bull. mar. biol. Lab. Woods Hole* **164**: 299–314.

Smith, V.J. & Söderhall, K. (1983b). Induction of degranulation and lysis of haemocytes in the freshwater crayfish, *Astacus astacus* by components of the prophenoloxidase activating system *in vitro. Cell Tissue Res.* **233**: 295–303

Smith, V.J., Söderhäll, K. & Hamilton, M. (1984). β 1,3-glucan induced cellular defence reactions in the shore crab, *Carcinus maenas. Comp. Biochem. Physiol.* **77A**: 635–9.

Söderhäll, K. (1981). Fungal cell wall β-1,3-glucans induce clotting and phenoloxidase attachment to foreign surfaces of crayfish hemocyte lysate. *Devl comp. Immunol.* **5**: 565–73.

Söderhäll, K. (1982). Prophenoloxidase activating system and melanization—a recognition mechanism of arthropods?—a review. *Devl comp. Immunol.* **6**: 601–11.

Söderhäll, K. (1985). The bee venom, melittin, induces lysis of arthropod granular cells and inhibits action of the prophenoloxidase activating system. *FEBS Lett.* **192**: 109–12.

Söderhäll, K. & Ajaxon, R. (1982). Effect of quinones and melanin on mycelial growth of *Aphanomyces* spp. and extracellular protease of *Aphanomyces astaci* a parasite on crayfish. *J. Invert. Path.* **39**: 105–9.

Söderhäll, K. & Häll, L. (1984). Lipopolysaccharide-induced activation of the prophenoloxidase activating system in crayfish haemocyte lysate. *Biochem. Biophys. Acta* **797**: 99–104.

Söderhäll, K. & Smith, V.J. (1983). Separation of the haemocyte populations of *Carcinus maenas* and other marine decapods, and prophenoloxidase distribution. *Devl comp. Immunol.* **7**: 229–39.

Söderhäll, K. & Smith, V.J. (1984). The prophenoloxidase activating system—a complement-like pathway in arthropods. In *Infection processes in fungi*: 160–7. (Eds Aist, J. & Roberts, D.W.). Rockefeller Foundation Study, Bellagio, Italy.

Söderhäll, K. & Smith, V.J. (In press). The prophenoloxidase activating cascade as a recognition and defense system in arthropods. In *Haemocytic and humoral immunity in arthropods*. (Ed. Gupta, A.P.). J. Wiley & Sons Ltd, New York.

Söderhäll, K., Smith, V.J. & Johansson, M. (In press). Exocytosis and uptake of bacteria by isolated haemocyte populations of crustaceans: Evidence for cell co-operation in the defence reactions of arthropods. *Cell Tissue Res.*

Söderhäll, K. & Unestam, T. (1979). Activation of crayfish serum prophenoloxidase: the specificity of cell wall glucan activation and activation by purified fungal glycoprotein. *Can. J. Microbiol.*, **25**: 404–16.

Söderhäll, K., Vey, A. & Ramstedt, M. (1984). Hemocyte lysate enhancement of fungal spore encapsulation by crayfish hemocytes. *Devl comp. Immunol.* **8**: 23–30.

Söderhäll, K., Wingren, A., Johansson, M. & Bertheussen, K. (1985). The cytotoxic reactions of haemocytes from the freshwater crayfish *Astacus astacus*. *Cell. Immunol.* **94**: 326–32.

Starkey, P.M. & Barrett, A.J. (1977). α_2 Macroglobulin, a physiological regulator of protease activity. In *Proteinases in mammalian cells and tissues*: 665–96. (Ed. Barrett, A.J.). Elsevier/North Holland Biochem. Press, Amsterdam.

Stoltz, D.B. & Cook, D.I. (1983). Inhibition of host phenoloxidase activity by parasitoid Hymenoptera. *Experientia* **39**: 1022–4.

Tyson, C.J. & Jenkin, C.R. (1974). Phagocytosis of bacteria *in vitro* by haemocytes from the crayfish (*Parachaeraps bicarinatus*). *Aust. J. exp. Biol. med. Sci.* **52**: 341–8.

Tyson, C.J., McKay, D. & Jenkin, C.R. (1974). Recognition of foreignness in the freshwater crayfish, *Parachaeraps bicarinatus*. *Contemp. Top. Immunobiol.* **4**: 159–66.

Unestam, T. (1981). Fungal diseases of freshwater and terrestrial Crustacea. In *Pathogenesis of invertebrate microbial diseases*: 485–510. (Ed. Davidson, E.W.). Allanheld, Osmun & Co., New Jersey.

Unestam, T. & Söderhäll, K. (1977). Soluble fragments from fungal cell walls elicit defence reactions in crayfish. *Nature, Lond.* **267**: 45–6.

Vey, A. & Vago, C. (1973). Protozoan and fungal diseases of *Austropotamobius pallipes* Lereboullet in France. *Freshwater Crayfish* **1**: 165–79.

Vranckx, R. & Durliat, M. (1981). Encapsulation of *Psorospermium haeckeli* by the haemocytes of *Astacus leptodactylus. Experientia* **37**: 40–2.

Yeaton, R.W. (1981). Invertebrate lectins: II. Diversity of specificity, biological synthesis and function in recognition. *Devl comp. Immunol.* **5**: 535–45.

Zlotkin, E., Gurevitz, M. & Shulov, A. (1973). The toxic effects of phenoloxidase from the haemolymph of tenebrionid beetles. *J. Insect Physiol.* **19**: 1057–65.

Symp. zool. Soc. Lond. (1986) No. 56 81–93

Lectins in molluscs and arthropods: their occurrence, origin and roles in immunity

Lothar RENWRANTZ

Zoologisches Institut und Zoologisches Museum der Universität Hamburg Martin-Luther-King-Platz 3, 2000 Hamburg 13, West Germany

Synopsis

Phagocytes of molluscs and arthropods detect and migrate towards invading micro-organisms. This reaction and the subsequent phagocytosis event require specific receptors on the haemocyte surface to recognize different types of foreign molecules either produced by or located on the membrane of the micro-invader. Apart from different sugar determinants, the only type of binding molecule so far identified on the membrane of haemocytes reveals carbohydrate specificity and may be characterized as a lectin. Information on the occurrence of haemocyte-bound lectins is summarized as well as indications of the presence of cell-bound lectins inside organs.

Soluble lectins also occur in the mucus, organs and body fluids of molluscs and arthropods. The detection of these molecules is discussed with special reference to the sensitivity of agglutination assays. It is demonstrated that lectin concentrations non-detectable by this test system nevertheless may reveal a high immunobiological activity in that they induce the recognition and subsequent uptake of foreign particles by haemocytes. Further results on this opsonizing effect of soluble lectins are discussed including a hypothetical explanation for the accelerated binding of a lectin/particle-complex to the haemocyte membrane in contrast to the reaction of free haemolymph lectin molecules.

Possible versatility of membrane-integrated recognition molecules of invertebrate haemocytes

It is apparent from numerous experimental observations that phagocytosis is essential for the protection of invertebrates from micro-organisms which have penetrated their tissues. Phagocytic invertebrate haemocytes therefore require the capacity to detect the presence of these invaders and to discriminate between self and non-self prior to ingestion of the foreign cells. Detection of and migration towards invading micro-organisms by phagocytes is based on the development of a concentration gradient of molecules released either by the micro-organisms or by tissue cells which are damaged in the

ZOOLOGICAL SYMPOSIUM No. 56
ISBN 0-19-854004-3

course of penetration. These chemotactic factors have to be recognized by phagocytes as well as the non-self properties of the micro-organism itself.

The chemical nature of factors chemotactic for invertebrate haemocytes is unknown. However, it is known that the chemotactic response of mammalian leucocytes is stimulated by a variety of polypeptides (Wilkinson 1981) and one may assume that peptides may also stimulate chemotaxis of invertebrate phagocytes. Polypeptides or more complex proteins also occur as membrane-integrated molecules of foreign cells. In addition, the surface of bacteria or cells is characterized by determinant carbohydrates which are part of polysaccharides or glycoproteins or glycolipids. They also may bear membrane-integrated lectins. Recognition of chemotactically active molecules or membrane components of micro-invaders requires an interaction between these molecules and complementary molecular structures on the surface of phagocytes. Consequently, we may not exclude the possibility that invertebrate haemocytes possess a versatility of receptors involved in the recognition of substances of different chemical nature.

Occurrence of haemocyte-bound lectins

Not much is known about membrane-integrated molecules of molluscan and arthropod haemocytes. Although different sugar determinants have been found (see Renwrantz 1983), the only successfully investigated cell surface receptor reveals carbohydrate binding specificity indicating its involvement in the recognition of carbohydrate determinants of foreign cells.

Amirante (1976) demonstrated that FITC-labelled antibodies raised in a rabbit against the humoral lectin of *Leucophaea maderae* bind to one haemocyte sub-population of this cockroach. A comparable cross-reaction between antibodies against serum lectin and the surface of haemocytes was reported for *Lymnaea stagnalis* by Van der Knaap, Boerrigter-Barendsen, Van der Hoeven & Sminia (1981). Haemocyte membranes of another mollusc, the oyster *Crassostrea virginica*, agglutinated indicator cells (Vasta, Sullivan, Cheng, Marchalonis & Warr 1982) which indicated the presence of a membrane-bound lectin. Its occurrence on the surface of haemocytes was subsequently demonstrated by Vasta, Cheng & Marchalonis (1984) using antibodies raised against the serum lectin of *Crassostrea*.

As in the oyster, antibodies specific for the lectin from the serum of *Mytilus edulis* bound to the surface of this mussel's haemocytes (Renwrantz & Stahmer 1983). Both the agglutinating activity of the serum lectin and the Ca^{++}-dependent attachment of yeast cells to the haemocyte surface were inhibited by bovine submaxillary gland mucin. This indicated combining sites of comparable specificity for both types of molecules, the serum and the haemocyte-bound lectin.

With respect to molluscan and arthropod haemocytes, no further results have been published on the occurrence of membrane-bound agglutinins. However, it should be mentioned that the excellent study of Coombe, Ey & Jenkin (1984a) on particle recognition by phagocytes from the colonial ascidian *Botrylloides leachii* contains proof of two different haemocyte-bound agglutinins. Furthermore, these authors demonstrated the secretion of a third agglutinin (HA2) by haemocytes which might explain one source of the serum agglutinins in invertebrates.

Apart from experiments which directly demonstrate the occurrence of lectins on the surface of haemocytes, we also have to consider findings which may indirectly indicate the presence of these recognition molecules. Attachment of target cells to arthropod haemocytes *in vitro* in the absence of serum factors (McKay & Jenkin 1970; Scott 1971) was interpreted to represent a non-specific phenomenon (Ratcliffe 1985), perhaps explicable by the hypothesis of Lackie (1981) that physicochemical properties, such as charge and wettability of foreign particles, may cause their adherence to the haemocyte surface. However, we also have to keep in mind that Ca^{++}-ions in the culture medium may activate surface-bound recognition molecules (McKay & Jenkin 1970; Renwrantz & Stahmer 1983) or that haemocytes in monolayers on slides may secrete agglutinin molecules.

Membrane-integrated lectins could also be involved in the recognition of saccharide moieties of different types of soluble foreign molecules, whereby haemocytes may become stimulated to release substances. In this connection, the best known example is the phenoloxidase system whose precursor molecules seem to be located inside granular cells of both insects (Schmit, Rowley & Ratcliffe 1977) and crustaceans (Söderhäll & Smith 1983). Degranulation of haemocytes and concomitant activation of prophenoloxidase is stimulated by different types of polysaccharides (Söderhäll 1982) or by lipopolysaccharides (Ratcliffe, Leonard & Rowley 1984) which are surface components of many micro-organisms. Since degranulation is stimulated by polysaccharides, haemocytes have to possess complementary combining sites on their surface, i.e. carbohydrate-specific molecules by which they recognize these foreign substances either in their solubilized form or as membrane-integrated components of micro-invaders.

Finally, with respect to membrane-bound lectins, we have to consider their occurrence on the surface of organ cells. Following the fate of foreign particles injected into different molluscan and arthropod species, different authors have described the accumulation of bacteria and non-self cells inside organs (Tyson & Jenkin 1973; Bayne 1973; Van der Knaap *et al.* 1981; Renwrantz, Schäncke, Harm, Erl, Liebsch & Gercken 1981; White & Ratcliffe 1981; Mullainadhan 1982). Histological investigation of these organs indicated that cells lining the haemal system bound bacteria or foreign cells (Tyson & Jenkin 1973; Van der Knaap *et al.* 1981; Renwrantz, Schäncke *et al.* 1981) and Van der Knaap *et al.* (1981) demonstrated, by immunocytochemical means, membrane-linked agglutinin at the surface of cells lining blood vessels of *Lymnaea stagnalis*. In *Helix*, the binding of foreign cells to organs could be inhibited by specific carbohydrates, supporting the conclusion that in gastropod snails membrane-associated lectins may also occur on the surface of tissue cells (Renwrantz, Schäncke *et al.* 1981).

Occurrence and origin of non-membrane-bound lectins

While indications or experimental proof of the existence of membrane-integrated lectins among arthropods and molluscs have only rarely been published, soluble lectins are known to occur in a variety of species belonging to different systematic groups (reviewed by Gold & Balding 1975; Yeaton 1981a; Ey & Jenkin 1982; Vasta & Marchalonis 1983; Ratcliffe, Rowley, Fitzgerald & Rhodes 1985). We may summarize that lectins have been found in the haemolymph of a variety of arthropods and molluscs and that agglutinating activity has also been detected in eggs and glands of the reproductive tract or in the mucus of different molluscan species.

Mucus lectins

The agglutinin in the slime from the surface of different cephalopod species was found to be secreted by epidermal gland cells (Marthy 1974). More recently, agglutinins were isolated from the mucus of the head-foot area of the gastropods *Arion empiricorum* (Habets, Vieth & Hermann 1979), *Helix aspersa* (Fountain & Campbell 1984) and *Achatina fulica* (Iguchi, Momoi, Egawa & Matsumota 1985). The mucus lectin of *Helix* is a protein, thought to be trapped within the mucopolysaccharides of the slime. It could originate from the protein glands, one of four functionally differing types of unicellular glands in the subepidermal connective tissue of the head-foot (Campion 1961).

Gland and egg lectins

The main source of lectins in pulmonate gastropods and their eggs is the albumin gland, a part of the sexual apparatus (Prokop, Schlesinger & Rackwitz 1965; Kilias, Schnitzler, Kothbauer, Stober & Prokop 1972). As this organ is in direct contact with the body fluid in the open circulatory system of the snail, it was suggested that lectin might be slowly released, thus causing the observed weak agglutinating activity of the pulmonate's haemolymph (see Ey & Jenkin 1982).

Haemolymph lectins

As far as investigated, gastropod haemolymph only contains a relatively low agglutinating activity often not detectable with untreated indicator cells.

Thus, according to Anderson & Good (1976), the haemolymph of *Otala lactea* does not contain an agglutinin. However, although the haemolymph of *Helix pomatia* does not clump untreated erythrocytes, enzyme-treated indicator cells are agglutinated (Reifenberg & Uhlenbruck 1971; Renwrantz 1979; Nielsen, Koch & Drachmann 1983). The agglutinating activity can also be enriched by ultrafiltration of haemocyanin-depleted *Helix* serum (Renwrantz 1979).

These examples clearly demonstrate that negative results of agglutination tests have to be very carefully evaluated. First, agglutination of indicator cells always depends on several factors, such as the presence of suitable lectin receptors, their number and their accessibility. Secondly, we have to be aware that agglutination tests are a very insensitive method for investigating the occurrence of lectins in biological fluids, a point which can be illustrated by the following example.

The agglutinin from the albumin gland of *Helix pomatia* (molecular weight of 8×10^4 daltons according to Hammarström, Westöö & Björk 1972) has a high affinity for the blood group A receptor, and only 5–10 mg lectin per litre are required for agglutination of red blood cells (Renwrantz, unpublished observation). A one molar solution of this lectin (= 8×10^4 g/litre or 8×10^{10} μg/litre) would contain 6.02×10^{23} molecules/litre. Consequently, a subagglutinating concentration of 1 μg lectin/litre would contain 7.5×10^{12} molecules/litre. Although this concentration of lectin molecules is still relatively high we would need a concentration 5,000 to 10,000 times higher to obtain first visible signs of agglutination of untreated human A erythrocytes. From this example, we can conclude that only relatively high lectin concentrations in invertebrate haemolymph are detectable by agglutination tests even by the use of optimal indicator cells!

In comparison to *Helix* haemolymph, high agglutinin activities occur in the serum of some other molluscs and a variety of arthropod species. However, lectins have only been purified from the haemolymph of the merostomates *Limulus polyphemus* (Marchalonis & Edelman 1968; Roche & Monsigny 1974), *Carcinoscorpius rotunda cauda* (Bishayee & Dorai 1980) and *Tachypleus tridentatus* (Shishikura & Sekiguchi 1983), from the haemolymph of the crustaceans *Homarus americanus* (Hall & Rowlands 1974a, b) and *Cancer antennarius* (Ravindranath, Higa, Cooper & Paulson 1985) and from the insects *Sarcophaga peregrina* (Komano, Mizuno & Natori 1980) and certain grasshoppers (Stebbins & Hapner 1985). With respect to molluscan haemolymph, lectins have been purified from the lamellibranchs *Crassostrea virginica* (Li & Flemming 1967; Acton, Bennett, Evans & Schrohenloher 1969). *Tridacna maxima* (Baldo, Sawyer, Stick & Uhlenbruck 1978), *Mytilus edulis* (Renwrantz & Stahmer 1983), from the gastropod *Biomphalaria glabrata* (Boswell & Bayne 1984) and from the cephalopod *Octopus vulgaris* (Rögener, Renwrantz & Uhlenbruck 1985).

The origin of serum lectins is largely unknown. Besides their possible release from albumin glands in gastropods, it seems likely that they are secreted by haemocytes (for review see Ratcliffe, Rowley *et al.* 1985).

Immunobiological importance of haemolymph lectins

Lectins precipitate soluble glycoconjugates and bind to determinant carbohydrate moieties of membrane glycoproteins or glycolipids thus causing agglutination of the respective cells. Consequently, lectins have been suggested to play a role, in immunobiological defence systems, as molecules which may cause clumping of viruses, bacteria and cellular micro-invaders or which may precipitate substances secreted by them (Uhlenbruck & Steinhausen 1977; Yeaton 1981b). However, agglutination of micro-organisms can only occur under the experimental condition where large numbers of particles are injected into an animal's circulation. Under natural conditions, it seems highly unlikely that infections may lead to a sufficient density of bacteria required for an agglutination reaction. Or, more concrete, even if 10,000 bacteria at a time should succeed in entering the circulation of an animal like *Helix pomatia*, their dilution by the approximately 10 ml of haemolymph would be so great that individual bacteria could barely maintain physical contact with each other, a prerequisite for their clumping by agglutinin molecules. On the other hand, we cannot exclude agglutination reactions between heterologous cells, i.e. between bacteria and haemocytes. In molluscs and arthropods, blood cells occur in a relatively high concentration, for example, 1×10^6 cells per ml of haemolymph, a density which may often increase by a factor of 10 during infections (Renwrantz, Schäncke *et al.* 1981). If lectin-induced binding of bacteria to the surface of haemocytes is followed by their ingestion, all requirements for calling the lectin an opsonin are fulfilled. Opsonizing molecules are defined as humoral factors which facilitate binding of foreign particles to and promote their ingestion by phagocytes.

About 20 years ago, Tripp (1966) incubated rabbit erythrocytes in the lectin-containing serum of the oyster *Crassostrea virginica* and demonstrated an enhanced uptake of these pretreated target cells by monolayers of *Crassostrea* haemocytes. Tripp correlated the increased phagocytic uptake with an opsonizing activity of lectin molecules adsorbed onto the erythrocyte's surface. Since then, this procedure has been adopted to study different animals by a variety of authors whose results mostly support the findings of Tripp (for review see Coombe *et al.* 1984b). Nevertheless, a general statement on the identity of haemolymph lectins and opsonins is not yet possible.

This is mainly due to the fact that purified agglutinins have rarely been used in the respective test systems.

A pure lectin has been tested in clearance studies with *Helix pomatia* (Harm & Renwrantz 1980). Injected human A erythrocytes adsorbed opsonin molecules from the haemolymph so that secondarily injected A erythrocytes were eliminated much more slowly from the circulation of the vineyard snail. However, if the second dose of erythrocytes was pretreated with a subagglutinating concentration of *Helix* albumin gland agglutinin, this blockade was reversed indicating the strong opsonizing effect of the lectin. Subsequently, in an *in vitro* test system, the opsonizing activity of the agglutinin purified from the haemolymph of *Mytilus edulis* was demonstrated (Renwrantz & Stahmer 1983). This lectin (10 μg/ml prepared in Ca^{++}-free buffer), increased the percentage of *Mytilus* haemocytes which phagocytosed yeast to about 60%, in comparison to the 5–8% which phagocytosed yeast in buffer alone. For vertebrates, it is known that an opsonin/particle-complex becomes coupled to specific opsonin receptor sites on the surface of phagocytes. With regard to invertebrate animals the nature of the receptor sites for opsonins is still essentially undetermined, although one may postulate that the sugar-specific combining sites of opsonizing humoral lectins bind to carbohydrates on the haemocyte surface. However, it also might be possible that a conformational change of opsonizing molecules is induced by their interaction with foreign particles so that a hitherto hidden combining site becomes exposed for a receptor on the haemocyte membrane (Coombe *et al.* 1984b). As different agglutinins of plant and invertebrate origin have been thoroughly investigated and are known to possess combining sites for complementary carbohydrates on the surface of cells, we recently chose heterologous lectins to induce recognition and phagocytosis of foreign cells by *Mytilus* haemocytes. We tried to clarify whether or not stimulation of phagocytosis by solubilized lectin molecules is dependent upon the binding of these molecules to carbohydrates on both effector and target cells (Mullainadhan & Renwrantz 1986). These phagocytosis experiments were performed in Ca^{++}-free media to avoid activation of the membrane-integrated lectin on the surface of *Mytilus* haemocytes (Renwrantz & Stahmer 1983).

The agglutinin from the albumin gland of *Helix pomatia* is known to react specifically with human blood group A erythrocytes (Uhlenbruck & Prokop 1966) and also binds to the surface of *Mytilus* haemocytes (Renwrantz, Daniels & Hansen1985). Therefore, fixed A-erythrocytes were used as target cells and pre-incubated in a subagglutinating concentration of *Helix* agglutinin (1 μg/ml). When *Mytilus* haemocyte monolayers on slides were covered with these presensitized target cells for 15 min, 27 ± 8.4% of the haemocytes were phagocytic in comparison to only 10 ± 4.9% in control monolayers which were exposed to non-presensitized target cells. When, instead of the target cells, the monolayers were pretreated with the agglutinin solution (30 μg/ml) and subsequently overlaid with A-erythrocytes, the percentage of phagocytosing haemocytes increased to 78 ± 5.6%. However, human O-erythrocytes which do not bind the *Helix* agglutinin were only phagocytosed by 6 ± 4.5% of the lectin-pretreated haemocytes. From these results we may draw the conclusion that, in our test system, lectin-induced phagocytosis only occurs when the lectin may bind to the phagocyte as well as to the target cell. For this reason the uptake of O-erythrocytes was not stimulated as they could not react with the blood group A specific *Helix* agglutinin bound to the surface of *Mytilus* haemocytes. This finding was further supported by the fact that Concanavalin A, which binds to human erythrocytes but not to *Mytilus* haemocytes (Renwrantz, Daniels *et al.* 1985), did not induce phagocytosis. However, two other lectins, wheat germ agglutinin and *Ricinus* 120, which, like *Helix* agglutinin, bind as well to *Mytilus* haemocytes as to human erythrocytes, also strongly increased phagocytosis.

The stimulating effect of the different lectins on phagocytosis could be inhibited by specific sugars, thus demonstrating the ability of carbohydrate-binding molecules to function as recognition molecules. Furthermore, our results support the assumption that carbohydrates on the surface of haemocytes represent the opsonin receptor sites to which an opsonin/particle-complex becomes coupled.

Opsonin/cell-interaction

As outlined above, opsonizing haemolymph agglutinins are thought first to bind to the surface of invading micro-organisms and subsequently to the surface of haemocytes, thus forming a bridge between phagocyte and target cell. However, it is still unknown why haemolymph agglutinins do not directly bind to their respective receptors on the phagocyte surface prior to contacting foreign cells. To explain this phenomenon, two frequently discussed assumptions were recently summarized by Coombe *et al.* (1984b: 247): 'Self cells may not bind uncomplexed opsonin. The adhesion of opsonin to foreign particles could induce a conformational change, however, and so generate or expose another opsonin-binding site for a receptor on the phagocytes.' Or, alternatively, 'an opsonin may bind with high affinity to some non-self particles but bind self cells with low affinity.' As yet, no experimental proof is available for a conformational change occurring in particle-bound agglutinins which increases their affinity for haemocyte surface receptors. However, with respect to the alternative suggestion, a theoretical explanation why opsonin-coated targets bind to the membrane of phagocytes with an increased avidity is possible.

If soluble serum lectin molecules only reveal a low affinity for the complementary receptor sites on the haemocyte's surface, the dissociation of the receptor-lectin complex would be very high, much higher than the association between receptors and lectin:

$$\text{receptors}\longleftrightarrow\text{lectin}\overset{\leftarrow}{\longrightarrow}\text{receptors} + \text{lectin} \qquad (1)$$

According to the law of mass action the following correlation exists between the *concentrations* of the three different components of Equation 1:

$$[\text{receptors}\longleftrightarrow\text{lectin}]\times\overrightarrow{k} = \overleftarrow{k} \times [\text{receptors}] \times [\text{lectin}] \qquad (2)$$

$$\text{or} \quad \frac{[\text{receptors}\longleftrightarrow\text{lectin}]}{[\text{receptors}] \times [\text{lectin}]} = \frac{\overleftarrow{k}}{\overrightarrow{k}} = Kc \qquad (3)$$

Equation 3 shows that an experimental increase in the concentration of lectin molecules (= increase of lectin density) also causes an increase in the concentration of the receptors⟷lectin complex, as *Kc* has to remain constant, i.e. more lectin molecules bind to the surface of a haemocyte. This fact demonstrates, with respect to the interaction of a lectin-coated (= opsonized) particle with a haemocyte, that if a large number of serum lectin molecules has accumulated on the surface of a micro-organism (= increase in lectin density), an increased number of these molecules would bind to the complementary receptor sites on the membrane of a haemocyte thus causing attachment of the target to the surface of the effector cell.

A priori, these considerations seem to be conclusive. But we have to recognize that the reaction between lectin-coated particles and the respective lectin receptors on the phagocyte's surface are more complex and include different additional parameters like, for example, receptor distribution and mobility. This phenomenon, however, may even improve the interaction. Thus binding of soluble lectin molecules to receptor sites on target or effector cells might stimulate receptor capping (Yoshino, Renwrantz & Cheng 1979) and thus cause an accumulation of lectin molecules in a relatively small area of the cell surface which would further increase the effect described above.

The role of humoral invertebrate lectins as part of a general self–non-self recognition system could be disputed because of a lack of diversity of their binding properties. Usually, the specificity of a lectin is defined in terms of the carbohydrate (monosaccharide or oligosaccharide) with the highest inhibiting effect on lectin-induced agglutination or precipitation. However, this does not mean that a lectin may not bind to a number of other carbohydrates with decreasing affinities. For example, the blood group A specific agglutinin from the albumin gland of *Helix pomatia* binds to a variety of carbohydrates as has been shown by different authors (for summary see Ey & Jenkin 1982). The best inhibitor of this lectin is 1-0-methyl-α-D-N-acetylgalactosamine which is twice as active as 1-0-phenyl-α-D-N-acetylgalactosamine or N-acetylgalactosamine and more than 12 times more active than N-acetylglucosamine or galactosamine. The *Helix* lectin also binds to non-reducing terminal α-D-galactose and to dextran. Explanations for the diverse molecular affinities of lectins have recently been discussed by Yeaton (1981b). They allow the conclusion that a 'graded' recognition of non-self can be expected from the binding properties of a single type of lectin. Moreover, we cannot exclude that invertebrate haemolymph usually contains more than one lectin.

To obtain further proof for the immunobiological role of lectins in body fluids of invertebrate animals, experiments with purified lectins are required in order to investigate precisely their specificities and to determine the affinity with which they bind to different carbohydrate structures on the surface of self and non-self cells.

References

Acton, R.T., Bennett, J.C., Evans, E.E. & Schrohenloher, R.E. (1969). Physical and chemical characterization of an oyster hemagglutinin. *J. biol. Chem.* **15**: 4128–45.

Amirante, G. (1976). Production of heteroagglutinins in haemocytes of *Leucophaea maderae* L. *Experientia* **32**: 526–8.

Anderson, R.S. & Good, R.A. (1976). Opsonic involvement in phagocytosis by mollusk hemocytes. *J. Invert. Path.* **27**: 57–64.

Baldo, B.A., Sawyer, W.H., Stick, R.V. & Uhlenbruck, G. (1978). Purification and characterization of a galactan-reactive agglutinin from the clam *Tridacna maxima* (Röding) and a study of its combining site. *Biochem. J.* **175**: 467–77.

Bayne, C.J. (1973). Molluscan internal defense mechanism: the fate of C^{14}-labelled bacteria in the land snail *Helix pomatia* (L). *J. comp. Physiol.* **86**: 17–25.

Bishayee, S. & Dorai, D.T. (1980). Isolation and characterization of a sialic acid binding lectin (Carcinoscorpin) from Indian horseshoe crab *Carcinoscorpius rotunda cauda*. *Biochim. biophys. Acta* **623**: 89–97.

Boswell, C.A. & Bayne, C.J. (1984). Isolation, characterization and functional assessment of a hemagglutinin from the plasma of *Biomphalaria glabrata*, intermediate host of *Schistosoma mansoni*. *Devl comp. Immunol.* **8**: 559–68.

Campion, M. (1961). The structure and function of the cutaneous glands in *Helix aspersa*. *Q. Jl microsc. Sci.* **102**: 195–216.

Coombe, D.E., Ey, P.L. & Jenkin, C.R. (1984a). Particle recognition by haemocytes from the colonial ascidian *Botrylloides leachii*: Evidence that the *B. leachii* HA-2 agglutinin is opsonic. *J. comp. Physiol.* **154B**: 509–21.

Coombe, D.E., Ey, P.L. & Jenkin, C.R. (1984b). Self/non-self recognition in invertebrates. *Q. Rev. Biol.* **59**: 231–55.

Dageförde, S., Schmücker, A. & Renwrantz, L. (1986). Capping of cell surface receptors on blood cells from the molluscs *Helix pomatia* (Gastropoda) and *Mytilus edulis* (Lamellibranchiata). *Eur. J. Cell Biol.* **41**: 113–20.

Ey, P.L. & Jenkin, C.R. (1982). Molecular basis of self/non-self discrimination in the Invertebrata. In *The reticuloendothelial system*: 321–91. (Eds Cohen, N. & Sigel, M.M.). Plenum Press, New York & London.

Fountain, D.W. & Campbell, B.A. (1984). A lectin isolated from mucus of *Helix aspersa*. *Comp. Biochem. Physiol.* **77B**: 419–25.

Gold, E.R. & Balding, P. (1975). *Receptor-specific proteins.* American Elsevier Publishing Company, Inc., New York.

Habets, L., Vieth, U.C. & Hermann, G. (1979). Isolation and new biological properties of *Arion empiricorum* lectin. *Biochim. biophys. Acta* **582**: 154–63.

Hall, J.L. & Rowlands, D.T. (1974a). Heterogeneity of lobster agglutinins. I. Purification and physicochemical characterization. *Biochemistry* **13**: 821–7.

Hall, J.L. & Rowlands, D.T. (1974b). Heterogeneity of lobster agglutinins. II. Specificity of agglutinin-erythrocyte binding. *Biochemistry* **13**: 828–32.

Hammarström, S., Westöö, A. & Björk, J. (1972). Subunit structure of *Helix pomatia* A hemagglutinin. *Scand. J. Immunol.* **1**: 295–309.

Harm, H. & Renwrantz, L. (1980). The inhibition of serum opsonins by a carbohydrate and the opsonizing effect of purified agglutinin on the clearance of nonself particles from the circulation of *Helix pomatia*. *J. Invert. Path.* **36**: 64–70.

Iguchi, S.M.M., Momoi, T., Egawa, K. & Matsumota, J.J. (1985). An N-acetylneuraminic acid-specific lectin from the body surface mucus of African giant snail. *Comp. Biochem. Physiol.* **81B**: 897–900.

Kilias, R., Schnitzler, S., Kothbauer, H., Stober, D. & Prokop, O. (1972). Further investigations on haemagglutinins in pulmonate snails. *Z. ImmunForsch. expl. Klin. Immun.* **144**: 157–66. (In German, English summary.)

Komano, H., Mizuno, D. & Natori, S. (1980). Purification of lectin induced in the haemolymph of *Sarcophaga peregrina* larvae on injury. *J. biol. Chem.* **255**: 2919–24.

Lackie, A.M. (1981). Immune recognition in insects. *Devl comp. Immunol.* **5**: 191–204.

Li, M.F. & Flemming, C. (1967). Hemagglutinins from oyster hemolymph. *Can. J. Zool.* **45**: 1225–34.

McKay, D. & Jenkin, C.R. (1970). Immunity in the invertebrates. The role of serum factors in phagocytosis of erythrocytes by haemocytes of the freshwater crayfish (*Parachaeraps bicarinatus*) *Aust. J. exp. Biol. med. Sci.* **48**: 139–50.

Marchalonis, J.J. & Edelman, G.M. (1968). Isolation and characterization of a hemagglutinin from *Limulus polyphemus*. *J. molec. Biol.* **32**: 453–65.

Marthy, H.-J. (1974). Nachweis und Bedeutung einer hämagglutinierenden Substanz aus der Haut von Cephalopoden. *Z. ImmunForsch. expl. Klin. Immun.* **148**: 225–34.

Mullainadhan, P. (1982). *Studies on the clearance of foreign substances from the hemolymph of* Scylla serrata *Forskal* (*Crustacea: Decapoda*). Ph. D. Thesis: Madras Univ., India.

Mullainadhan, P. & Renwrantz, L. (1986). Lectin-dependent recognition of foreign cells by hemocytes of the mussel, *Mytilus edulis*. *Immunobiology* **171**: 263–73.

Nielsen, H.E., Koch, C. & Drachmann, O. (1983). Non-respiratory haemolymph proteins in the vineyard snail *Helix pomatia*. Changes after phagocytosis *in vivo*. *Devl comp. Immunol.* 7: 413–22.

Prokop, O., Schlesinger, D. & Rackwitz, A. (1965). Über eine thermostabile "antibody-like substance" (Anti-A_{hel}) bei *Helix pomatia* und deren Herkunft. *Z. ImmunForsch. expl. Klin. Immun.* **129**: 402–12.

Ratcliffe, N.A. (1985). Invertebrate immunity—a primer for the non-specialist. *Immunol. Lett.* **10**: 253–70.

Ratcliffe, N.A., Leonard, C. & Rowley, A.F. (1984). Prophenoloxidase-activation: Non-self recognition and cell cooperation in insect immunity. *Science, N.Y.* **226**: 557–9.

Ratcliffe, N.A., Rowley, A.F., Fitzgerald, S.W. & Rhodes, C.P. (1985). Invertebrate immunity—Basic concepts and recent advances. *Int. Rev. Cytol.* **97**: 183–349.

Ravindranath, M.H., Higa, H.H., Cooper, E.L. & Paulson, J.C. (1985). Purification and characterization of an O-acetylsialic acid-specific lectin from a marine crab *Cancer antennarius*. *J. biol. Chem.* **260**: 8850–6.

Reifenberg, U. & Uhlenbruck, G. (1971). Über ein Agglutinin in der Hämolymphe von *Helix pomatia*. *Immun-Information* **1**: 14–15.

Renwrantz, L. (1979). Eine Untersuchung molekularer und zellulärer Bestandteile der Hämolymphe von *Helix pomatia* unter besonderer Berücksichtigung immunbiologisch aktiver Komponenten. *Zool. Jb.* (*Allg. Zool.*) **83**: 283–333.

Renwrantz, L. (1983). Involvement of agglutinins (lectins) in invertebrate defense reactions: the immuno-biological importance of carbohydrate-specific binding molecules. *Devl comp. Immunol.* 7: 603–8.

Renwrantz, L., Daniels, J. & Hansen, P.-D. (1985). Lectin-binding to hemocytes of *Mytilus edulis*. *Devl comp. Immunol.* **9**: 203–10.

Renwrantz, L., Schäncke, W., Harm, H., Erl, H., Liebsch, H. & Gercken, J. (1981). Discriminative ability and function of the immunobiological recognition system of the snail *Helix pomatia*. *J. comp. Physiol.* **141**: 477–88.

Renwrantz, L. & Stahmer, A. (1983). Opsonizing properties of an isolated hemolymph agglutinin and demonstration of lectin-like recognition molecules at the surface of hemocytes from *Mytilus edulis*. *J. comp. Physiol.* **149**: 535–46.

Roche, A.C. & Monsigny, M. (1974). Purification and properties of limulin: A lectin (agglutinin) from hemolymph of *Limulus polyphemus*. *Biochim. biophys. Acta* **371**: 242–54.

Rögener, W., Renwrantz, L. & Uhlenbruck, G. (1985). Isolation and characterization of a lectin from the hemolymph of the cephalopod *Octopus vulgaris* (Lam.) inhibited by α-D-lactose and N-acetyl-lactosamine. *Devl comp. Immunol.* **9**: 605–16.

Schmit, A.R., Rowley, A.F. & Ratcliffe, N.A. (1977). The role of *Galleria mellonella* hemocytes in melanin formation. *J. Invert. Path.* **29**: 232–4.

Scott, M.T. (1971). Recognition of foreignness in invertebrates. II. *In vitro* studies of cockroach phagocytic haemocytes. *Immunology* **21**: 817–28.

Shishikura, F. & Sekiguchi, K. (1983). Agglutinins in the horseshoe crab *hemolymph: Purification of a potent agglutinin of horse erythrocytes from the* hemolymph of *Tachypleus tridentatus*, the Japanese horseshoe crab. *J. Biochem.* **93**: 1539–46.

Söderhäll, K. (1982). Prophenoloxidase activating system and melanization—a recognition mechanism of arthropods? A review. *Devl comp. Immunol.* **6**: 601–11.

Söderhäll, K. & Smith, V.J. (1983). Separation of the haemocyte populations of *Carcinus maenas* and other marine decapods, and prophenoloxidase distribution. *Devl comp. Immunol.* 7: 229–39.

Stebbins, M.R. & Hapner, K.D. (1985). Preparation and properties of hemagglutinin from haemolymph of Acrididae (grasshoppers). *Insect Biochem.* **15**: 451–62.

Tripp, M.R. (1966). Hemagglutinin in the blood of an oyster *Crassostrea virginica. J. Invert. Path.* **8**: 478–84.

Tyson, C.J. & Jenkin, C.R. (1973). The importance of opsonic factors in the removal of bacteria from the circulation of the crayfish (*Parachaeraps bicarinatus*). *Aust. J. exp. Biol. med. Sci.* **51**: 609–15.

Uhlenbruck, G. & Steinhausen, G. (1977). Tridacnins: Symbiosis-profit or defense-purpose? *Devl comp. Immunol.* **1**: 183–92.

Uhlenbruck, G. & Prokop, O. (1966). An agglutinin from *Helix pomatia* which reacts with terminal N-acetyl-D-galactosamine. *Vox Sang.* **11**: 519–26.

Van der Knaap, W.P.W., Boerrigter-Barendsen, L.H., van den Hoeven, D.S.P. & Sminia, T. (1981). Immunocytochemical demonstration of a humoral defence factor in blood cells (amoebocytes) of the pond snail, *Lymnaea stagnalis. Cell Tissue Res.* **219**: 291–6.

Vasta, G.R., Cheng, T.C. & Marchalonis, J.J. (1984). A lectin on the hemocyte membrane of the oyster (*Crassostrea virginica*). *Cell. Immunol.* **88**: 475–88.

Vasta, G.R. & Marchalonis, J.J. (1983). Humoral recognition factors in the Arthropoda. The specificity of Chelicerata serum lectins. *Am. Zool.* **23**: 157–71.

Vasta, G.R., Sullivan, J.T., Cheng, T.C., Marchalonis, J.J. & Warr, G.W. (1982). A cell membrane-associated lectin of the oyster hemocyte. *J. Invert. Path.* **40**: 367–77.

White, K.N. & Ratcliffe, N.A. (1981). Crustacean internal defense mechanisms: clearance and distribution of injected bacteria by the shore crab *Carcinus maenas* (L.). In *Developmental and comparative immunology* **1**: 153–8. (Ed. Solomon, J.B.). Pergamon Press, Elmsford, N.Y.

Wilkinson, P.C. (1981). Peptide and protein chemotactic factors and their recognition by neutrophil leucocytes. *Symp. Soc. Exp. Biol.* **12**: 53–72.

Yeaton, R.W. (1981a). Invertebrate lectins: I. Occurrence. *Devl comp. Immunol.* **5**: 391–402.

Yeaton, R.W. (1981b). Invertebrate lectins: II. Diversity of specificity, biological synthesis and function in recognition. *Devl comp. Immunol.* **5**: 535–45.

Yoshino, T.P., Renwrantz, L. & Cheng, T.C. (1979). Binding and redistribution of surface membrane receptors for Concanavalin A on oyster hemocytes. *J. exp. Zool.* **207**: 439–50.

Symp. zool. Soc. Lond. (1986) No. 56 95–115

Interference—immunity of mosquitoes to bunyavirus superinfection

David H. L. BISHOP
N.E.R.C. Institute of Virology, Mansfield Road, Oxford OX1 3SR

and Barry J. BEATY
Department of Microbiology, College of Veterinary Medicine and Biomedical Studies Colorado State University, Fort Collins, CO 80523, U.S.A.

Synopsis

It has been shown that *Aedes triseriatus* mosquitoes are permissive for the replication of La Crosse, LAC, virus, a member of the California serogroup of bunyaviruses (family Bunyaviridae). Naturally or experimentally infected *Ae. triseriatus* are able to transmit LAC virus to vertebrates following virus replication in the mosquito species. Infected female mosquitoes may also pass virus transovarially to their offspring thereby allowing the virus to overwinter in the eggs of the invertebrate host. In addition, transovarially infected male mosquitoes can venereally transmit virus to female mosquitoes. Since bunyaviruses have a segmented RNA genome, it is possible that dual infections of a mosquito could yield novel recombinant viruses through the mechanism of RNA segment reassortment. The importance of such an attribute to the evolutionary capability of bunyaviruses has to be considered in relation to vector-virus interactions (vector preferences, virus interference, immunity etc.). Recombination has been demonstrated for genetically compatible bunyaviruses in experiments involving dual, simultaneous infections of mosquitoes with temperature-sensitive (*ts*) mutants of different bunyaviruses and by the recovery of wild-type, intertypic, reassortant virus progeny. However, when viruses were introduced successively into a female mosquito using different intervals of infection, the superinfecting virus did not appear to elicit recombination when the inoculation interval was longer than one to two days after the first virus infection. By employing alternative genetic procedures, it has been shown that virus superinfection was unsuccessful (after the first two days) when the superinfecting virus belonged to the same bunyavirus gene pool as the initial virus. Superinfection was, however, successful when a genetically compatible second arbovirus was introduced within the first two days, or at any time if the second virus belonged to a different bunyavirus gene pool, or to another virus family. The limitations that have been observed for successful superinfection upon the evolution potential of bunyaviruses in mosquito vectors are discussed.

ZOOLOGICAL SYMPOSIUM No. 56
ISBN 0-19-854004-3

Introduction

The intent of this review is to describe and discuss the data that demonstrate interference to virus superinfection in an experimental bunyavirus—*Aedes triseriatus* mosquito system in relation to its potential impact upon virus transmission by the invertebrate host as well as its limitation on virus evolution. Up to the present time the subject of the arthropod host response to a virus infection has been inadequately studied. No evidence has been reported for an acquired, antigen-specific, immune response (either humoral or cell-mediated) that can be compared to the immune response of vertebrates to virus infections. The role of pre-existing humoral or cellular reactions (as discussed by other participants in this symposium) to a virus infection is not clear either with regard to the role of phagocytic cells, or the elimination of virus-infected cells, or with regard to the removal of free virus (virus liberated from infected cells or virus that has not been able to gain entry into cells). For further information on resistance mechanisms of arthropods to viruses, the reader is referred to earlier reviews edited by Maramorosch & Shope (1975).

The Bunyaviridae

The majority of viruses that have been assigned to the Bunyaviridae on the basis of common morphological features have been grouped by serological and biochemical tests into four genera (*Bunyavirus*, *Nairovirus*, *Phlebovirus* and *Uukuvirus*; see Berge, Shope & Work 1970; Berge, Chamberlain, Shope & Work 1971; Berge 1975; Karabatsos 1978; Bishop & Shope 1979; Bishop, Calisher *et al.* 1980). A fifth genus (*Hantavirus*, including Hantaan virus and related species) has been proposed (McCormick, Palmer, Sasso & Kiley 1982; Schmaljohn & Dalrymple 1983; Schmaljohn, Hasty, Harrison & Dalrymple 1983). There are also many viruses that have yet to be assigned to a genus (see Bishop, Calisher *et al.* 1980); their inclusion in the family is based solely on morphological and morphogenetic features by comparison with recognized members of the family. The diseases elicited in vertebrate hosts range from inapparent infections to neurological disorders, haemorrhages and visceral organ infections depending on the virus and host species (Bishop & Shope 1979).

Other than the Hantaan-related viruses, almost all of the viruses assigned to the Bunyaviridae are considered to be arthropod-borne viruses (arboviruses; i.e., in addition to replication in vertebrates, they are believed to replicate in and to be transmitted by arthropods, although formal demonstration of this postulate has only been obtained for a few members). By contrast, Hantaan and related viruses do not appear to involve arthropods in their transmission cycles. Often only specific species

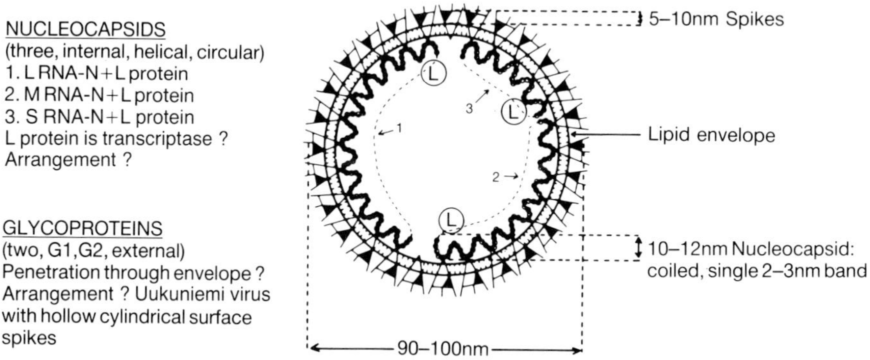

Fig. 1. Schematic bunyavirus particle

of mosquitoes (or gnats, ticks, phlebotomines and tabanids) are permissive for the replication of a particular Bunyaviridae member. The molecular bases for such vector-virus specificities are not known. In general, no major effect of virus infection is evident in the behaviour and longevity of the infected arthropods. Some changes in the feeding habits, slight reductions in the longevity of the arthropod, and reduced yields of viable eggs have been reported in laboratory studies as discussed by Turell & LeDuc (1983). The significance of such observations to the natural infection course is not clear.

The structural and genetic properties of the Bunyaviridae

In brief, Bunyaviridae members are spherical viruses (*c.* 100 nm in diameter), enveloped in lipid and in possession of an external layer of glycoproteins (see Fig. 1). The viral glycoproteins consist of two protein species that are designated in relation to their relative sizes as G1 and G2. Two or three glycoproteins have been reported for Hazara virus, a member of the *Nairovirus* genus (Clerx, Casals & Bishop 1981; Foulke, Rosato & French 1981). Bunyaviridae viruses have internal components consisting of three nucleocapsids. Electron microscopic analyses of the nucleocapsids have revealed that they consist of coiled strands, 2–3 nm in diameter, that are occasionally observed to be circular and sometimes supercoiled with diameters of between 7 and 12 nm. Each nucleocapsid consists of a single species of RNA (large, L, medium, M, and small, S), nucleoprotein (N) and minor quantities of a large protein that has a size of $180–200 \times 10^3$ Daltons (the L protein). The viral RNA species of several representative members of the family have been characterized by a variety of procedures and have been shown to consist of three molecules of single-stranded RNA with distinct sizes and sequences. Fuller descriptions of the characteristics of members of the Bunyaviridae, their isolation and molecular properties can be found in the review by Bishop & Shope (1979). Most of the genetic, molecular and vector transmission studies that have so far been reported have employed *Bunyavirus* genus members as discussed below.

Some 151 virus serotypes, subtypes and varieties have been described as bunyaviruses (Table 1). The viruses have been placed into 16 serogroups to reflect the results of serological comparisons (see Bishop & Shope 1979). In general, members of each serogroup are serologically more closely related to each other by particular tests (e.g., neutralization, haemagglutination inhibition) than to members of other serogroups although, depending on the test (e.g., complement fixation analyses), distant serological relationships have been demonstrated between representative members of different bunyavirus serogroups (see Bishop & Shope 1979; Klimas, Ushijima, Clerx-van Haaster & Bishop 1981). These latter observations, plus the common biochemical features of the serogroups, form the rationale for the inclusion of all 16 serogroups into the *Bunyavirus* genus. No serological relationships have been detected between bunyaviruses and members of other Bunyaviridae genera.

DNA cloning studies have revealed that the S RNA species of snowshoe hare (SSH) bunyavirus (see Table 1) is of the order of 3×10^5 Daltons (Bishop, Gould, Akashi & Clerx-van Haaster 1982). From sequence analyses it has been determined that the S RNA of SSH bunyavirus codes for two proteins that are read from overlapping reading frames in viral-complementary S mRNA sequences. These proteins are the viral N protein (26×10^3 Daltons) and a non-structural protein (NS_S, 10×10^3 Daltons). Protein and genetic analyses have confirmed these coding assignments as well as the existence of the S coded N and NS_S proteins (Gentsch, Wynne, Clewley, Shope & Bishop 1977; Gentsch & Bishop 1978; Fuller & Bishop 1982; Fuller, Bhown & Bishop 1983). The results of cloning the M RNA species of SSH virus have indicated that its size is 1.5×10^6 Daltons (Eshita & Bishop 1984). From these analyses, the SSH M RNA has been shown to code for a precursor to the viral glycoproteins (162×10^3 Daltons) in a viral-complementary mRNA sequence. Genetic and molecular analyses have demonstrated that the bunyavirus M RNA codes for both G1 (115×10^3 Daltons), G2 (38×10^3 Daltons) and a second non-structural protein, NS_M, that has been estimated to be of the order of 15×10^3 Daltons (Gentsch & Bishop 1979; Fuller & Bishop 1982). Presumably these proteins come from the glycoprotein precursor, although their order and mode of derivation from that polypeptide are not known. As expected for the external glycoproteins of viruses, the M coded gene products elicit neutralizing antibodies and are the principal determinants of the virulence of bunyaviruses in animals (Gentsch, Rozhon, Klimas, El Said, Shope & Bishop 1980; Shope, Rozhon & Bishop 1981; Shope, Tignor, Jacoby, Watson, Rozhon & Bishop 1982). The glycoproteins are also important in determining whether a virus can establish a productive infection in a particular mosquito species (Beaty, Holterman, Tabachnick, Shope, Rozhon & Bishop 1981; Beaty, Miller, Shope, Rozhon & Bishop 1982). The size of the SSH (or other bunyavirus) L RNA has not been determined by DNA cloning; however, it is estimated to be of the order of 3×10^6 Daltons (Bishop & Shope 1979). The SSH L RNA is believed to code for the $180–200 \times 10^3$ Dalton L protein that has been identified in virus preparations (the putative transcriptase-replicase), although formal proof of that postulate has not been reported.

As judged by gel electrophoresis, the sizes of the viral RNA and proteins of some 30–40 other bunyaviruses that have been analysed are similar to those of SSH virus (Clewley, Gentsch & Bishop 1977; Bishop & Shope 1979; El Said *et al.* 1979; Ushijima, Clerx-van Haaster & Bishop 1981; Klimas, Thompson, Calisher, Clark, Grimstad & Bishop 1981; Klimas, Ushijima *et al.* 1981). In summary, bunyaviruses, which are mostly transmitted by mosquitoes or gnats, have RNA sizes of around 3.3×10^5 Daltons (S), 1.5×10^6 Daltons (M), and (estimated) 3×10^6 Daltons (L). These RNA species code for the $20–25 \times 10^3$ Dalton N and 10×10^3 Dalton NS_S proteins (S RNA), the $110–120 \times 10^3$ Dalton G1, $30–40 \times 10^3$ Dalton G2 and 15×10^3 Dalton NS_M proteins (M RNA) and, presumably, the $180–200 \times 10^3$ Dalton L protein (L RNA). The structural features of members of the other Bunyaviridae genera are distinct from those of bunyaviruses (Bishop, Calisher *et al.* 1980).

Bunyavirus evolution

Bunyavirus evolution by genetic drift

Evidence has been presented that for LAC virus no two virus isolates recovered from nature have identical genome sequences as evidenced by RNA oligonucleotide fingerprinting, or by RNA sequencing (El Said *et al.* 1979; Klimas, Thompson *et al.* 1981; Clerx-van Haaster, Akashi, Auperin & Bishop 1982). This observation applies to viruses isolated from the same place but at different times (Fig. 2), or at the same time but different places. However, by such procedures most of the LAC virus isolates can be shown to be closely related to each other, albeit they are also distinguishable. No doubt identical LAC viruses could be isolated (e.g. from different siblings of an infected female mosquito); however, the diversity that has been seen is taken as evidence for genetic evolution of the virus through, principally, the accumulation of point mutations.

Table 1. Proposed serological classification of viruses of Family Bunyaviridae, Genus *Bunyavirus*[a].

Anopheles A Group	*Bwamba Group*	*Gamboa Group*	*Simbu Group*
Anopheles A	Bwamba	Gamboa	Simbu
CoAr 3624[b]	Pongola	Pueblo Viejo (75–2621[b])	Akabane
ColAn 57389[b]	*C Group*	Alajuela[b]	Yaba–7[b]
Las Maloyas	Caraparu	San Juan (78V2441[b],	Manzanilla
Lukuni	Caraparu (BeH5546[b],	75V–2374[b])	Ingwavuma
Trombetas[b]	Trinidad[b])		Inini
Tacaiuma	Ossa	*Guama Group*	Mermet
H–32580[b]	Apeu	Guama	Buttonwillow
SPAr 2317[b] (Virgin River)	Vinces	Ananindeua	Nola
CoAr 1071[b] (CoAr 3627[b])	Bruconha[b]	Moju	Oropouche
	Madrid	Mahogany Hammock	Facey's Paddock[b]
Anopheles B Group	Marituba	Bertioga	Utinga
Anopheles B	Murutucu	Cananeia	Utive[b]
Boraceia	Restan	Guaratuba	Sabo
	Nepuyo (63U11[b])	Itimirim	Tinaroo
Bunyamwera Group	Gumbo Limbo	Mirim	Sathuperi (Douglas)
Bunyamwera	Oriboca	Bimiti	Shamonda
Batai (Calovo)	Itaqui	Catu	Sango
Birao		Timboteua	Peaton
Cache Valley (Tlacotalpan)	*California Group*		Shuni
Maguari (CbaAr 426[b])	California encephalitis	*Koongol Group*	Aino (Kaikalur, Samford[b])
Playas	Inkoo	Koongol	Thimiri
Xingu[b]	La Crosse (snowshoe hare)	Wongal	
Germiston	San Angelo		*Tete Group*
Ilesha	Tahyna (Lumbo[b])	*Minatitlan Group*	Tete
Lokern	Melao	Minatitlan	Bahig
Northway	Keystone	Palestina	Matruh
Santa Rosa	Jamestown Canyon (South River[b],		Tsuruse
Shokwe[b]	Jerry Slough)	*Olifantsvlei Group*	Batama
Tensaw	Serra do Navio	Olifantsvlei (Bobia)	
Kairi	trivittatus	Botambi	*Turlock Group*
Main Drain	Guaroa		Turlock

Wyeomyia		*Patois Group*	Lednice
Anhembi (BeAr 314206[b], BeA- 328208[b])	*Capim Group*	Patois	Umbre
Macaua	Capim	Abras	M'Poko
Sororoca	Acara	Babahoyo	Yaba–1[b]
Taiassui[b]	Moriche	Shark River	
	Benevides	Zegla	
	BushBush	Pahayokee	
	Benfica		
	GU71U344[b]		
	Juan Diaz		
	Guajara (GU71U350[b])		

[a] Viruses are classified in three steps indicated by degrees of indentation—complex, virus, and subtype; viruses in parentheses are varieties.
[b] These viruses are not in the published or working *International Catalogue of Arboviruses* (Berge 1975; Karabatsos 1978).

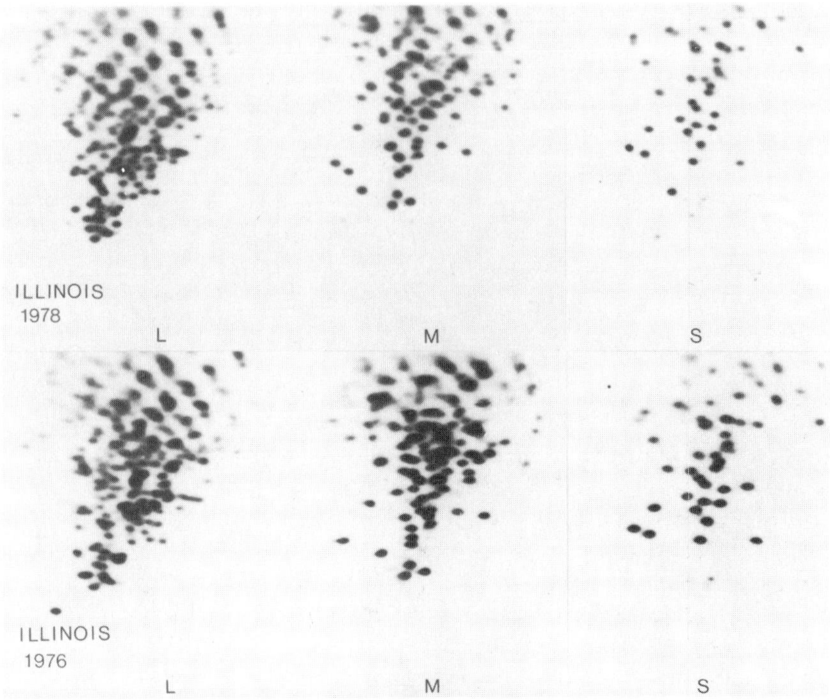

Fig. 2. Oligonucleotide fingerprints of the viral RNA species of two La Crosse virus isolates demonstrating minor sequence variations. The L, M and S RNA species of two La Crosse virus isolates obtained from mosquitoes collected in the same site but at different times (1976, 1978) were recovered from virus-infected cells grown in the presence of radioactive phosphorus. The RNA species were purified, digested to completion with ribonuclease T1 (hydrolysing the RNA at guanosine residues) and the products resolved by two-dimensional gel electrophoresis. The first dimension was run from left to right, the second from bottom to top. By this means the unique sequence largest ($A_nC_mU_oG$) oligonucleotides were located in the lower half of the gels and could be identified by autoradiography. Comparison of the two sets of fingerprints indicates extensive homology, but also sequence differences (for further details see Clewley *et al.* 1977; Klimas, Thompson *et al.* 1981).

Bunyavirus evolution by genetic recombination (reassortment)

In view of the observation that bunyaviruses have a segmented RNA genome with RNA segments coding for different gene products, the possibility that recombinant viruses may be generated in dual virus infections by RNA segment reassortment or by other mechanisms has been investigated. Genetic studies using mammalian tissue cultures and *ts* mutants of SSH virus have confirmed that intratypic wild-type recombinant viruses can be formed from *ts* mutants representing different RNA species (Gentsch & Bishop 1976; Gentsch, Wynne *et al.* 1977; Gentsch, Robeson & Bishop

1979). Similar results have been reported for LAC and other bunyaviruses. For example, LAC M RNA mutant *ts* I–16 and LAC L RNA mutant *ts* II–5 on co-infection of the same cell yielded recombinant wild-type viruses that, unlike the parent *ts* viruses which only gave plaques at 33°C, produced plaques at both 39.8°C (40°C) and 33°C.

The limited intertypic genetic recombination studies that have been reported for California, Bunyamwera and Group C bunyaviruses suggest that many members of a bunyavirus serogroup are genetically compatible (i.e. capable of RNA segment reassortment; see Gentsch, Wynne *et al.* 1977; Gentsch, Robeson *et al.* 1979; Gentsch, Rozhon *et al.* 1980; Gentsch & Bishop 1978, 1979; Iroegbu & Pringle 1981a,b; Rozhon, Gensemer, Shope & Bishop 1981; Shope, Rozhon *et al.* 1981; Pringle & Iroegbu 1982; Bishop, Fuller & Akashi 1983). However, viruses representing different bunyavirus serogroups do not appear to be capable of generating recombinant viruses, also not all viruses assigned to a serogroup are genetically interactive. Whether this conclusion extends to all bunyavirus serogroups, or to members of serogroups of other genera, remains to be determined. For SSH and LAC viruses, all the possible genotypes of progeny viruses (2^3, i.e., 8) have been isolated from dual virus infections in tissue culture, but none from crosses involving SSH and the Group C Oriboca and Caraparu viruses (unpublished data). Reassortant viruses have been obtained between LAC, SSH, California encephalitis, trivitattus, Lumbo and Tahyna viruses (all members of the California group), although none has been detected between these viruses and Guaroa virus (a serologically distant California group virus), Oriboca, or Caraparu (Group C viruses), or with several members of the Bunyamwera serogroup.

A derivatory question from these *in vitro* experimental observations is whether bunyaviruses in their natural environment evolve through genetic recombination (reassortment). This question has to be considered in relation to the preferred vertebrate and invertebrate hosts, the viral determinants of permissive infections and the opportunities afforded for dual virus infection (see below). In brief, though, direct evidence for naturally occurring reassortant viruses has been obtained. Thus, RNA genome fingerprint analyses of LAC virus field isolates have provided evidence for intratypic recombinant (reassortant) LAC viruses (Klimas, Thompson *et al.* 1981). Reassortant viruses have also been identified among field isolates of members of the Patois serogroup of bunyaviruses (Ushijima *et al.* 1981).

Bunyavirus infections of mosquito species

Natural bunyavirus-vector relationships

The natural relationship between LAC virus and *Ae. triseriatus* mosquitoes has been extensively studied. This mosquito species has been demonstrated to be an efficient oral (Watts, Morris, Wright, DeFoliart & Hanson 1972), transovarial (Pantuwatana, Thompson, Watts, Yuill & Hanson 1974; Watts, Grimstad, DeFoliart, Yuill & Hanson 1973; Watts, Pantuwatana, DeFoliart, Yuill & Thompson, 1973), and venereal (Thompson & Beaty 1977) transmitter of LAC virus. In temperate regions of the United States, LAC virus overwinters in diapaused *Ae. triseriatus* eggs (Watts, Thompson, Yuill, DeFoliart & Hanson 1974; Beaty & Thompson 1975). Immunofluorescence techniques were used to determine the virogenesis of LAC in *Ae. triseriatus* and to derive anatomical explanations of the unique vector-virus interactions (transovarial and venereal transmission) observed in this system. Subsequent to oral infection, virus antigen was first detected in the pyloric portion of the midgut, six days post-infection. By ten days the virus had disseminated from the midgut and antigen was detected in most secondary organ systems, including ovaries and salivary glands. It was observed that LAC virus infection was virtually pantropic in the arthropod with most organ systems containing large quantities of virus antigen. Detection of virus antigen in ovarian follicles and in accessory sex gland fluid provided anatomical explanations for the observed transovarial and venereal transmission, respectively (Beaty & Thompson 1976, 1977).

Although serologically closely related, each of the California serogroup bunyaviruses has a distinct epizootiology often involving a select vector species and a particular preferred, but not exclusive, vertebrate host. For example, in the United States, trivittatus virus is closely associated with an *Ae. trivittatus*-cottontail rabbit feeding cycle. Keystone virus is associated with an *Ae. atlanticus*-squirrel cycle. SSH virus (which is serologically almost indistinguishable from LAC virus), is associated with an *Ae. canadensis* and *Ae. communis* group-snowshoe hare cycle, while LAC virus in the midwest of the United States, and elsewhere, is associated with an *Ae. triseriatus*-chipmunk-tree squirrel cycle (Sudia, Newhouse, Calisher & Chamberlain 1971; Pantuwatana, Thompson, Watts & Hanson 1972; Le Duc 1979).

California serogroup viruses have, however, been isolated on occasion from alternative vectors in nature. An example is LAC virus which has been isolated occasionally from *Ae. canadensis* and *Ae. communis* group mosquitoes, as well as from tabanids. SSH virus has been isolated infrequently from *Ae. triseriatus* mosquitoes. In addition, at least six of the California group viruses have been isolated from *Ae. vexans*. These observations pose the question of whether there are differences in the efficiencies of bunyavirus infection for the different vector species.

Viral determinants of permissive bunyavirus infections of mosquitoes

Since all the possible genotype combinations of LAC and SSH virus reassortants are available from *in vitro* (cell culture) experiments (Gentsch, Wynne *et al.* 1977; Gentsch, Robeson *et al.* 1979; Rozhon *et al.* 1981) and because *Ae. triseriatus* mosquitoes are not the normal vectors of SSH virus, the question of the viral determinants for a permissive replication of LAC and SSH virus in that arthropod species has been investigated by employing either LAC, or SSH virus, or LAC–SSH reassortant viruses to infect *Ae. triseriatus* mosquitoes. The results obtained from several studies are summarized in Table 2 (Beaty, Holterman *et al.* 1981; Beaty, Miller *et al.* 1982). The data have been interpreted to indicate that the LAC viral M RNA gene products (the glycoprotein species) are the principal determinants of the efficiency of LAC virus, and of reassortants containing a LAC M RNA, both to establish a disseminated infection and to be transmitted by *Ae. triseriatus* mosquitoes. By contrast, viruses with a SSH M RNA were inefficiently transmitted. Although attenuating mutations in other LAC RNA species may affect the LAC M gene property, as determined with an attenuated LAC/LAC/SSH reassortant virus (Table 2 and Rozhon *et al.* 1981), the major viral determinants of efficient vector transmission appear to be the LAC viral glycoproteins (Beaty, Holterman *et al.* 1981; Beaty, Miller *et al.* 1982).

Bunyavirus recombination in mosquitoes after intrathoracic infection

From the bunyavirus isolation data it can be concluded that, even though the viruses are distinct epizootiologically, many are sympatric throughout much of their respective ranges, theoretically providing ample opportunity for dual virus infection of vector species to occur in nature. In order to experimentally investigate the question of whether dual California group virus infection of mosquitoes would result in intertypic recombinant virus formation, dual infections of laboratory stocks of colonized *Ae. triseriatus* mosquitoes were undertaken using intrathoracic inoculation of *ts* mutants of LAC and SSH viruses. The study yielded evidence for wild-type SSH-LAC reassortant virus formation and for the transmission of the recombinant viruses to a vertebrate host (Table 3; Beaty, Rozhon, Gensemer & Bishop 1981). In parenthesis, one combination of mutants (SSH II–21 × LAC I–20) did not yield the expected reassortant progeny. Similar results were obtained in tissue culture (Gentsch, Wynne *et al.* 1977). It was subsequently shown that in tissue culture the expected reassortants are only infrequently obtained (Rozhon *et al.* 1981). In summary, it has been demonstrated that at least certain intertypic bunyaviruses can be generated in the arthropod host.

Bunyavirus recombination in mosquitoes after oral infection

Intrathoracic infection is not a normal route of infection, therefore studies were conducted to determine if *Ae. triseriatus* mosquitoes would yield recombinant viruses if they were allowed to acquire the appropriate viruses by the oral route either simultaneously or by interrupted feeding. Mosquitoes were allowed to partially engorge on blood-virus mixtures containing either LAC *ts* I–16, or LAC *ts* II–5, or wild-type LAC virus (to serve as three control experiments), or mixtures of LAC *ts* I–16 and wild-type LAC virus, or LAC *ts* I–16 and LAC *ts* II–5 viruses. As shown in Table 4, after 14 days of incubation in the mosquitoes, wild-type viruses were only obtained from the mosquitoes that fed on blood containing the two *ts* mutants, or from those that received the mutant and wild-type viruses. As expected, only mutant viruses were obtained from the controls that ingested a single *ts* mutant; likewise wild-type virus was recovered from the mosquitoes that only received the wild-type virus inoculum (data not shown). The results of the simultaneous *ts* mutant infections indicated therefore that recombination had occurred in the dually infected mosquitoes.

For the interrupted feeding studies, mosquitoes were allowed to partially engorge on blood meals containing LAC *ts* I–16 and two hours later permitted to engorge to completion on meals containing LAC *ts* II–5 (Table 4). Wild-type viruses were detected in the mosquitoes after 14 days of incubation. As expected, mosquitoes that received wild-type LAC through the interrupted feeding protocol also yielded wild-type virus (Table 4). Thus

Table 2. The role of the M RNA in infection, dissemination and transmission of LAC and SSH parent and SSH-LAC reassortant viruses by *Aedes triseriatus* mosquitoes.

Virus L/M/S/ genotype	% Disseminated infection[a]	% Transmission[b]
LAC/LAC/LAC	100	100
SSH/LAC/LAC	97	96
SSH/LAC/SSH	97	90
LAC/LAC/SSH	12	64
SSH/SSH/SSH	17	33
LAC/SSH/LAC	42	42
LAC/SSH/SSH	8	36
SSH/SSH/LAC	29	31

[a] Mosquitoes were allowed to engorge on blood-virus mixtures containing either wild-type LAC, or SSH, or LAC-SSH reassortants of the indicated L/M/S RNA genotypes. After 14 days of extrinsic incubation the mosquitoes were then analysed. Viral antigen in mosquito tissues was identified by immunofluorescence and the % disseminated infection was scored on the basis of the numbers of mosquitoes in which antigen was observed in all tissues, divided by those for which antigen was only located in midgut cells. The LAC/LAC/SSH data used in the dissemination analyses are probably atypical since they involved a reassortant that subsequent analyses (Rozhon *et al.* 1981) demonstrated carried a silent attenuating L defect. In each of the analyses, representing the data from several experiments, between 13 and 45 individual mosquitoes were used (for details see Beaty, Miller *et al.* 1982).

[b] The % transmission was scored on the basis of the numbers of individual mosquitoes that transmitted virus and induced disease in the suckling mice (moribund or dead mice) following intrathoracic inoculation of 10^3 plaque-forming units (pfu) of virus and after ten days (average) of extrinsic incubation, divided by the number of feeding mosquitoes that exhibited disseminated infections when the engorged mosquitoes were sacrificed and analysed by immunofluorescence. Viruses were inoculated intrathoracically to bypass the mesenteron and thereby preclude variables associated with midgut passage. In each of the analyses, representing the data from several experiments, between 14 and 60 mosquitoes were used (for details see Beaty, Rozhon *et al.* 1981).

Table 3. Viruses recovered from intrathoracic dually infected *Aedes triseriatus* mosquitoes and from mice on which the infected mosquitoes were allowed to feed[a].

	Virus cross		Mosquitoes		Mice	
			%*ts*	%*wt*	%*ts*	%*wt*
SSH I–1	x	SSH II–22	95	5	90	10
LAC I–20	x	LAC II–4	45	55	20	80
SSH I–1	x	LAC II–5[b]	35	65	55	45
SSH II–21	x	LAC I–20[c]	100	0	100	0

[a] Viruses recovered from dually infected mosquitoes that had been inoculated with *ts* viruses of LAC or SSH virus representing different RNA segments (mutants SSH I–1, SSH II–22, etc.), or viruses obtained from derived moribund and dead mice, were plated on BHK–21 cells at 33°C, virus plaques picked and reassayed at both 33°C and 39.8°C to score for temperature-sensitive (*ts*) and wild-type (*wt*) viruses. The results for each cross represent the averages of analyses of several mosquitoes recovered after 7, 14 and 21 days post-inoculation, or of one or more mice obtained after the 7, 14 and 21 day mosquito feedings (for details see Beaty, Bishop *et al.* 1983).
[b] For different progeny *wt* virus clones, both virus-induced intracellular polypeptides and RNA oligonucleotide fingerprint analyses indicated that the expected SSH/LAC/SSH reassortants were present (Gentsch, Wynne *et al.* 1977; Beaty, Rozhon *et al.* 1981).
[c] The *wt* progeny that would be expected from this cross are rarely obtained even in tissue culture, presumably owing to inefficient gene product interactions of the heterologous viral RNA species (Gentsch, Robeson *et al.* 1979; Rozhon *et al.* 1981).

Table 4. Dual infection of *Aedes triseriatus* mosquitoes and generation of recombinant viruses[a].

Infection protocol	Infection rates			
	33°C assay		40°C assay	
Simultaneous				
LAC *ts* I–16 + LAC *wt*	15/15	(100%)	15/15	(100%)
LAC *ts* I–16 + LAC *ts* II–5	8/8	(100%)	2/8	(25%)
Interrupted feeding				
LAC *ts* I–16 then LAC *wt*	19/19	(100%)	18/19	(95%)
LAC *ts* I–16 then LAC *ts* II–5	20/20	(100%)	4/20	(20%)

[a] Infection rates are expressed as the number of mosquitoes that were found to contain virus (> 10 pfu) as detected by plaque assay at 33°C or 40°C, divided by the number tested. Virus-blood meals on which the mosquitoes were initially allowed to feed (partially for the interrupted feeding protocol) contained 6.5–7.3 logs of each of the indicated viruses per ml. At 2 h post-ingestion the mosquitoes in the interrupted feeding experiment were allowed to engorge to completion on blood-virus mixtures containing 6.5–7.3 logs per ml of the second virus (LAC *ts* II–5 or LAC wild-type, *wt*, virus). All mosquitoes were held for 14 days, triturated, then assayed for virus. For further details see Beaty, Sundin *et al.* (in press).

analyses of the mosquitoes that were superinfected through interrupted feeding protocol showed that recombinant viruses were produced by this method.

Since bunyavirus RNA segment reassortment would only be epidemiologically significant if the newly formed recombinant viruses were transmitted to a vertebrate host, mosquitoes from the above experiment were permitted to feed (after 14 days of incubation) on groups of five to seven baby mice. Brains were extracted from the resulting moribund or dead mice and assayed for virus at the *ts* mutant permissive (33°C) and non-permissive (40°C) temperatures. Viruses producing plaques at 40°C (i.e., with a wild-type phenotype) were isolated from the mice on which mosquitoes were allowed to feed that had previously ingested LAC *ts* I–16 followed two hours later by wild-type LAC virus, indicating that prior infection by the mutant did not preclude transmission of the superinfecting virus. Also wild-type virus was recovered from the mice that were substrates for feeding by mosquitoes that had been initially infected with LAC *ts* I–16 then superinfected with LAC *ts* II–5. These results indicated therefore that recombinant viruses could be recovered from a vertebrate host on which a dually infected mosquito had fed.

The preceding experiments indicated a low level of recombinant virus formation (Table 4) in both the simultaneous and interrupted feeding protocols. The reason for this observation may be trivial (e.g., the particular experimental conditions that were employed), or may be related to factors such as the relative numbers of the infecting viruses. The effect of these and other factors, such as the incubation time, the number of gonadotrophic cycles, or the use of alternative vectors, upon the production of recombinant bunyaviruses has yet to be determined.

Bunyavirus interference in mosquito species

In nature, the opportunities for simultaneous infection of vector species involving the ingestion of two or more viruses in a single blood meal are probably severely limited by the acute (short-term) character of bunyavirus infections in a vertebrate host. In most instances, critical viraemia threshold titres for infection of mosquitoes are present for only a few days duration before the virus is cleared. Infection and production of maximum viraemia in a particular vertebrate host by two viruses simultaneously is probably relatively rare, although it may occur. It can be argued that the opportunities for sequential infection of a vector species are more probable if several blood meals are taken by the female mosquito. It is known that *Ae. triseriatus* females may ingest several blood meals during their lifetime, thereby allowing the possibility for dual virus infection to occur by that route (DeFoliart 1983). Other means of dual virus infection are possible, for instance when an infected male inseminates and infects a previously infected female, or when an infected mosquito that has acquired virus transovarially from its mother ingests a blood meal containing an alternate virus.

A series of experiments were therefore conducted to analyse the potential for bunyavirus superinfection of *Ae. triseriatus* and to determine possible temporal and phylogenetic constraints on these phenomena. In initial experiments, mosquitoes were inoculated intrathoracically with a *ts* mutant of LAC virus (*ts* II–5, an L RNA mutant) and three, seven, or 14 days later were superinfected by the same route with SSH *ts* I–1 (an M RNA mutant). After further extrinsic incubation the mosquitoes were triturated in tissue culture medium and the homogenates assayed at 33°C and 40°C to quantify

Table 5. Parenteral superinfection of *Aedes triseriatus* mosquitoes previously inoculated with a La Crosse (LAC) *ts* mutant virus[a].

Virus inoculum Day 0	Day 7	Geometric mean titre (log pfu/mosquito) 33°C assay	40°C assay	Log difference
LAC I–16	None	4.5 ± 0.3	< 1.0	
None	LAC *wt*	4.5 ± 0.4	4.2 ± 0.2	
LAC I–16	LAC *wt*	4.4 ± 0.1	< 1.0	> 3.2
None	SSH *wt*	3.3 ± 0.5	3.0 ± 0.5	
LAC I–16	SSH *wt*	4.1 ± 0.4	< 1.0	> 2.0
None	TAH *wt*	4.1 ± 0.6	3.3 ± 0.5	
LAC I–16	TAH *wt*	4.1 ± 0.4	< 1.0	> 2.3
None	TVT *wt*	4.6 ± 0.3	3.7 ± 0.2	
LAC I–16	TVT *wt*	3.8 ± 0.2	< 1.0	> 2.7
None	WN *wt*	5.1 ± 0.3	4.4 ± 0.2	
LAC I–16	WN *wt*	5.2 ± 0.3	4.4 ± 0.4	0
None	VSV *wt*	4.9 ± 0.3	4.0 ± 0.4	
LAC I–16	VSV *wt*	4.5 ± 0.2	3.4 ± 0.7	0.6

[a] Groups of at least four mosquitoes were inoculated intrathoracically with 1.4 log plaque-forming units (pfu) of a LAC *ts* mutant representing the M RNA segment (LAC I–16) and superinfected 7 days later by inoculation with 2–4 log pfu of wild-type (*wt*) viruses representing LAC, or other California group bunyaviruses (snowshoe hare, SSH; Tahyna, TAH; trivittatus, TVT), or a flavivirus (West Nile, WN), or a rhabdovirus (vesicular stomatitis virus, VSV). After a further 7 days of extrinsic incubation, the presence of wild-type and *ts* viruses in the mosquitoes was determined by plaque assays. For further details see Beaty, Bishop *et al.* (1983).

the numbers of *ts* and reassortant wild-type viruses, respectively. Despite the presence of 10^{3-5} *ts* viruses, no wild-type viruses were detected (Beaty, Bishop, Gay & Fuller 1983). The reverse virus inoculation schedule (i.e. SSH *ts* I–1 followed by LAC *ts* II–5) gave similar results, suggesting that genetic interaction between the two viruses had been inhibited.

Reassortment can be an event that occurs at a low frequency; therefore, in order to examine the interference phenomenon with a more sensitive procedure, mosquitoes were inoculated intrathoracically with a *ts* mutant of LAC virus and subsequently challenged seven days later by intrathoracic inoculation of either homologous or heterologous wild-type virus. Since wild-type virus progeny can easily be quantified by *in vitro* plaque assays (at 40°C), the ability of the superinfecting virus to replicate in the mosquito could be determined. The results of the analyses are shown in Table 5. They indicated that the mosquitoes were resistant to superinfection with related California group viruses but not to viruses of other families (Table 5), or to viruses representing other bunyavirus gene pools (Beaty, Bishop *et al.* 1983). The data can be interpreted in terms of an interference phenomenon that is specific to viruses that are members of a gene pool.

Intrathoracic inoculation is not a natural route of infection for mosquitoes. Nonetheless if mosquitoes became resistant to superinfection by natural routes of infection, the opportunities for dual infection of vectors and,

Table 6. Interference to LAC virus oral superinfection of *Aedes triseriatus* mosquitoes[a].

Time until ingestion of challenge virus	Infection rates 33°C assay		40°C assay	
Simultaneous	15/15	(100%)	15/15	(100%)
30 min	8/8	(100%)	8/8	(100%)
2 h	19/19	(100%)	18/19	(95%)
4 h	7/7	(100%)	7/7	(100%)
1 day	18/18	(100%)	11/18	(60%)
2 days	11/11	(100%)	3/11	(27%)
7 days	6/6	(100%)	0/6	(0%)
14 days	5/5	(100%)	0/5	(0%)
21 days	3/3	(100%)	0/3	(0%)
28 days	4/4	(100%)	0/4	(0%)

[a] Infection rates are expressed as the number of mosquitoes that were found to contain virus (> 10 pfu) as detected by plaque assay at 33°C or 40°C, divided by the number tested. Virus blood meals on which the mosquitoes were initially allowed to feed (partially) contained 6.5–7.8 logs of LAC *ts* mutant I–16 per ml. At the indicated times post-ingestion the mosquitoes were allowed to engorge to completion on blood-virus mixtures containing 7–7.8 logs per ml of the challenge wild-type virus. All mosquitoes were held for 14 days after the second meal, triturated, then assayed for virus. For further details see Beaty, Sundin *et al.* (in press).

for bunyaviruses, virus evolution through RNA segment reassortment, would be limited. In nature, many mosquito species exhibit a behaviour pattern called interrupted feeding. If the defensive reaction of a host causes the mosquito to interrupt its feeding, the vector may complete engorgement at a later time on an alternative host. Thus mosquitoes may ingest blood meals from two different vertebrate hosts that are viraemic with two different viruses in a period of time brief enough to preclude interference. In light of these considerations, an experiment was conducted to determine when interference to oral superinfection occurs in mosquitoes (Beaty, Sundin, Chandler & Bishop in press).

In order to determine if mosquitoes could be superinfected by the oral route, *Ae. triseriatus* mosquitoes were permitted to ingest a partial or complete blood meal containing a LAC *ts* mutant virus (LAC I–16). At predetermined times post-feeding, the mosquitoes were permitted to engorge to repletion on a blood meal containing wild-type LAC virus. One cohort of mosquitoes ingested a meal containing both LAC *ts* I–16 and wild-type virus. As shown in Table 6, mosquitoes that received the wild-type virus challenge in the first 24 hours replicated the superinfecting virus. Mosquitoes that ingested wild-type virus after 48 hours were resistant to superinfection (Beaty, Sundin *et al.* in press). Control groups of mosquitoes that received only *ts* virus yielded only *ts* progeny viruses (i.e., the virus was phenotypically stable; Beaty, Sundin *et al.* in press). It was concluded from these studies and the earlier experiment (Table 4) demonstrating that reassortment could occur in mosquitoes superinfected two hours after the initial virus infection, that with the greater elapsed time between the two blood meals the mosquitoes became refractory to homologous virus superinfection.

Conclusions

The molecular basis for the interference phenomenon remains to be determined. Possible mechanisms of interference that require further investigation include the removal, or blocking by gene products of the initial virus, of particular cellular receptors that are used for virus infection of a cell, the induction (from the initial inoculum virus) of defective interfering viruses that prevent the replication of a second, genetically interactive, virus, or the establishment of a host immune mechanism. Using the available genetic tools further experimentation is required to explore these possibilities and to determine whether interference is operative in transovarially infected mosquitoes.

Whatever the mechanism, the experimental observation of virus interference between genetically permissive viruses, if indeed it has a counterpart in nature, may restrict the ability of a bunyavirus to evolve by RNA segment reassortment. That reassortment occurs has been shown by analyses of natural virus isolates (Klimas, Thompson *et al.* 1981; Ushijima *et al.* 1981). However, of some 25 virus isolates that have been analysed, only one intratypic LAC recombinant virus has been identified, which may be interpreted to indicate that recombination is a rare event. It is not clear at this time whether interference is a viral or host-mediated phenomenon. In view of the importance of this subject, research is continuing.

Acknowledgements

This research was supported by grants AI 15400 and AI 19688 from the U.S. Public Health Service.

References

Beaty, B.J., Bishop, D.H.L., Gay, M. & Fuller, F. (1983). Interference between bunyaviruses in *Aedes triseriatus* mosquitoes. *Virology* **127**: 83–90.

Beaty, B.J., Holterman, M., Tabachnick, W., Shope, R.E., Rozhon, E.J. & Bishop, D.H.L. (1981). Molecular basis of bunyavirus transmission by mosquitoes: Role of the middle-sized RNA segment. *Science, N.Y.* **211**: 1433–5.

Beaty, B.J., Miller, B.R., Shope, R.E., Rozhon, E.J. & Bishop, D.H.L. (1982). Molecular basis of bunyavirus *per os* infection of mosquitoes: Role of the middle-sized RNA segment. *Proc. natn. Acad. Sci. U.S.A.* **79**: 1295–7.

Beaty, B.J., Rozhon, E.J., Gensemer, P. & Bishop, D.H.L. (1981). Formation of reassortant bunyaviruses in dually-infected mosquitoes. *Virology* **111**: 662–5.

Beaty, B.J., Sundin, D.R., Chandler, L. & Bishop, D.H.L. (In press). Evolution of bunyaviruses via genome segment reassortment in dually-infected (per os) *Aedes triseriatus* mosquitoes. *Science, N.Y.*

Beaty, B.J. & Thompson, W.H. (1975). Emergence of La Crosse virus from endemic foci: Fluorescent antibody studies of overwintered *Aedes triseriatus. Am. J. trop. Med. Hyg.* **24**: 685–91.

Beaty, B.J. & Thompson, W.H. (1976). Delineation of La Crosse virus in developmental stages of transovarially infected *Aedes triseriatus. Am. J. trop. Med. Hyg.* **25**: 505–12.

Beaty, B.J. & Thompson, W.H. (1977). Tropisms of La Crosse virus in *Aedes triseriatus* following infective blood meals. *J. med. Ent. Honolulu.* **14**: 499–503.

Berge, T.O. (1975). *International catalogue of arboviruses.* DHEW, Atlanta. (*DHEW Publication* No. (CDC) 75–8301.)

Berge, T.O., Chamberlain, R.W., Shope, R.E. & Work, T.H. (1971). The subcommittee on information exchange of the American committee on arthropod-borne viruses: Catalogue of arthropod-borne and selected vertebrate viruses of the world. *Am. J. trop. Med. Hyg.* **20**: 1018–50.

Berge, T.O., Shope, R.E. & Work, T.H. (1970). The subcommittee on information exchange of the American committee on arthropod-borne viruses: Catalogue of arthropod-borne viruses of the world. *Am. J. trop. Med. Hyg.* **19**: 1079–1160.

Bishop, D.H.L., Calisher, C., Casals, J., Chumakov, M.P., Gaidamovich, S.Ya., Hannoun, C., Lvov, D.K., Marshall, I.D., Oker-Blom, N., Pettersson, R.F., Porterfield, J.S., Russell, P.K., Shope, R.E. & Westaway, E.G. (1980). Bunyaviridae. *Intervirology* **14**: 125–43.

Bishop, D.H.L., Fuller, F.J. & Akashi, H. (1983). Coding assignments of the RNA genome segments of California serogroup viruses. *Progr. clin. biol. Res.* **123**: 107–17.

Bishop, D.H.L., Gould, K.G., Akashi, H. & Clerx-van Haaster, C.M. (1982). The complete sequence and coding content of snowshoe hare bunyavirus small (S) viral RNA species. *Nucl. Acids Res.* **10**: 3703–13.

Bishop, D.H.L. & Shope, R.E. (1979). Bunyaviridae. *Compr. Virol.* **14**: 1–156.

Clerx, J.P.M., Casals, J. & Bishop, D.H.L. (1981). Structural characteristics of nairoviruses (Genus *Nairovirus*, Bunyaviridae). *J.gen. Virol.* **55**: 165–78.

Clerx-van Haaster, C.M., Akashi, H., Auperin, D.D. & Bishop, D.H.L. (1982). Nucleotide sequence analyses and predicted coding of bunyavirus genome RNA species. *J. Virol.* **41**: 119–28.

Clewley, J., Gentsch, J. & Bishop, D.H.L. (1977). Three unique viral RNA species of snowshoe hare and La Crosse bunyaviruses. *J. Virol.* **22**: 459–68.

DeFoliart, G.R. (1983). *Aedes triseriatus*: Vector biology in relationship to the persistence of La Crosse virus in endemic foci. *Progr. clin. biol. Res.* **123**: 89–104.

El Said, L.H., Vorndam, V., Gentsch, J.R., Clewley, J.P., Calisher, C.H., Klimas, R.A., Thompson, W.H., Grayson, M., Trent, D.W. & Bishop, D.H.L. (1979). A comparison of La Crosse virus isolates obtained from different ecological niches and an analysis of the structural components of California encephalitis serogroup viruses and other bunyaviruses. *Am. J. trop. Med. Hyg.* **28**: 364–86.

Eshita, Y. & Bishop, D.H.L. (1984). The complete sequence of the medium RNA of snowshoe hare bunyavirus reveals the presence of internal hydrophobic domains in the viral glycoprotein. *Virology* **137**: 227–40.

Foulke, R.W., Rosato, R.R. & French, G.R. (1981). Structural polypeptides of Hazara virus, *J. gen. Virol.*. **53**: 169–72.

Fuller, F., Bhown, A.S. & Bishop, D.H.L. (1983). Bunyavirus nucleoprotein, N, and a non-structural protein, NS_S, are coded by overlapping reading frames in the S RNA. *J. gen. Virol.* **64**: 1705–14.

Fuller, F. & Bishop, D.H.L. (1982). Identification of viral coded non-structural polypeptides in bunyavirus infected cells. *J. Virol.* **41**: 643–8.

Gentsch, J. & Bishop, D.H.L. (1976). Recombination and complementation between temperature-sensitive mutants of the bunyavirus, snowshoe hare virus. *J. Virol.* **20**: 351–4.

Gentsch, J. & Bishop, D.H.L. (1978). Small viral RNA segment of bunyaviruses codes for viral nucleocapsid protein. *J. Virol.* **28**: 417–9.

Gentsch, J.R. & Bishop, D.H.L. (1979). M viral RNA segment of bunyaviruses codes for two unique glycoproteins: G1 and G2. *J. Virol.* **30**: 767–76.

Gentsch, J.R., Robeson, G. & Bishop, D.H.L. (1979). Recombination between snowshoe hare and La Crosse bunyaviruses. *J. Virol.* **31**: 707–17.

Gentsch, J.R., Rozhon, E.J., Klimas, R.A., El Said, L.H., Shope, R.E. & Bishop, D.H.L. (1980). Evidence from recombinant bunyavirus studies that the M RNA gene products elicit neutralizing antibodies. *Virology* **102**: 190–204.

Gentsch, J., Wynne, L.R., Clewley, J.P., Shope, R.E. & Bishop, D.H.L. (1977). Formation of recombinants between snowshoe hare and La Crosse bunyaviruses. *J. Virol.* **24**: 893–902.

Iroegbu, C.U. & Pringle, C.R. (1981a). Genetic interactions among viruses of the Bunyamwera complex. *J. Virol.* **37**: 383–94.

Iroegbu, C.U. & Pringle, C.R. (1981b). Genetics of the Bunyamwera complex. In *The replication of negative strand viruses*: 159–65. (Eds Bishop, D.H.L. & Compans, R.W.) Elsevier, New York.

Karabatsos, N. (1978). Supplement to international catalogue of arboviruses including certain other viruses of vertebrates. *Am. J. trop. Med. Hyg.* **27**: 372–3.

Klimas, R.A., Thompson, W.H., Calisher, C.H., Clark, G.G., Grimstad, P.R. & Bishop, D.H.L. (1981). Genotypic varieties of La Crosse virus isolated from different geographic regions of the continental United States and evidence for a naturally occurring intertypic recombinant La Crosse virus. *Am. J. Epidem.* **114**: 112–31.

Klimas, R.A., Ushijima, H., Clerx-van Haaster, C.M. & Bishop, D.H.L. (1981). Radioimmune assays and molecular studies that place Anopheles B and Turlock serogroup viruses in the *Bunyavirus* genus (Bunyaviridae). *Am. J. trop. Med. Hyg.* **30**: 876–87.

Le Duc, J. (1979). The ecology of California group viruses. *J. med. Ent. Honolulu* **16**: 1–17.

Maramorosch, K. & Shope, R.E. (Eds). (1975). *Invertebrate immunity: Mechanisms of invertebrate vector-parasite relations.* Academic Press, New York.

McCormick, J.G., Palmer, E.L., Sasso, D.R. & Kiley, M.P. (1982). Morphological identification of the agent of Korean hemorrhagic fever (Hantaan virus) as a member of Bunyaviridae. *Lancet* **1982(i)**: 765–8.

Pantuwatana, S., Thompson, W.H., Watts, D.M. & Hanson, R.P. (1972). Experimental infection of chipmunks and squirrels with La Crosse and trivittatus viruses and biological transmission of La Crosse by *Aedes triseriatus. Am. J. trop. Med. Hyg.* **21**: 476–81.

Pantuwatana, S., Thompson, W.H., Watts, D.M., Yuill, T.M. & Hanson, R.P. (1974). Isolation of La Crosse virus from field collected *Aedes triseriatus* larvae. *Am. J. trop. Med. Hyg.* **23**: 246–50.

Pringle, C.R. & Iroegbu, C.U. (1982). A mutant identifying a third recombination group in a bunyavirus. *J. Virol.* **42**: 873–9.

Rozhon, E.J., Gensemer, P., Shope, R.E. & Bishop, D.H.L. (1981). Attenuation of virulence of a bunyavirus involving an L RNA defect and isolation of LAC/SSH/LAC and LAC/SSH/SSH reassortants. *Virology* **111**: 125–38.

Schmaljohn, C.S. & Dalrymple, J.M. (1983). Analysis of Hantaan virus RNA: Evidence for a new genus of Bunyaviridae. *Virology* **131**: 482–91.

Schmaljohn, C.S., Hasty, S.E., Harrison, S.A. & Dalrymple, J.M. (1983). Characterization of Hantaan virions, the prototype virus of hemorrhagic fever with renal syndrome. *J. infect. Dis.* **148**: 1005–12.

Shope, R.E., Rozhon, E.J. & Bishop, D.H.L. (1981). Role of the middle-sized bunyavirus RNA segment in mouse virulence. *Virology* **114**: 273–6.

Shope, R.E., Tignor, G.H., Jacoby, R.O., Watson, H., Rozhon, E.J. & Bishop, D.H.L. (1982). Pathogenicity analyses of reassortant bunyaviruses: coding assignments. In *International symposium on tropical arboviruses and hemorrhagic fever*: 135–46. (Ed. Pinheiro, F.). Impresso Nat. Fund. Sci. Dev. Tech., Belem, Brazil.

Sudia, W.D., Newhouse, V.F., Calisher, C.H. & Chamberlain, R.W. (1971). California group arboviruses: Isolations from mosquitoes in North America. *Mosquito News* **31**: 576–600.

Thompson, W. & Beaty, B. (1977). Venereal transmission of La Crosse (California encephalitis) arbovirus in *Aedes triseriatus* mosquitoes. *Science, N. Y.* **196**: 530–1.

Turell, M.J. & LeDuc, J.W. (1983). The role of mosquitoes in the natural history of California serogroup viruses. *Progr. clin. biol. Res.* **123**: 43–55.

Ushijima, H., Clerx-van Haaster, C.M. & Bishop, D.H.L. (1981). Analyses of Patois group bunyaviruses: evidence for naturally occurring recombinant bunyaviruses and existence of viral coded nonstructural proteins induced in bunyavirus infected cells. *Virology* **110**: 318–32.

Watts, D.M., Grimstad, P.R., DeFoliart, G.R., Yuill, T.M. & Hanson, R.P. (1973). Laboratory transmission of La Crosse encephalitis virus by several species of mosquito. *J. med. Ent. Honolulu* **10**: 583–6.

Watts, D.M., Morris, C.D., Wright, R.E., DeFoliart, G.R. & Hanson, R.P. (1972).

Transmission of La Crosse virus (California encephalitis group) by the mosquito *Aedes triseriatus. J. med. Ent. Honolulu* **9**: 125–7.

Watts, D.M., Pantuwatana, S., DeFoliart, G.R., Yuill, T.M. & Thompson, W.H. (1973). Transovarial transmission of La Crosse virus (California encephalitis group) in the mosquito, *Aedes triseriatus. Science, N.Y.* **182**: 1140–1.

Watts, D.M., Thompson, W.H., Yuill, T.M., DeFoliart, G.R. & Hanson, R.P. (1974). Over-wintering of La Crosse virus in *Aedes triseriatus. Am. J. trop. Med. Hyg.* **23**: 694–700.

Symp. zool. Soc. Lond. (1986) No. 56 117–144

Insect immunity to Trypanosomatidae

D.H. MOLYNEUX
G. TAKLE[1]
E.A. IBRAHIM
and G.A. INGRAM

Department of Biological Sciences, University of Salford, Salford M5 4WT

[1]*Department of Zoology, University of Glasgow, Glasgow G12 8QQ*

Synopsis

The relationships between Trypanosomatidae, which include the important pathogens *Trypanosoma* and *Leishmania*, and various species of arthropod which act as vectors or hosts are briefly reviewed. Although there have been many studies on parasite/vector biology, comparatively little attention has been focused on the responses of the vectors to Trypanosomatidae.

The haemocytes of the major vectors *Glossina* and *Rhodnius* have been characterized and studies on the haemocytic responses have shown that inoculation of flagellates provokes reduced total haemocyte counts in most vector/flagellate systems. Phagocytosis of flagellates by plasmatocytes does occur but not all species of flagellate are phagocytosed by the plasmatocytes of the same species of insect. Phagocytosis of *Leishmania hertigi* by locust plasmatocytes is described; the process involves initial formation of filopodial extensions, entrapment and vacuole formation which internalizes the parasite.

The study of model systems as well as of the vectors has revealed the existence of various humoral control mechanisms which include parasite agglutinins in locust and cockroach haemolymph, specific factors which reduce *Trypanosoma brucei* motility in *Glossina* and haemagglutinins in *Glossina* haemolymph. Prior injection of trypanosomatids can induce increased agglutinin and lysozyme levels in some systems but no evidence of induced responses in natural vectors is yet available. Different insects display a variable capacity to control different flagellates as *Glossina*, for example, is rapidly killed by a fulminating infection of *Crithidia* whereas *Trypanosoma brucei* is controlled. In addition, evidence for the role of other parasites (e.g. mermithids) in suppressing the ability to respond is demonstrated by inoculation of flagellates into mermithid-infected locusts. Nematode-infected locusts inevitably die from massive flagellate infections, but in locusts not infected with mermithids flagellate numbers are always controlled. Recent studies on *Rhodnius* and *Glossina* have indicated that lectins from various tissues may have a role in determining parasite-vector specificity; tissue extracts show different parasite agglutinating activities which can be inhibited at low concentration by certain carbohydrates, indicating that parasite agglutination is mediated by surface receptor molecules capable of recognizing these carbohydrate determinants.

ZOOLOGICAL SYMPOSIUM No. 56
ISBN 0-19-854004-3

Introduction

The relationship between Trypanosomatidae and insects has been a subject of detailed study since the recognition that tsetse flies, *Glossina* (Diptera: Glossinidae), and Triatominae were vectors of pathogenic human trypanosomes (*Trypanosoma brucei* group and *T. cruzi* respectively) and *Glossina* was the vector of the pathogenic animal trypanosomes (*T. vivax*, *T. congolense*). Subsequent to these findings the role of the phlebotomine sandflies (genera *Phlebotomus*, *Lutzomyia* and *Psychodopygus*) as vectors of human *Leishmania* was also elucidated. Studies on the epidemiology of these important diseases have involved studies on the infection rates, modes of transmission and bloodmeal preferences of vectors and the host-parasite relationship within the vector. Complementary but valuable studies on the life cycles of other trypanosomes in a variety of vectors have also been undertaken (see Table 1 for a list of trypanosomes, vectors and hosts).

Monoxenous Trypanosomatidae of the genera *Crithidia*, *Leptomonas*, *Blastocrithidia*, *Herpetomonas* and rarely *Rhynchoidomonas* (Wallace 1966, 1979) are parasites of insects, particularly the orders Hemiptera, Siphonaptera and Diptera. Work on monogenetic flagellates has been directed at elucidation of life cycles, description of species and studies of host-parasite relationships although they are increasingly used as models in cell biology. Despite the many studies on the Trypanosomatidae little attention has been paid to the insect's response to the invasion. Jordan (1976) stated that there was no information on the immune response of *Glossina* to trypanosomes and although this statement is no longer valid, there are still considerable gaps in our knowledge as to the nature and extent of these responses. In view of the diverse nature of the organisms and the variety of vectors and mammalian hosts the range of reactions to invasion by Trypanosomatidae will remain an important and fascinating study. This chapter reviews our current knowledge and reports recent work on the immune response of insects to Trypanosomatidae and hopefully indicates lines of research which may prove productive.

Life-cycles of Trypanosomatidae in insects

Digenetic parasites

Trypanosoma

Protozoans of the genus *Trypanosoma* are tissue and blood parasites transmitted by insects, occasionally by ticks or mites and less frequently directly by coitus (*T. equiperdum*) or mechanically by insects in which no cyclical development takes place and where the insect acts as a 'flying syringe' (e.g. the transmission of *T. evansi* of camels by tabanids). During the cycles physiological changes occur within the insect and these changes are accompanied by morphological and configurational changes of the organelles of the parasites. Although the major site of development of trypanosomes in insects is the gut, there is occasionally invasion of the haemolymph and salivary glands. Transmission is usually the result of contamination by the infective form (the metacyclic trypanosomes) in the faeces of the vector (Stercoraria), more rarely by the bite of the insect when the metacyclics are present in the foregut or salivary glands (Salivaria) (see Hoare 1964). *T. rangeli*, formerly classified in the *Herpetosoma* subgenus, is now placed by Anez (1982) in the subgenus *Tejeraia*. This parasite has morphological similarities to *Herpetosoma* trypanosomes but invades the haemocoele and the salivary glands, infection being transmitted by the bite of *Rhodnius prolixus*. This organism has provided a useful model for studies on interactions between trypanosomes and insect defence reactions as it invades the haemocoele naturally and causes pathogenicity in bugs (see pp. 133 and 135).

Leishmania

Leishmania parasites are transmitted by sandflies; three genera are known to be vectors—*Phlebotomus* in the Old World and *Lutzomyia* and *Psychodopygus* in the neotropics. Lainson & Shaw (1979) have recently characterized *Leishmania* infections in sandflies into Suprapylaria and Peripylaria; the majority of *Leishmania* are Suprapylaria with development in the midgut and foregut only. Peripylarian *Leishmania* (*L. braziliensis* spp.) develop initially in the pylorus and ileum of sandflies and later in the midgut and foregut. Transmission of *Leishmania* is by the bite of an infected sandfly.

Endotrypanum

Endotrypanum is an intra-erythrocytic trypanosomatid found only in sloths (genera *Bradypus* and *Choloepus*) and transmitted by sandflies (Shaw 1969, 1981).

Monoxenous parasites of insects

Wallace (1966, 1979) has extensively reviewed the biology of the genera—*Leptomonas*, *Blastocrithidia*, *Crithidia*, *Herpetomonas* and *Rhynchoidomonas*. They are usually confined to the gut of insects and transmission is by contamination of the environment and subsequent acquisition of free-living parasites or cysts.

Midgut cell invasion and natural haemocoelic infections of Trypanosomatidae

Insects

The vast majority of trypanosomatids reside in the gut and rarely invade the haemocoele, the gut acting as an effective barrier to the parasite. Nonetheless

Table 1. Range of selected *Trypanosoma* species, hosts and vectors.

Trypanosoma species and/or subgenus	Host	Vector	Author
Fish trypanosomes			
Trypanosoma tinca	*Tinca tinca*, tench	*Hemiclepsis marginata*, leech	Needham (1969)
T. murmanensis	*Gadus morhua*, Atlantic cod	*Myzobdella*, leech	Khan (1974)
T. raiae	*Raia* sp., ray	*Pontobdella muricata*, leech	Robertson (1910)
Amphibian trypanosomes			
T. canadensis	*Rana pipiens*, frog	*Placobdella* sp., leech	Woo (1969a)
T. bufophlebotomi	*Bufo boreas halophilus*, toad	*Lutzomyia vexatrix*, sandfly	Ayala (1971)
Reptilian trypanosomes			
T. chrysemidis	*Chrysemys picta*, turtle	*Placobdella*, leech	Woo (1969b)
T. grayi	*Crocodilus niloticus*, crocodile	*Glossina fuscipes*, tsetse	Hoare (1931a,b)
T. boueti	*Mabuya striata*, skink	*Sergentomyia bedfordi*, sandfly	Ashford, Bray & Foster (1973)
Avian trypanosomes			
T. macfiei	*Serinus canaria*, canary	*Dermanyssus gallinae*, mite	MacFie & Thomson (1929) Cotton (1970)
T. corvi	Corvidae	*Ornithomyia avicularia*, flatfly	Baker (1956)
'*T. avium*'	Several Canadian species	*Aedes*, mosquitoes	Bennet (1961)
T. numidae	*Numida mitrata*, Guinea fowl	*Simulium*, blackflies	Fallis, Jacobson & Raybould (1973)

Mammalian trypanosomes			
Herpetosoma			
T. lewisi	*Rattus*	*Nosopsyllus fasciatus*	Molyneux (1969)
Megatrypanum			
T. theileri	Bovidae	Tabanids, ticks	Wells (1976)
Schizotrypanum			
T. cruzi	Man Many mammalian reservoir hosts	Triatomines	Hoare (1972)
Trypanozoon			
T. brucei	Man Reservoir hosts	*Glossina*	Hoare (1972)
Nannomonas			
T. congolense	Cattle Wild herbivores	*Glossina*	Hoare (1972)
Duttonella			
T. vivax	Cattle Wild herbivores	*Glossina*	Hoare (1972)

Table 2. Invasion of arthropod gut cells by Trypanosomatidae.

Organism	Vector	Author(s)
T. lewisi	*Nosopsyllus fasciatus*	Minchin & Thomson (1911, 1915)
	Ctenocephalides canis	Nöller (1912)
		Molyneux (1969)
T.b. rhodesiense	*Glossina m. morsitans*	Evans & Ellis (1983)
	G. palpalis	
T. congolense	*G. pallidipes*	Kaddu & Mutinga (1980a, b)
T. cruzi	*Ornithodorus moubata*	See Rodhain (1942a, b)
T. lewisi	"	"
T. pipistrelli	"	"
L. mexicana	"	Lelijveld (1966)
T. cruzi	*Triatoma megista*	De Fario & Cruz (1927)
L.m. amazonensis	*Lutzomyia longipalpis*	Molyneux, Killick-Kendrick & Ashford (1975)
	Lutzomyia flaviscutellata	Molyneux, Ryan, Lainson & Shaw (in press)
L.b. braziliensis	*Psychodopygus wellcomei*	Killick-Kendrick, Lainson, Leaney, Ward & Shaw (1977)

it is evident that invasion of midgut cells does take place but the significance of this invasion is not clear. Table 2 lists occasions where invasion of midgut cells has been documented.

From Table 2 it is clear that invasion of the midgut epithelium by trypanosomatids is not uncommon; the pathological and physiological consequences of such invasions on the vector are unknown. There are, however, some situations where the gut is regularly penetrated; *T. rangeli* invades the haemocoele of *R. prolixus* where some strains are pathogenic. Watkins (1971a,b) suggested that this was achieved by the secretion of enzymes which altered the basal lamina of the gut cells in *R. prolixus*.

Penetration of *T. brucei* through the midgut cells and basal lamina of *Glossina* also occurs but its frequency and significance remain unknown (Evans & Ellis 1975, 1983). It does appear that in both *T. brucei* and *T. rangeli* there are considerable differences in the ability of strains to penetrate the midgut barrier in their respective vectors. The list of natural haemocoelic infections of flagellates in insects and arachnids is given in Table 3.

Foster (1964) described haemocoelic infections of *Glossina palpalis* and *pallicera* in Liberia which he attributed to a non-pathogenic trypanosome possibly of non-mammalian origin. Mshelbwala (1972) experimentally infected *Glossina morsitans*, *palpalis* and *tachinoides* with *T. brucei* and found infections in the haemocoele, a finding confirmed by Otieno (1973). Otieno & Darji (1979) then found three out of 955 *G. pallidipes* with haemocoelic infections of *T. brucei*. Monoxenous flagellates of the genus *Herpetomonas* have been recorded from the haemocoele of Hymenoptera and Diptera on a few occasions (Kramer 1961; Smirnoff & Lipa 1970; Bailey & Brooks 1972a, b; Smirnoff 1974). *H. swainei*, a parasite of the sawfly, *Neodiprion swainei*, was considered by Smirnoff (1974) as a potential biological

Table 3. Haemocoelic infections of Trypanosomatidae in arthropods (excluding infections induced by intrahaemocoelic inoculations).

Organism	Vector	Author(s)
T. rangeli	*R. prolixus*	Watkins (1971a, b)
		D'Alessandro (1976)
Trypanosoma sp.	*G. palpalis*	Foster (1963, 1964)
		Croft, Kuzoe, Ryan & Molyneux (1984)
Trypanozoon	*G. pallidipes*	Otieno & Darji (1979)
T. macfiei	*Dermanyssus gallinae*	MacFie & Thomson (1929)
		Cotton (1970)
T. theileri	*Boophilus decoloratus*	Burgdorfer *et al.* (1973)
	Rhipicephalus pulchellus	
Trypanosoma sp.	*Ixodes ricinus*	Rehacek *et al.* (1974)
		Aeschlimann *et al.* (1979)
Trypanosoma sp.	*Amblyomma americanum*	Krinsky & Burgdorfer (1976)
Herpetomonas swainei	*Neodiprion swainei*	Smirnoff & Lipa (1970)
H. muscarum	*Musca domestica*	Kramer (1961)
	Hippolates pusio	Bailey & Brooks (1972a, b)
Leishmania tropica	*Phlebotomus sergenti*	Adler & Theodor (1929)

control agent. *Herpetomonas* infections which invade the haemocoele cause pathogenicity but whether this is due to the parasite itself or concomitant bacterial infection is not known.

Arachnids

There have been a number of reports of various genera of ticks (both of Ixodidae and Argasidae) with trypanosome infections in haemolymph. Burgdorfer, Schmidt & Hoogstraal (1973) reported 'enormous masses' of epimastigotes of *Trypanosoma theileri* in one out of 69 *Boophilus decoloratus* and 19 out of 258 *Rhipicephalus pulchellus* from Ethiopian cattle. Parasites were associated with plasmatocytes and were believed to be located intracellularly. Intracellular localization was also observed in cells of the ovaries, salivary glands and Malpighian tubules, and the connective tissue and muscles were also invaded. O'Farrell (1913), Carpano (1932) and Arifdzhanov & Nikitina (1961) earlier reported *T. theileri* in ticks; however, the presence of trypomastigotes in the salivary glands of *R. pulchellus* prompted Burgdorfer *et al.* (1973) to suggest that *T. theileri* could be transmitted by bite. Krinsky & Burgdorfer (1976) also found a *Megatrypanum* species similar to *T. theileri* in the haemocoele of *Amblyomma americanum* collected from *Odocoileus virginianus* (white-tailed deer) and Shastri & Despande (1981) reported *T. theileri* infections in *Hyalomma a. anatolicum*. There is evidence that in Europe *Ixodes ricinus* harbours an unidentified trypanosome in the haemolymph (Rehacek, Sixl & Sebek 1974; Aeschlimann, Burgdorfer, Matile, Peter & Wyler 1979). Mites (*Dermanyssus gallinae*) are vectors of avian trypanosomes and development of massive infections of the haemocoele has been described (MacFie & Thomson 1929; Cotton 1970). No details of the life cycles or host-parasite relationship of these interesting infections are known.

Behaviour of trypanosomatids in, and response of, model systems

Cellular responses

Several authors have inoculated trypanosomatids into the haemocoele of insects (Zotta 1921; Glaser 1922; Shortt 1923; Ivanoff 1925; Hoare 1938; Linder 1960). Wide variations in the ability of inoculated insects to control such infections have been reported. Ivanoff (1925) found that trypanosomes survived for several days in the haemolymph of *Galleria mellonella* and Hoare (1938) observed 100 per cent survival of *T. cruzi* in *G. mellonella* larvae, though in contrast Linder (1960) reported *T. cruzi* did not develop in the same insect. Linder also described the development of other flagellate species in *G. mellonella* and showed that the behaviour of the different species of flagellate in the same insect was highly variable. She observed phagocytosis of flagellates and reduction in total haemocyte counts following inoculation. Differential haemocyte counts revealed prohaemocytes were reduced whilst the proportion of plasmatocytes rose.

Schmittner & McGhee (1970) investigated the effects of intrahaemocoelic inoculation of six *Crithidia* species into *Drosophila viridis*, *Acheta domestica* and *Tenebrio molitor*. All three mounted a cellular defence reaction but none was able to clear circulating parasites completely from the haemolymph. Parasites thrived and multiplied in the body cavities of the insects and in most cases death occurred; *Crithidia fasciculata* was particularly virulent. Schmittner & McGhee (1970) also reported variations in ability of the haemocytes of the different insects to phagocytose the parasites and form nodules.

These results have provided a baseline for studies on other model systems which are more easily manipulable and have sufficient haemolymph to allow investigations of haemocyte kinetics and serum agglutinin levels. We have used *Schistocerca gregaria* and *Periplaneta americana* in the studies described below (see also Ingram, East & Molyneux 1983, 1984).

We have used as parasites cultured *T. brucei* procyclics (stock EATRO1125), *L. hertigi* and *C. fasciculata*; inoculation of these parasites into locusts has been studied in relation to the mechanisms of clearance and control. The mechanisms of plasmatocyte/*L. hertigi* interaction also provide a model for investigating phagocytosis of trypanosomatids. Inoculation of these flagellates into *S. gregaria* provoked a rapid reduction in the total haemocyte count over a 48h period post-inoculation, when compared with sham-inoculated and culture-medium-inoculated controls; recovery to nor-

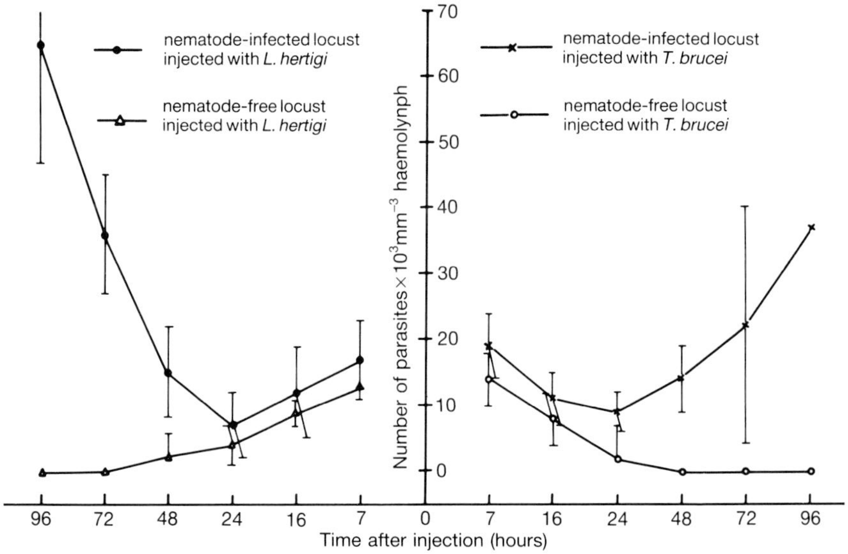

Fig. 1. Mean number of *T. brucei* and *L. hertigi* in nematode-free and nematode-infected *S. gregaria* at various times after injection of trypanosomatids.

mal levels was achieved after 72–96h post-inoculation. The cell type most noticeably reduced was the plasmatocyte; the proportion of plasmatocytes dropped from 50 per cent to 30 per cent within 24h.

In parallel with the changes observed in total haemocyte counts following inoculation of flagellates there was a rapid clearance of flagellates from the haemocoele. Around 75 per cent of locusts examined 24h after inoculation of *T. brucei* were free of detectable parasites in the haemolymph although *C. fasciculata* and *L. hertigi* took longer to clear (see Fig. 1).

Observations on locusts following inoculation of *T. brucei* revealed that a proportion of them were not clearing the parasites within the normal 24h period. These locusts were found to be infected with *Mermis nigrescens*. Locusts infected with *M. nigrescens* had significantly lower total haemocyte counts (see Table 4) and differential haemocyte counts revealed significant reduction in the numbers of plasmatocytes and granular cells and increases in coagulocytes and prohaemocytes compared with uninfected ones. A lower rate of phagocytosis of flagellates in mermithid-infected locusts was also observed.

In locusts not infected with mermithids, there was rapid nodule formation following injection of flagellates. *L. hertigi* was most rapidly and effectively phagocytosed but *C. fasciculata* and *T. brucei* appeared to be only entrapped, by means of filopodia-like extensions from the plasmatocytes. Concomitant with nodule formation, plasmatocytes either phagocytosed or entrapped parasites, and clumps of cells could be recovered 12–16h post-

Table 4. Total haemocyte counts (THC) × 10^{-5} mm^3 and agglutinating activity of haemolymph against Trypanosomatidae of *Mermis*-free and *Mermis*-infected *S. gregaria*. (Male and female locusts showed no significant differences in each group and results are pooled.)

Haemolymph	THC ± SE	*T. brucei*	*L. hertigi*
Mermis-free	16.40 ± 3.14	2^{-8}–2^{-9}	2^{-8}–2^{-9}
Mermis-infected	5.89 ± 1.89	2^{-4}–2^{-5}	0–2^{-2}

injection having attached to various tissues. Nodules reached maximum size within 48h with a mean diameter of 30μm. Parasites associated with such nodules were inactive and not viable. In mermithid-infected locusts, however, no clearance of flagellates occurred; instead, there was a rapid increase in the numbers of flagellates (*T. brucei* and *L. hertigi*) between two and six days post-inoculation at which time the locusts died (Fig. 1). No mortality is usually associated with mermithid infection of locusts alone (Rutherford & Webster 1978) or with the inoculation of trypanosomatids alone and this appears to be a clear example of an immunosuppression by the nematode of the capacity of the locust to react to challenge inoculation of flagellates. (See also section on humoral response.)

No studies have been published to date on the mechanism of phagocytosis by insect plasmatocytes of trypanosomatid flagellates, although the mechanism of phagocytosis and mammalian phagocyte-flagellate interactions have been intensively studied (Thorne & Blackwell 1983), as have plasmatocyte-bacteria interactions (Rowley & Ratcliffe 1976). Using the locust plasmatocyte/*L. hertigi* model, the mechanism of internalization of *L. hertigi* promastigotes has been studied. Contact between parasite and plasmatocyte stimulates formation of cytoplasmic extensions which engulf the parasites and internalize them in a parasitophorous vacuole (Figs 2, 3). Such extensions form multimembranous whorls; parasites rapidly lose their typical structural integrity, round up and can become disrupted. Several parasites can be phagocytosed by a single plasmatocyte and more than a single parasite can be found in a parasitophorous vacuole (Figs 2, 3). The vacuoles are surrounded by granular inclusions but little release of such material into the vacuole is seen (Figs 2, 3). Cytochalasin B (at 10 $\mu g.ml^{-1}$) reduced phagocytosis of living *L. hertigi* and their internalization (Fig 4) but no effect was observed on the attachment of the promastigotes to the plasmatocytes. In addition, histochemical studies using nitroblue tetrazolium have indicated that phagocytosis of *L. hertigi* by locust plasmatocytes does not stimulate an oxidative burst when living, heat-killed or formalized *L. hertigi* are phagocytosed; the lack of an oxidative burst in insect haemolymph was first reported by Anderson, Holmes & Good (1973).

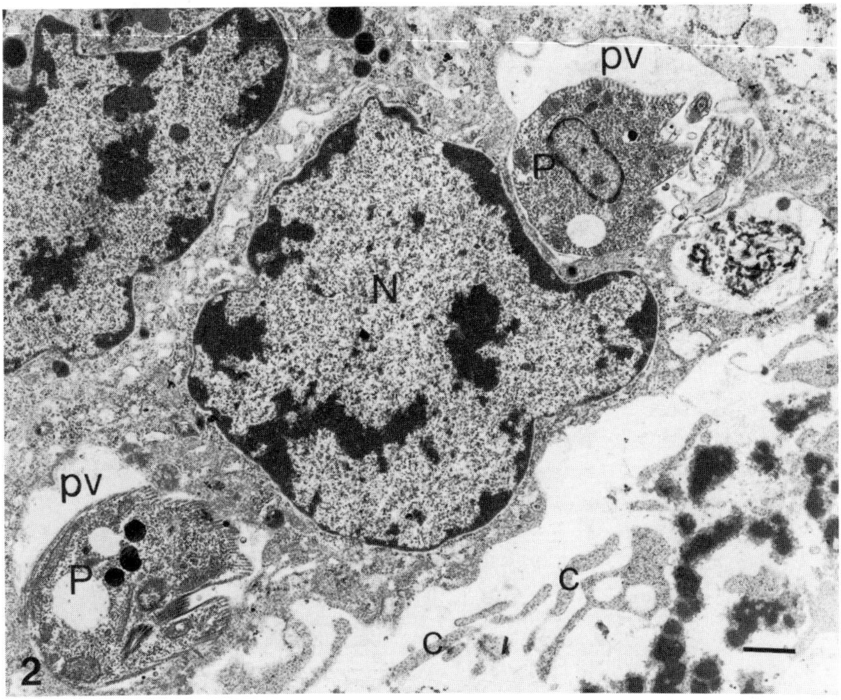

Fig. 2. Locust plasmatocyte containing phagocytosed *L. hertigi* promastigotes (P) within phagosome or parasitophorous vacuole (pv), prepared for electron microscopy 14h after intrahaemocoelic injection of *L. hertigi*. Nucleus (N) of plasmatocyte occupies high proportion of cell. Cytoplasmic extensions or filopodia (c) in lower part of micrograph are involved in entrapment of parasites prior to phagocytosis. Scale bar = 1 µm.

Humoral responses in model systems

Haemolymph of insects contains molecules which agglutinate mammalian erythrocytes (haemagglutinins), opsonins which stimulate phagocytosis, and bactericidins (including lysozyme). The role of serum agglutinins in internal defence has been discussed earlier in this volume by Renwrantz. We have used locusts and cockroaches to determine whether trypanosomatids can be agglutinated by serum and to ascertain if any induced responses could be detected to challenge by flagellates.

These studies (Ingram *et al.* 1983, 1984) using procyclic cultures of *T. brucei*, *L. hertigi* promastigotes and *C. fasciculata* choanomastigotes revealed that high titres of parasite agglutinin activity were present against *T. brucei* and *L. hertigi*. The mean titres ranged from 2^{-4} to 2^{-13}; cockroaches in general displayed higher titres against both parasites and there was no significant variation in values detected between either sex examined. Levels of agglutinins to *C. fasciculata* in locust and cockroach haemolymph were only up to 2^{-3}. *Calliphora erythrocephala* haemolymph did not

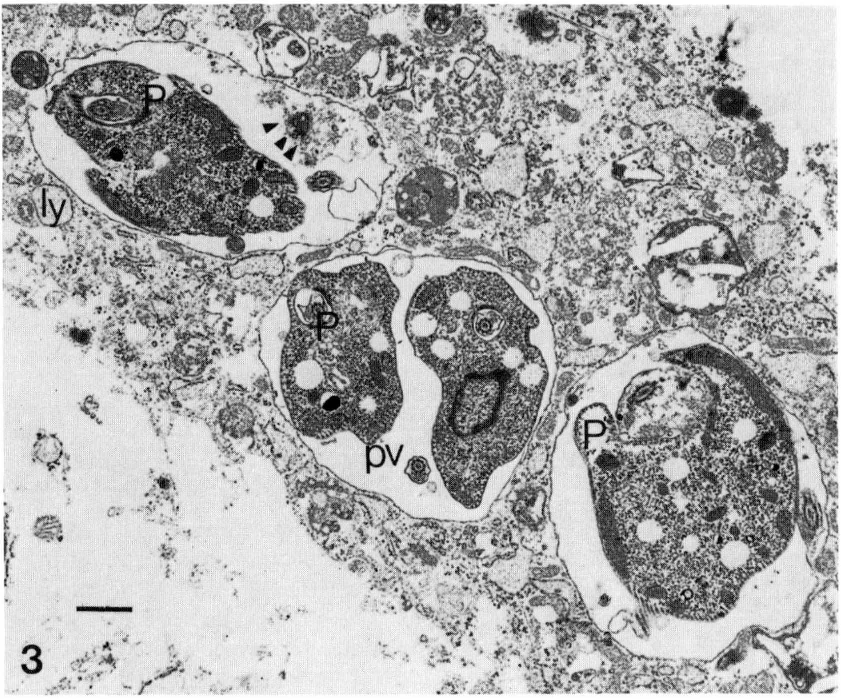

Fig. 3. Same subject as Fig. 2 showing three phagosomes containing parasites and two parasites within a single vacuole. Lysosomes are adjacent to vacuoles (ly) and one phagosome shows diffuse material present of non-parasite origin (arrowed). Other abbreviations as for Fig. 2. Scale bar = 1 μm.

demonstrate any agglutinating ability against these flagellates (Ingram *et al.* 1983). Pre-adsorption of haemolymph with *T. brucei* or *L. hertigi* abolishes or negates agglutinin activity, suggesting the agglutinin in both locusts and cockroaches has a similar degree of specificity for both parasites. Physico-chemical treatment of locust and cockroach haemolymph suggests the flagellate agglutinins are proteins or glycoproteins.

Leishmania hertigi inoculation of locusts and cockroaches stimulates a significant increase in agglutinin levels when compared with controls. However, inoculation of *T. brucei* into locusts resulted in an initial decrease in agglutinin titres after which titres returned to normal levels, compared with cockroaches in which significant increases in agglutinin levels were observed. These studies demonstrated for the first time the capacity of trypanosomatids to induce humoral responses in insects. Lysozyme levels in locust and cockroach haemolymph were also measured after inoculation of *T. brucei* and *L. hertigi*, and were found to increase in all injected insects except in *T. brucei*-injected locusts where levels were reduced (Ingram *et al.* 1984).

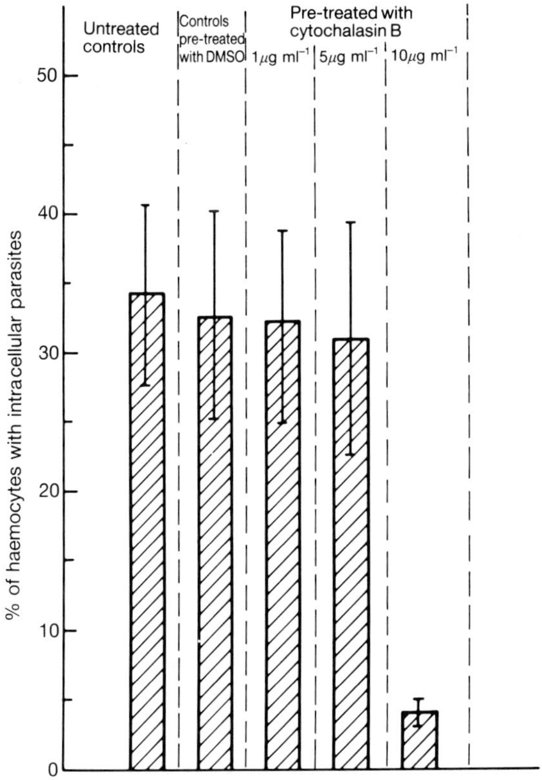

Fig. 4. Effect of cytochalasin B on uptake of living *L. hertigi* by locust haemocyte monolayers pre-treated for 60 min with varying concentrations of the drug.

In studies on the serum agglutinins of mermithid-infected locusts it was found that the presence of the nematode almost completely destroyed the agglutinating ability of the haemolymph against both *T. brucei* and *L. hertigi* (see Table 4). This finding, in association with the observation reported earlier that nematode-infected locusts succumb to these flagellate infections, suggests that locust agglutinins have a role in controlling such infections.

Immune responses of natural vectors

The haemocytes

Haemocytes of *Glossina*

Extensive information is available on the form and function of insect blood cells (Gupta 1979; Rowley & Ratcliffe 1981). The haemocytes of *Glossina* have only recently been studied; East, Molyneux & Hillen (1980) first described the haemocytes of *Glossina* and categorized several cell types, including a curious haemocyte, the spindle cell, which made up a very high proportion of the tsetse haemocyte population (Fig. 5). Spindle cells are characterized by an axial thickening observable under the phase contrast microscope and found under the electron microscope to be a longitudinally arranged axis of microtubules (Fig. 6); these cells are usually observed to be devoid of a nucleus (Kaaya & Ratcliffe 1982). *Glossina* plasmatocytes closely resemble the phagocytic plasmatocytes of other insects (Fig. 7). Kaaya & Ratcliffe (1982) reported that *Glossina* plasmatocytes ingest the spindle cells, an observation which accounts for the decline in number of spindle cells after the adult has emerged from the pupa. *Glossina* also contain the putative progenitor cells, prohaemocytes, which make up 1–4% of the haemocyte population (Fig. 8), granulocytes (Fig. 9), nephrocytes and adipohaemocytes. Prohaemocytes and adipohaemocytes are rare cell types and nephrocytes are normally in fixed locations (Kaaya & Ratcliffe 1982). The spindle cells are a unique cell type whose function remains unknown, but it has been suggested that whatever role they play is a temporary one as they appear in the late pupal stage and decline in numbers as the adult ages (Kaaya & Otieno 1981; Kaaya & Ratcliffe 1982). All species of *Glossina* examined to date, both laboratory-reared and wild-caught flies, possess these cells. The plasmatocytes of *Glossina* are phagocytic and have been shown to phagocytose bacilli (*B. cereus*) *in vivo* (Poinar, Wassink, Leegwater-Van Der Linden & Van Der Geest 1979; Kaaya 1980) and certain trypanosomatids. Experimental intrahaemocoelic inoculations with a variety of flagellates demonstrate that procyclic trypomastigotes of *T. brucei* are not phagocytosed by *Glossina* cells whilst the smaller *L. hertigi* promastigotes and *C. fasciculata* cultured choanomastigotes are. The spindle cells and plasmatocytes were involved in phagocytosis. Inoculation of all flagellates resulted in a reduction of total haemocyte counts within 48h; no such reduction was observed in sham-inoculated controls. However, *Glossina* species (*G. morsitans*, *palpalis* and *tachinoides*) are unable to control haemocoelic infections of *C. fasciculata* despite the ability of the plasmatocytes particularly to phagocytose this flagellate. Most deaths occurred on Days 6 and 7 whilst no mortality was observed in control flies. Reduced doses of 100–200 *C. fasciculata* delayed death until Days 10–15.

Haemocytes of *Rhodnius* and other reduviid bugs

Four to six major haemocyte types have been identified in *R. prolixus* by a number of authors (Lai-Fook 1970; Jones 1967; Price & Ratcliffe 1974). There may be additional haemocyte types but these are probably variations of the four main categories (Jones 1965). In general, plasmatocytes formed the greatest proportion of the haemocyte population; granulocyte numbers varied widely and in some instances exceeded the numbers of plasmatocytes. The other two haemocyte types, prohaemocytes and oenocytoids, were in the minority (Jones 1967).

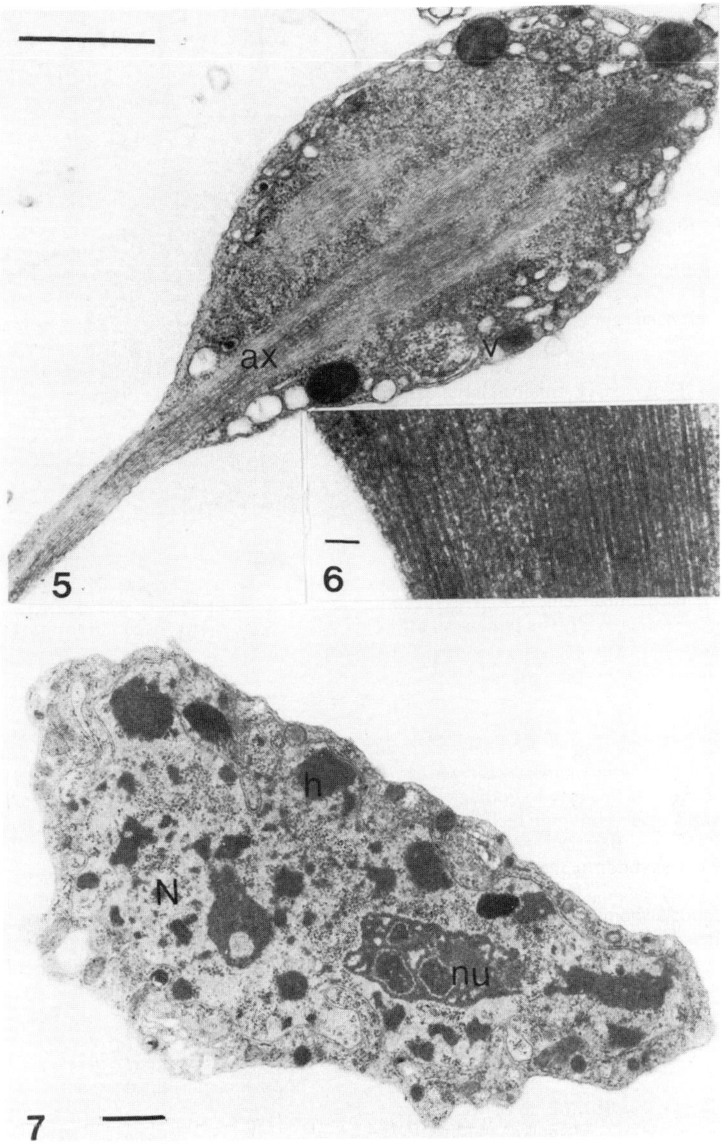

Fig. 5. Electron micrograph of spindle cell of *Glossina* showing polar extension of axial thickening (ax) of microtubules. Note absence of nucleus; presence of peripheral vacuoles (v), and ribosomes in cytoplasm; dense peripheral bodies have been described by earlier workers as mitochondria. Scale bar = 1 μm.

Fig. 6. Higher-power micrograph of detail of axial bundle of microtubules of a spindle cell of *Glossina*. Scale bar = 1 μm.

Fig. 7. Plasmatocyte of *Glossina* showing large irregularly shaped nucleus (N) with nucleolus (nu) and large concentrations of heterochromatin. The peripheral cytoplasm contains ribosomes, vacuoles and mitochondria. Scale bar = 1 μm.

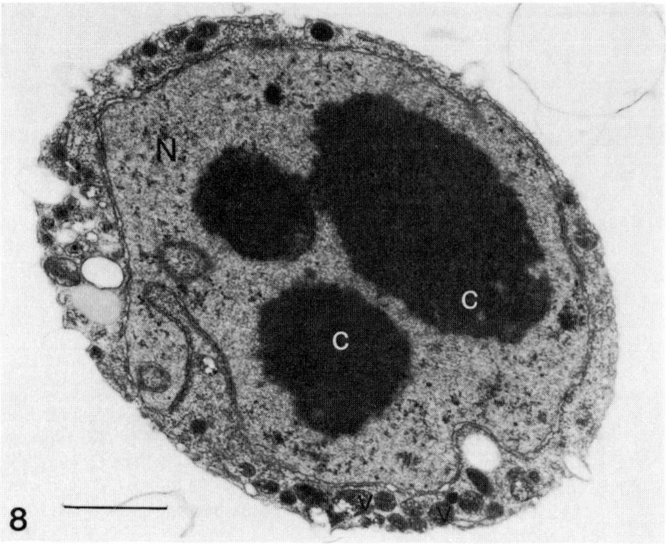

Fig. 8. *Glossina* prohaemocyte, witsh prominent chromatin blocks (c), nucleus (N) occupying large proportion of cell. The peripheral area of cytoplasm contains vacuoles of several different types. Scale bar = 1 μm.

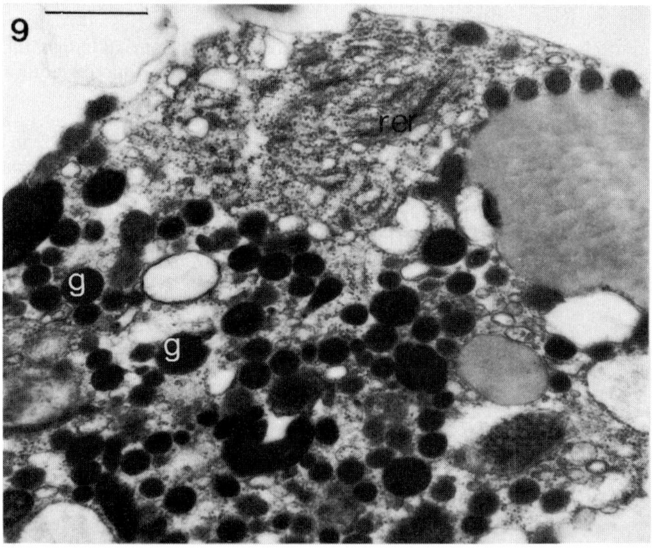

Fig. 9. TEM section of part of a *Glossina* granulocyte showing an area of ribosomes and rough endoplasmic reticulum (rer) with concentration of mature amorphous dense granules (g). Note regular arrangement of granules around the periphery of a large electron-lucent inclusion body. Scale bar = 0.5 μm.

Price & Ratcliffe (1974) identified prohaemocytes, plasmatocytes, granular cells and coagulocytes in *R. prolixus*, whilst Drif (1983) using light- and transmission-electron microscopy of *R. prolixus* and *Triatoma infestans* haemocytes observed oenocytoids in addition to the latter four haemocyte types. The largest of the South American reduviid bugs, *Dipetalogaster maximus*, has four haemocyte types: (i) spindle-shaped granular cells (20%), (ii) small round cells (60%), (iii) large spreading plasmatocytes (10%), and (iv) coagulocytes (10%) (G. Takle, unpubl.). Of interest here is the relatively low number of plasmatocytes, since these cells are the main cells responsible for the phagocytosis of intrahaemocoelic microorganisms.

The total haemocyte counts (THCs) for 4th/5th instar *Rhodnius* derived from the haemocyte concentration and the haemolymph volume (Núñez 1962) indicate that, compared with the model species *Periplaneta* and *Schistocerca*, *Rhodnius* has very few haemocytes ($1.46 \pm 1.14 \times 10^4$ haemocytes per bug compared with 8.5×10^6 per *Periplaneta* and 3.3×10^6 per *Schistocerca* (Lackie, Takle & Tetley 1985). These differences in haemocyte numbers may explain why *Rhodnius* is susceptible to *T. rangeli* whilst cockroaches and locusts are not. Interestingly, THCs from another reduviid species, *Triatoma infestans*, are higher than those from *Rhodnius* and *Triatoma* is resistant to *T. rangeli* (Zeledón & de Monge 1966). The numbers of phagocytic plasmatocytes may give a more representative idea of the possible susceptibility of insects to trypanosome infections, since these cells, as well as being phagocytic, are also involved in the formation of nodules and capsules.

When *T. rangeli* enters the haemocoele of *Rhodnius*, either by experimental intrahaemocoelic injection (Tobie 1968) or by penetration of the gut wall after the bug has ingested an infected bloodmeal (Anez 1981), multiplication can occur both intra- and extracellularly (D'Alessandro 1976) before the parasite enters the salivary glands (Ellis, Evans & Stamford 1980). Following injection of 10^4 *T. rangeli* into 4th/5th instars of *Rhodnius* a rapid initial decrease in the trypanosome/haemocyte ratio occurs (Fig. 10). This could be a result of an increase in the number of haemocytes (Zeledón & de Monge 1966). However, trypanosome numbers gradually increase by two orders of magnitude in the haemolymph of *Rhodnius* over about 10 days after injection, after which *T. rangeli* infection can be fatal for the bugs. The susceptible reduviids do mount a limited haemocytic response as small nodules of haemocytes associated with trypanosomes can be found in infected haemolymph. Since very low numbers of *T. rangeli* can be pathogenic (Zeledón & de Monge 1966) a deficiency in the activity of the bug's cellular defence system may explain its susceptibility.

Recently haemocyte and substratum surface charge has been implicated as a control for haemocyte/substratum adhesion (Lackie 1983; Takle &

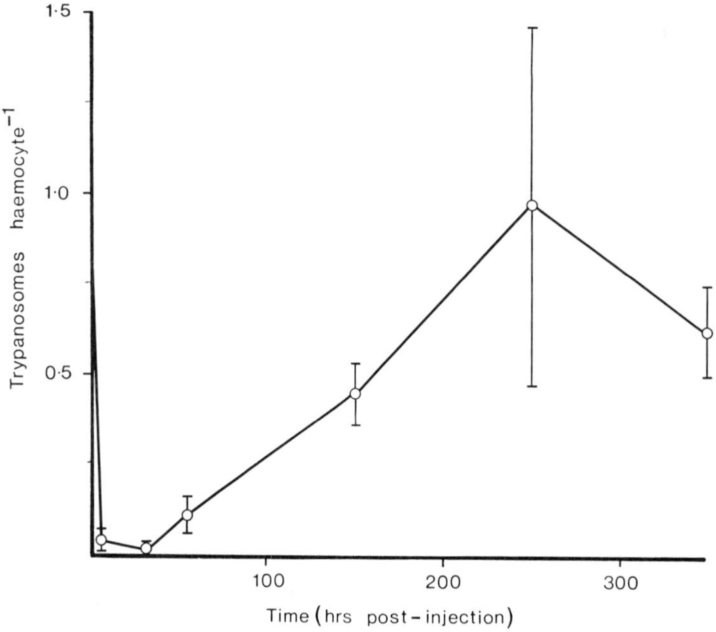

Fig. 10. Development of *T. rangeli* infection in *Rhodnius*. Trypanosome: haemocyte ratio in the haemolymph during the 11 days following an injection of 10^4 trypanosomes. Each point is the mean ± s.d. of 20 measurements from each of 5 insects.

Lackie 1985). The possible involvement of surface charge in the reduviid haemocyte/*T. rangeli* system was thus examined. *T. rangeli* culture forms are significantly less negatively charged than haemocytes of both *Dipetalogaster* and *Rhodnius*. Since *Schistocerca* haemocytes have the same surface charge as *Dipetalogaster* and *Rhodnius* haemocytes (Takle & Lackie 1985), yet *Schistocerca* can easily clear *T. rangeli*, it was concluded that haemocyte surface charge *per se* cannot determine whether an insect will be susceptible to *T. rangeli* infection.

In conclusion, at first inspection the data suggest that the numbers of circulating haemocytes appear to play a role in the pathogenicity of *T. rangeli* for the reduviid bugs. However, although *Periplaneta* and *Schistocerca* contain 200–500 times more haemocytes than *Rhodnius* (Lackie *et al.* 1985), they can very effectively clear injected doses of *T. rangeli* 1000 times higher than those causing infection in *Rhodnius*. Also very low doses of *T. rangeli* are sufficient to bring about an infection in *Rhodnius* (Zeledón & de Monge 1966), so low haemocyte numbers alone do not render *Rhodnius* susceptible to *T. rangeli*. Surface charge of reduviid haemocytes also does not seem to determine the response of *Dipetalogaster* and *Rhodnius* to *T. rangeli* and it is therefore likely that the presence of *T. rangeli*-specific haemolymph molecules may be closely involved in specifying whether a particular insect species is capable of clearing infections of this trypanosome (see next section).

Recent work (Bitkowska, Dzbenski, Szadziewska & Wegner 1982) suggests that infection of *T. cruzi* in the gut of *T. infestans* inhibited xenograft rejection when compared with uninfected bugs. A similar effect was achieved by inoculation of material from *T. cruzi* cultures. Haemocytic encapsulation of auto- and allografts is weak whereas xenografts stimulate a strong encapsulation response. This clearly requires more intense investigation but the suggestion that exoantigens of *T. cruzi* are responsible for suppression of the response requires confirmation: however, the figures given by Bitkowska *et al.* (1982) do suggest that a difference between the response of infected and uninfected bugs to implants does occur.

Humoral control of trypanosome infections

Glossina

The occurrence of haemocoelic infection of trypanosomes in *Glossina* was referred to on pp. 120–123. The significance of such infections in the life cycle of *Trypanozoon* has been discussed by Evans & Ellis (1983).

Studies by Croft, East & Molyneux (1982) showed that when the cell-free haemolymph of *Glossina* was incubated *in vitro* with procyclics of *T. brucei*, the parasites were immobilized at dilutions as low as 2^{-9} within 1–2h. When *C. fasciculata*, *L. hertigi* and the epimastigotes of a bat trypanosome (*Trypanosoma dionisii*) were incubated under the same conditions there was only a slight reduction in motility of *T. dionisii* but no reduction in the observed motility of *L. hertigi* and *C. fasciculata*. The thermosensitive non-inducible antitrypanosomal factor was present in both male and female teneral (i.e. tsetse which had recently emerged from puparia and not yet taken a bloodmeal) and non-teneral flies. Studies on the effects of *G.m. morsitans* serum on *T. congolense* bloodstream and procyclic forms and *T. vivax* bloodstream forms showed a similar capacity to immobilize these pathogenic trypanosomes (East, Molyneux, Maudlin & Dukes 1983). The results suggest that should *T. congolense* and *T. vivax* ever reach the haemolymph of *Glossina* they will be rapidly destroyed by these as yet uncharacterized factors in tsetse haemolymph.

Studies by Pereira, Andrade & Ribeiro (1981) (see p. 136) prompted the assessment of haemagglutinating activity of *Glossina* gut extracts and haemolymph against various red blood cell types. *Glossina* serum showed no agglutinating activity against *T. brucei*, *L. hertigi* or *C. fasciculata* but gut and salivary gland extracts and haemolymph showed variable levels of haemagglutinating activity when tested against four human blood group erythrocytes and calf, guinea pig and chicken erythrocytes (Ibrahim, Ingram & Molyneux 1984) (see Table 5). Carbohydrate inhibition studies demonstrated that the different red cell agglutination reactions could be inhibited specifically by a variety of sugar molecules at various concentrations, suggesting that such reactions are mediated through the presence of specific lectins in the various parts of the gut and haemolymph.

Rhodnius

Pereira, Andrade *et al.* (1981) demonstrated the presence of lectins in *R. prolixus* following earlier studies which had identified lectin receptors as markers characteristic of the developmental stages of *T. cruzi* (Pereira, Loures, Villalta & Andrade 1980). Pereira, Andrade *et al.* (1981) demonstrated that lectins of different activities were found in crop, midgut and haemolymph; such lectin receptors were present in the epimastigote stage of *T. cruzi* but not in the trypomastigotes which were not agglutinated. Several other trypanosomatid flagellates (*Leishmania donovani*, *L. mexicana amazonensis*, *Crithidia deanei*, *Herpetomonas samuelpessoai*, *Phytomonas davidii*) did not agglutinate in the presence of *R. prolixus* lectins. Pereira, Andrade *et al.* (1981) also reported that several species of *Triatoma* and *Dipetalogaster maximus* and *Panstrongylus megistus* had similar lectins. These studies demonstrated the presence of lectins specific for the insect stage of *T. cruzi* in a vector but also the absence of any cross-reactivity of such a lectin with other trypanosomatid flagellates. Different regions of the gut clearly have different specificities which indicate, in view of our knowledge of the distribution of parasites in the guts of insects and evidence that lectin-like molecules are present in insect guts (Peters, Kolb & Kolb-Bachofen 1983), that lectins could act as determinants in the selection of the site of development of trypanosomatids in the guts of the vector.

Discussion

From the available information from studies on both the natural insect hosts of Trypanosomatidae as well as in model systems there is clear evidence of an ability to respond to and control such infections by both cellular and humoral mechanisms. In both natural vectors and models there are considerable differences in the ability of individual insect species to respond to different but closely related organisms. This is illustrated by the ability of *Glossina* to clear salivarian trypanosomes yet its inability to marshal a response to the flagellate *Crithidia*; in addition there are varying degrees of pathogenicity of strains of *T. rangeli* in *R. prolixus* suggesting that such differences in the immunogenicity of parasite strains are mediated by very limited differences in molecular structure of the invading organisms. It remains to be demonstrated at what site or location of the parasite such differences occur.

Table 5. Agglutinogenic activity of *Glossina* (*morsitans* and *austeni*) haemolymph and midgut and hindgut extracts against various human and animal red blood cells and *T. brucei* procyclics.

Agglutinin Source	OR_1-	OR_2+	ARh+	BRh−	ARh−	Calf	Guinea Pig	Chicken	*T. brucei* procyclics
Control (PBS[a])	0	0	0	0	0				
Haemolymph	2^{-7}	2^{-9}	2^{-9}	2^{-10}	2^{-6}	2^{-6}	2^{-6}	2^{-14}	0
Haemolymph of 5 day infection with *C. fasciculata*	2^{-10}	2^{-9}	2^{-10}	2^{-10}	2^{-10}	NOT DONE			
Midgut extract	2^{-3}	2^{-4}	0	0	0	2^{-6}	2^{-6}	2^{-4}	2^{-7}
Hindgut extract	2^{-5}	2^{-5}	2^{-6}	0	0	2^{-6}	2^{-6}	2^{-6}	2^{-8}

OR_1+ — subtype of ORh + with OR_2 genotype CDe/CDe surface antigens.
OR_2+ — subtype of ORh + with OR_2 genotype cDE/cDE surface antigens.
[a] PBS — phosphate buffered saline pH 7.4.
No reaction with donkey, horse and sheep.

Trypanosomatid invasion of the haemocoele in the systems studied to date initiates a reduction in total haemocyte counts. This response, however, is usually short-lived and is likely to be due to the sequestration of the parasites by haemocytes within nodules. Phagocytosis of flagellates by plasmatocytes does occur but is not universally exhibited; indeed, experimental studies show again that levels of specificity exist in the ability of insect plasmatocytes to phagocytose trypanosomatids. Where phagocytosis does occur the process is similar to that observed when parasites are phagocytosed by mammalian phagocytic cells. However, plasmatocyte/flagellate phagocytosis does not, on the basis of our observations, stimulate a respiratory burst. That the process is, however, an active one is demonstrated by its inhibition by cytochalasin B in the locust plasmatocyte/*Leishmania* system referred to in this paper.

Although evidence of induced responses in a natural vector is not yet available the finding from the locust model indicates that an induction of elevated levels of agglutinins can occur as well as enhanced lysozyme secretion following flagellate inoculation. The presence of concurrent infections such as mermithids in locusts can compromise the immune mechanisms, permitting trypanosomatid infections which would normally be controlled to become pathogenic. Conversely, *T. cruzi*-infected *Triatoma* have a reduced ability to reject xenografts compared with uninfected bugs. Both these observations parallel phenomena observed in mammalian hosts infected with *Trypanosoma* and *Leishmania*, where on the one hand immunodepressed hosts succumb more rapidly, whilst on the other the parasites themselves cause immunodepression.

It is hoped that this review has highlighted areas of study for the future on the relationship between trypanosomatids and their vectors, enabling a more detailed understanding of the basis of the interactions described to be elucidated, thus furthering our knowledge of parasite host specificity and epidemiology.

Acknowledgements

This work was in part supported by a grant from the UNDP/World Bank/ WHO Special Programme for Research and Training in Tropical Diseases and the King Faisal Foundation. GT was in receipt of an MRC Studentship.

References

Adler, S. & Theodor, O. (1929). Attempts to transmit *Leishmania tropica* by bite: The transmission of *L. tropica* by *Phlebotomus sergenti*. *Ann. trop. Med. Parasit.* 23: 1–16.

Aeschlimann, A., Burgdorfer, W., Matile, H., Peter, O. & Wyler, R. (1979). Aspects nouveaux du rôle de vecteur joué par *Ixodes ricinus* L. en Suisse. Note préliminaire. *Acta trop.* **36**: 181–91.

Anderson, R.S., Holmes, B. & Good, R.A. (1973). Comparative biochemistry of phagocytizing insect haemocytes. *Comp. Biochem. Physiol.* **46B**: 595–602.

Anez, N. (1981). *Trypanosomatidae of Venezuela with special reference to* Trypanosoma rangeli *and* Leishmania garnhami. PhD Thesis: University of London.

Anez, N. (1982). Studies on *Trypanosoma rangeli* Tejera, 1920. III. Direct transmission of *Trypanosoma rangeli* between triatomine bugs. *Ann. trop. Med. Parasit.* **76**: 641–7.

Arifdzhanov, K.A. & Nikitina, R.E. (1961). Detection of *Crithidia hyalomma* (O'Farrell 1913) in *Hyalomma a. anatolicum* (Koch 1844) ticks. *Zool. Zhur.* **40**: 20–4.

Ashford, R.W., Bray, M.A. & Foster, W.A. (1973). Observations on *Trypanosoma boueti* (Protozoa) parasitic in the skink *Mabuya striata* (Reptilia) and the sandfly *Sergentomyia bedfordi* in Ethiopia. *J. Zool., Lond.* **171**: 285–92.

Ayala, S.C. (1971). Trypanosomes in wild California sandflies, and extrinsic stages of *Trypanosoma bufophlebotomi*. *J. Protozool.* **18**: 433–6.

Bailey, C.H. & Brooks, W.M. (1972a). Histological observations on larvae of the eye gnat, *Hippolates pusio* (Diptera: Chloropidae), infected with the flagellate *Herpetomonas muscarum*. *J. Invert. Path.* **19**: 342–53.

Bailey, C.H. & Brooks, W.M. (1972b). Effects of *Herpetomonas muscarum* on development and longevity of the eye gnat, *Hippolates pusio* (Diptera: Chloropidae). *J. Invert. Path.* **20**: 31–6.

Baker, J.R. (1956). Studies on *Trypanosoma avium* Danilewsky 1885. III. Life cycle in vertebrate and invertebrate hosts. *Parasitology* **46**: 335–52.

Bennett, G.F. (1961). On the specificity and transmission of some avian trypanosomes. *Can. J. Zool.* **39**: 17–33.

Bitkowska, E., Dzbenski, T.H., Szadziewska, M. & Wegner, Z. (1982). Inhibition of xenograft rejection reaction in the bug *Triatoma infestans* during infection with a protozoan, *Trypanosoma cruzi*. *J. Invert. Path.* **40**: 186–9.

Burgdorfer, W., Schmidt, M.L. & Hoogstraal, H. (1973). Detection of *Trypanosoma theileri* in Ethiopian cattle ticks. *Acta trop.* **30**: 340–6.

Carpano, M. (1932). Localisations du *Trypanosoma theileri* dans les organes internes des bovins. Son cycle évolutif. *Ann. Parasit. hum. comp.* **10**: 305–22.

Cotton, T.D. (1970). A life cycle study of *Trypanosoma macfiei*, a natural haemoflagellate of canaries (*Serinus canarius*). *J. Parasit.* **56**(4) Section **2** (1): 63. (Abstract).

Croft, S.L., East, J.S. & Molyneux, D.H. (1982). Anti-trypanosomal factor in the haemolymph of *Glossina*. *Acta trop.* **39**: 293–302.

Croft, S.L., Kuzoe, F.A.S., Ryan, L. & Molyneux, D.H. (1984). Trypanosome infection rates of *Glossina* spp. (Diptera: Glossinidae) in transitional forest-savanna near Bouaflé, Ivory Coast. *Tropenmed. Parasit.* **35**: 247–50.

De Fario, G. & Cruz, O. (1927). Sur l'éxistence d'un stade évolutif intracellulaire du *Trypanosoma cruzi* dans la *Triatoma megista* Burm. *C.r. Séanc. Soc. Biol.* **97**: 1355–7.

D'Alessandro, A. (1976). Biology of *Trypanosoma* (*Herpetosoma*) *rangeli* Tejera,

1920. In *Biology of the Kinetoplastida* **1**: 328–403. (Eds Lumsden, W.H.R. & Evans, D.A.). Academic Press, London, New York & San Francisco.

Drif, L. (1983). *Etude des cellules circulantes d'insectes vecteurs hématophages: ultrastructure et fonctions*. PhD Thesis: Université des Sciences et Techniques du Languedoc, France.

East, J., Molyneux, D.H. & Hillen, N. (1980). Haemocytes of *Glossina*. *Ann. trop. Med. Parasit.* **74**: 471–4.

East, J., Molyneux, D.H., Maudlin, I. & Dukes, P. (1983). Effect of *Glossina* haemolymph on salivarian trypanosomes *in vitro*. *Ann. trop. Med. Parasit.* **77**: 97–9.

Ellis, D.S., Evans, D.A. & Stamford, S. (1980). The penetration of the salivary glands of *Rhodnius prolixus* by *Trypanosoma rangeli*. *Z. Parasitenk.* **62**: 63–74.

Evans, D.A. & Ellis, D.S. (1975). Penetration of mid-gut cells of *Glossina morsitans morsitans* by *Trypanosoma brucei rhodesiense*. *Nature, Lond.* **258**: 231–3.

Evans, D.A. & Ellis, D.S. (1983). Recent observations on the behaviour of certain trypanosomes within their insect hosts. *Adv. Parasit.* **22**: 1–42.

Fallis, A.M., Jacobson, R.L. & Raybould, J.N. (1973). Experimental transmission of *Trypanosoma numidae* Wenyon to guinea fowl and chickens in Tanzania. *J. Protozool.* **20**: 436–7.

Foster, R. (1963). Infection of *Glossina* spp. Weidemann 1830 (Diptera) and domestic stock with *Trypanosoma* spp. Gruby 1843 (Protozoa) in Liberia. *Ann. trop. Med. Parasit.* **57**: 383–96.

Foster, R. (1964). An unusual protozoal infection of tsetse flies (*Glossina* Weidemann 1830 spp.) in West Africa. *J. Protozool.* **11**: 100–6.

Glaser, R.W. (1922). *Herpetomonas muscae-domesticae*, its behaviour and effect in laboratory animals. *J. Parasit.* **8**: 99–108.

Gupta, A.P. (Ed.) (1979). *Insect haemocytes: Development, forms, functions and techniques*. Cambridge University Press, Cambridge, London etc.

Hoare, C.A. (1931a). The peritrophic membrane of *Glossina* and its bearing upon the life-cycle of *Trypanosoma grayi*. *Trans. R. Soc. trop. Med. Hyg.* **25**: 57–64.

Hoare, C.A. (1931b). Studies on *Trypanosoma grayi*. III. Life-cycle in the tsetse-fly and in the crocodile. *Parasitology* **23**: 449–84.

Hoare, C.A. (1938). Miscellanea Protistologica. II. Development of mammalian trypanosomes in the body-cavity of caterpillars. *Trans. R. Soc. trop. Med. Hyg.* **32**: 8.

Hoare, C.A. (1964). Morphological and taxonomic studies on mammalian trypanosomes. X. Revision of the systematics. *J. Protozool.* **11**: 200–7.

Hoare, C.A. (1972). *The trypanosomes of mammals: A zoological monograph*. Blackwell, Oxford.

Ibrahim, E.A.R., Ingram, G.A. & Molyneux, D.H. (1984). Haemagglutinins and parasite agglutinins in haemolymph and gut of *Glossina*. *Tropenmed. Parasit.* **35**: 151–6.

Ingram, G.A., East, J. & Molyneux, D.H. (1983). Agglutinins of *Trypanosoma*, *Leishmania* and *Crithidia* in insect haemolymph. *Devl comp. Immunol.* **7**: 649–52.

Ingram, G.A., East, J. & Molyneux, D.H. (1984). Naturally occurring agglutinins against trypanosomatid flagellates in the haemolymph of insects. *Parasitology* **89**: 435–51.

Ivanoff, E. (1925). Un nouveau mode de conservation et d'envoi des trypanosomes et des spirochètes dans des larves de *Galleria mellonella*. *C.R. hebd. Séanc. Acad. Sci., Paris* (D) **181**: 230–2.

Jones, J.C. (1965). The haemocytes of *Rhodnius prolixus* Stål. *Biol. Bull. mar. biol. lab. Woods Hole* **129**: 282–94.

Jones, J.C. (1967). Normal differential counts of haemocytes in relation to ecdysis and feeding in *Rhodnius*. *J. Insect Physiol.* **13**: 1133–41.

Jordan, A.M. (1976). Tsetse flies as vectors of trypanosomes. *Vet. Parasit.* **2**: 143–52.

Kaaya, G.P. (1980). Defence reactions of some medical vectors to pathogens and other foreign materials. *International Centre for Insect Physiology and Ecology, Nairobi, Annual Report* **8**: 111–113.

Kaaya, G.P. & Otieno, L.H. (1981). Haemocytes of *Glossina*. I. Morphological classification and the pattern of change with age of the flies. *Insect Sci. Appl.* **2**: 175–80.

Kaaya, G.P. & Ratcliffe, N.A. (1982). Comparative study of hemocytes and associated cells of some medically important dipterans. *J. Morph.* **173**: 351–65.

Kaddu, J.B. & Mutinga, M.J. (1980a). *Trypanosoma* (*Nannomonas*) *congolense* in the basement lamina of the anterior midgut cells of *Glossina pallidipes*. *Acta trop.* **37**: 91–2.

Kaddu, J.B. & Mutinga, M.J. (1980b). *Trypanosoma* (*Nannomonas*) *congolense* in the anterior midgut cells of *Glossina pallidipes*. *Ann. trop. Med. Parasit.* **74**: 255–6.

Khan, R.A. (1974). Transmission and development of the trypanosome of the Atlantic cod by a marine leech. In *Proceedings of the Third International Congress of Parasitology* **3**: 1608. Facta, Vienna. (Abstract.)

Killick-Kendrick, R., Lainson, R., Leaney, A.J., Ward, R.D. & Shaw, J.J. (1977). Promastigotes of *L.b. braziliensis* in the gut wall of a natural vector, *Psychodopygus wellcomei*. *Trans. R. Soc. trop. Med. Hyg.* **71**: 381.

Kramer, J.P. (1961). *Herpetomonas muscarum* (Leidy) in the haemocoele of larvae *Musca domestica* L. *Ent. News* **72**: 165–6.

Krinsky, W.L. & Burgdorfer, W. (1976). Trypanosomes in *Amblyomma americanum* from Oklahoma. *J. Parasit.* **62**: 824–5.

Lackie, A.M. (1983). Effect of substratum wettability and charge on adhesion *in vitro* and encapsulation *in vivo* by insect haemocytes. *J. Cell Sci.* **63**: 181–90.

Lackie, A.M., Takle, G.B & Tetley, L. (1985). Haemocytic encapsulation in the locust *Schistocerca gregaria* (Orthoptera) and in the cockroach *Periplaneta americana* (Dictyoptera). *Cell Tissue Res.* **240**: 343–51.

Lai–Fook, L. (1970). Haemocytes in the repair of wounds in an insect (*Rhodnius prolixus*). *J. Morph.* **130**: 297–313.

Lainson, R. & Shaw, J.J. (1979). The role of animals in the epidemiology of South American leishmaniasis. In *Biology of the Kinetoplastida* **2**: 1–116. (Eds Lumsden, W.H.R. & Evans, D.A.). Academic Press, London, New York, San Francisco.

Lelijveld, J.L.M. (1966). Leishmania mexicana *in ticks*. Unpublished DAPE Thesis: University of London.

Linder, J. (1960). Behaviour of various Trypanosomatidae in larvae of the bee-moth, *Galleria mellonella*. *Bull. Res. Coun. Israel* **8E**: 128–34.

MacFie, J.W.S. & Thompson, J.G. (1929). A trypanosome of the canary (*Serinus canarius* Koch). *Trans. R. Soc. trop. Med. Hyg.* **23**: 185–91.

Minchin, E.A. & Thomson, J.D. (1911). On the occurrence of an intracellular stage in the development of *Trypanosoma lewisi* in the rat-flea. (Prel. Note.) *Br. med. J.* **1911**(**ii**): 361–4.

Minchin, E.A. & Thomson, J.D. (1915). The rat-trypanosome, *Trypanosoma lewisi*, in its relation to the rat-flea, *Ceratophyllus fasciatus*. *Q. Jl microsc. Sci.* **60**: 463–92.

Molyneux, D.H. (1969). Intracellular stages of *Trypanosoma lewisi* in fleas and attempts to find such stages in other trypanosome species. *Parasitology* **59**: 737–44.

Molyneux, D.H., Killick-Kendrick, R. & Ashford, R.W. (1975). *Leishmania* in phlebotomid sandflies. III. The ultrastructure of *Leishmania mexicana amazonensis* in the midgut and pharynx of *Lutzomyia longipalpis*. *Proc. R. Soc.* (B) **190**: 341–57.

Molyneux, D.H., Ryan, L., Lainson, R. & Shaw, J.J. (In press). The *Leishmania*-sandfly interface. Proceedings of a symposium on *Leishmania* evolution. Montpellier, 1984. *Ann. Parasit. hum. comp.*

Mshelbwala, A.S. (1972). *Trypanosoma brucei* in the haemocoel of tsetse flies. *Trans. R. Soc. trop. Med. Hyg.* **66**: 637–43.

Needham, E.A. (1969). *Protozoa parasitic in fish*. Ph.D. Thesis: University of London.

Nöller, W. (1912). Die Übertragungsweise der Rattentrypanosomen durch Flohe. *Arch. Protistenk.* **25**: 386–424.

Núñez, J.A. (1962). Regulation of water economy in *Rhodnius prolixus*. *Nature, Lond.* **194**: 704.

O'Farrell, W.R. (1913). Hereditary infection, with special reference to its occurrence in *Hyalomma aegyptium* infected with *Crithidia hyalommae*. *Ann. trop. Med. Parasit.* **7**: 545–62.

Otieno, L.H. (1973). *Trypanosoma* (*Trypanozoon*) *brucei* in the haemolymph of experimentally infected young *Glossina morsitans*. *Trans. R. Soc. trop. Med. Hyg.* **72**: 622–6.

Otieno, L.H. & Darji, N. (1979). An assessment of trypanosome infections of wild *Glossina pallidipes* Austen using fly dissection, salivation and mouse inoculation methods. In OUA/STRC International Scientific Council for Trypanosomiasis Research and Control (ISCTRC). 15th Meeting, Banjul, The Gambia, 15–30 April 1977. *Publ. OAU/STRC* No. 110: 273–8.

Pereira, M.E.A., Andrade, A.F.B. & Ribeiro, J.M.C. (1981). Lectins of distinct specificity in *Rhodnius prolixus* interact selectively with *Trypanosoma cruzi*. *Science, N.Y.* **211**: 597–600.

Pereira, M.E., Loures, M.A., Villalta, F. & Andrade, A.F.B. (1980). Lectin receptors as markers for *Trypanosoma cruzi*. Developmental stages and a study of the interaction of wheat germ agglutinin with sialic acid residues on epimastigote cells. *J. exp. Med.* **152**: 1375–92.

Peters, W., Kolb, H. & Kolb-Bachofen, V. (1983). Evidence for a sugar receptor (lectin) in the peritrophic membrane of the blowfly larva, *Calliphora erythrocephala* Mg. (Diptera). *J. Insect Physiol.* **29**: 275–80.

Poinar, G.O., Jr., Wassink, H.J.M., Leegwater-Van Der Linden, M.E. & Van Der Geest, L.P.S. (1979). *Serratia marcescens* as a pathogen of tsetse flies. *Acta trop.* **36**: 223–7.

Price, C.D. & Ratcliffe, N.A. (1974). A reappraisal of insect haemocyte classification by the examination of blood from fifteen insect orders. *Z. Zellforsch. mikrosk. Anat.* **147**: 537–49.

Rehacek, V.J., Sixl, W. & Sebek, Z. (1974). Trypanosomen in der Haemolymphe von Zecken. *Mitt. Abt. Zool. Bot. Landesmus. Joanneum.* **3**: 33.

Robertson, M. (1910). Further notes on a trypanosome from the gut of *Pontobdella muricata. Q. Jl microsc. Sci.* **54**: 119–41.

Rodhain, J. (1942a). Au sujet du développement intracellulaire de *Trypanosoma lewisi* chez *Ornithodorus moubata. Acta biol. belg.* **2**: 413–5.

Rodhain, J. (1942b). Au sujet du développement intracellulaire de *Trypanosoma pipistrelli* (Chatton et Courrier) chez *Ornithodorus moubata. Acta biol. belg.* **2**: 416–20.

Rowley, A.F. & Ratcliffe, N.A. (1976). An ultrastructural study of the *in vitro* phagocytosis of *Escherichia coli* by the haemocytes of *Calliphora erythrocephala. J. Ultrastruct. Res.* **55**: 193–202.

Rowley, A.F. & Ratcliffe, N.A. (1981). Insects. In *Invertebrate blood cells* **2**: 421–88. (Eds Ratcliffe, N.A. & Rowley, A.F.). Academic Press, London, New York, Toronto, Sydney, San Francisco.

Rutherford, T.A. & Webster, J.M. (1978). Some effects of *Mermis nigrescens* on the hemolymph of *Schistocerca gregaria. Can. J. Zool.* **56**: 339–47.

Schmittner, S.M. & McGhee, R.B. (1970). Host specificity of various species of *Crithidia* Leger. *J. Parasit.* **56**: 684–93.

Shastri, U.V. & Despande, P.D. (1981). *Hyalomma anatolicum anatolicum* (Koch 1844) as a possible vector for transmission of *Trypanosoma theileri*, Laveran, 1902 in cattle. *Vet. Parasit.* **9**: 151–5.

Shaw, J.J. (1969). *The haemoflagellates of sloths.* (London School of Hygiene and Tropical Medicine Memoir No. 13: 1–127.) H.K. Lewis, London.

Shaw, J.J. (1981). The behaviour of *Endotrypanum schaudinni* (Kinetoplastida: Trypanosomatidae) in three species of laboratory bred Neotropical sandflies (Diptera: Psychodidae) and its influence on the classification of the genus *Leishmania.* In *Parasitological topics*: 232–42. (Ed. Canning, E.U.). Allen Press, Kansas. (Soc. Protozool. Spec. Publ. No. 1.)

Shortt, H.E. (1923). Record of Kala-azar research work carried out at the King Edward VII Memorial Pasteur Institute, Shillong during 1922. *Ind. J. med. Res.* **10**: 1150–68.

Smirnoff, W.A. (1974). Reduction de la viabilité et de la fecondité de *Neodiprion swainei* (Hymenoptera: Tenthredinidae) par le flagellate *Herpetomonas swainei. Phytoprotection* **55**: 64–6.

Smirnoff, W.A. & Lipa, J.J. (1970). *Herpetomonas swainei* sp.n. a new flagellate parasite of *Neodiprion swainei* (Hymenoptera: Tenthredinidae). *J. Invert. Path.* **16**: 187–95.

Takle, G. & Lackie, A.M. (1985). Surface charge of insect haemocytes, examined using cell electrophoresis and cationized ferritin binding. *J. Cell Sci.* **75**: 207–14.

Thorne, K.J.I. & Blackwell, J.M. (1983). Cell-mediated killing of Protozoa. *Adv. Parasit.* **22**: 43–151.

Tobie, E.J. (1968). Fate of some culture flagellates in the hemocoel of *Rhodnius prolixus*. *J. Parasit.* **54**: 1040–6.

Wallace, F.G. (1966). The trypanosomatid parasites of insects and arachnids. *Expl Parasit.* **18**: 124–93.

Wallace, F.G. (1979). Biology of the Kinetoplastida of arthropods. In *Biology of the Kinetoplastida* **2**: 213–40. (Eds Lumsden, W.H.R. & Evans, D.A.). Academic Press, London, New York, San Francisco.

Watkins, R. (1971a). Histology of *Rhodnius prolixus* infected with *Trypanosoma rangeli*. *J. Invert. Path.* **17**: 59–66.

Watkins, R. (1971b). *Trypanosoma rangeli*: effect on excretion in *Rhodnius prolixus*. *J. Invert. Path.* **17**: 67–71.

Wells, E.A. (1976). Subgenus *Megatrypanum*. In *Biology of the Kinetoplastida* **1**: 257–84. (Eds Lumsden, W.H.R. & Evans, D.A.). Academic Press, London, New York, San Francisco.

Woo, P.T.K. (1969a). The development of *Trypanosoma canadensis* of *Rana pipiens* in *Placobdella* sp. *Can. J. Zool.* **47**: 1257–9.

Woo, P.T.K. (1969b). The life cycle of *Trypanosoma chrysemydis*. *Can. J. Zool.* **47**: 1139–51.

Zeledón, R. & de Monge, E. (1966). Natural immunity of the bug *Triatoma infestans* to the Protozoan *Trypanosoma rangeli*. *J. Invert. Path.* **8**: 420–4.

Zotta, G. (1921). Sur la transmission expérimentale du *Leptomonas pyrrhocoris* Z. chez des insectes divers. *C. r. Séanc. Soc. Biol.* **85**: 135–7.

Symp. zool. Soc. Lond. (1986) No. 56 145–160

Immune mechanisms and mosquito-filarial worm relationships

Bruce M. CHRISTENSEN

Department of Veterinary Science, University of Wisconsin, 1655 Linden Drive, Madison, Wisconsin 53706, USA

Synopsis

A review of the literature regarding the melanization response of adult mosquitoes against filarioid nematodes indicates that various species and strains of mosquitoes differ somewhat in their immune capabilities. Studies also show that the ability to mount a melanization response decreases with the age of the host and is significantly reduced in male mosquitoes. Although data concerning haemocyte populations and their classification in adult mosquitoes are equivocal, ultrastructural studies verify that haemocytes are actively involved in the melanization response. Mosquito species that support filarial worm development, as well as those that are refractory, possess the inherent ability to destroy microfilariae by melanization. However, data suggest microfilariae obtain protection from immune recognition by exposure to the midgut in compatible mosquito hosts and that developing larvae have the ability to actively suppress the melanization response; therefore, it is proposed that immune evasion tactics on the part of the parasite play a key role in determining successful mosquito-filarial worm associations.

Introduction

Knowledge concerning encapsulation and/or melanization reactions of insects against metazoan parasites has increased significantly in the last 30 years. Unfortunately, little of this research has been concerned with the immune capabilities of adult mosquitoes. The paucity of information available on defence mechanisms of mosquitoes is striking when one considers the public health importance of these insects.

The majority of early literature dealing with adult mosquito defence reactions is primarily descriptive and deals almost exclusively with incidental observations of melanized microfilariae (mff) or larvae of filarioid nematodes. This melanization response was first thought to occur only around dead parasites (Brug 1932; Hu 1939a), but later studies by Hu (1939b), Highby (1943) and Kartman (1953a) demonstrated that the response was also made against living filarial worms. Although numerous reports of mela-

ZOOLOGICAL SYMPOSIUM No. 56
ISBN 0-19-854004-3

nization of filarial worms by mosquito hosts have been published (Sambon 1902; Newton, Wright & Pratt, 1945; Kartman 1953a,b; Lavoipierre 1958; Esslinger 1962; Intermill 1973; Oothuman, Simpson & Laurence 1974; Nayar & Sauerman 1975; Lindemann 1977; Christensen 1981; Ho, Yap & Sin 1982; Bradley & Nayar 1985; Yamamoto, Kobayashi, Ogura, Tsuruoka & Chigusa 1985), the reaction has long been considered a response that occurs in mosquitoes that are refractory to filarial worm development, or one that occurs so seldom that it is of no consequence to the vector-parasite system.

It is now well accepted that in certain mosquito-filarial parasite systems the survival of the vector for the extrinsic incubation period of the parasite is seriously threatened when the intensity of the infection is excessive (Kershaw, Lavoipierre & Chalmers 1953; Christensen 1978). It also has been suggested that the melanization reaction of *Aedes trivittatus* against *Dirofilaria immitis* might function in reducing the parasite load in heavily infected mosquitoes, thereby helping to ensure the survival of both the parasite and the vector (Christensen 1981). However, experimental data are needed to substantiate these preliminary observations.

This chapter is concerned with the defence capabilities of adult mosquitoes against filarioid nematodes, and with the mechanisms of immune evasion that are relevant to parasite survival in compatible vectors.

Barriers to the host haemocoel

Nearly all mff face three potential mechanical barriers upon entry into a mosquito vector: (1) the bucco-pharyngeal armature, (2) the peritrophic membrane and (3) the midgut epithelial cells. Certain species of *Dirofilaria* develop in and emerge from the Malpighian tubules and thus avoid both the peritrophic membrane and midgut wall. Schacher (1973) has provided an excellent summary of known species of filarioid nematodes, their definitive and intermediate hosts, and the developmental site of filarioid larvae within their invertebrate vectors.

Coluzzi & Trabucchi (1968) investigated the mechanism by which *Dirofiliaria repens* mff were rapidly destroyed in the midgut of *Anopheles* and *Culex* mosquitoes. They tested 18 species of mosquitoes and found physically damaged mff in those hosts that possessed a well-developed pharyngeal armature. They suggested that mff damaged by the armature during ingestion were more susceptible to digestion by the mosquito. Mff introduced into the gut through the anus, however, were undamaged, and even completely resistant species became infected. When these mosquitoes ingested smaller mff (*Eufilaria sergenti*), only a very few were found damaged. Subsequent studies by other researchers have supported these findings (Bryan, Oothuman, Andrews & McGreevy 1974).

The peritrophic membrane has often been considered a barrier that might hinder parasite infections in mosquitoes; however, it has not been shown to limit the migration of mff from the midgut into the haemocoel. Esslinger (1962) and Ewert (1965) reported that *Brugia pahangi* migrated through the midgut before the peritrophic membrane had hardened, and recently, Christensen & Sutherland (1984) found that 60% of the *B. pahangi* mff ingested by *Aedes aegypti* black-eyed Liverpool strain (LVP) migrated out of the midgut within 2.5 h following ingestion. Those mff that failed to penetrate within this time period usually remained within the midgut. Similar findings were reported for *B. malayi* in both *A. aegypti* (Ramachandran 1966) and *Mansonia longipalpis* (Wharton 1957). Orihel (1975), in his review of the peritrophic membrane as a potential barrier to infection in various arthropod hosts, concluded that the membrane does not hinder migration of mff.

It has been assumed for a number of years that migration of mff from the midgut is significantly reduced if the ingested blood meal undergoes rapid clot formation (Kartman 1953a). Recently, however, it was suggested that progressive movement of *B. malayi* mff might be aided by the clotting of ingested blood (Mather, Hamill, Ribeiro, Rossignol & Spielman unpublished abstract). These authors reported accelerated midgut penetration by mff when *A. aegypti* were subjected to pre-feeding ablation of their salivary ducts, thereby providing a more rapid coagulation of the ingested blood meal. These results are in contrast to recent studies by Sutherland, Christensen & Lasee (1986), who showed a significant increase in the percentage of *B. pahangi* mff penetrating *A. aegypti* LVP midguts *in vitro* when mosquitoes were exposed to infective blood containing anticoagulants. They demonstrated that a midgut barrier does exist in *A. trivittatus* that significantly limits migration of *B. pahangi* mff. They suggested that there is a physiological difference in the midgut environment of this species compared to *A. aegypti*, and microscopical examinations indicated that inhibition of microfilarial migration was not caused by structural differences in the composition of the midgut wall.

If mff circumvent these mechanical barriers and gain access to the haemocoel, they must then contend with the cellular and humoral responses of the mosquito host. The innate capabilities of mosquitoes to respond and destroy invading mff, and the means whereby mff avoid or inhibit this response are the subjects of the remainder of this chapter.

Immune components of haemolymph

It generally has been accepted that circulating haemocytes are not very abundant in nematoceran dipterans, and several studies with larval stages have reported humoral melanization of parasites without haemocyte involvement (see review by Poinar 1974). Data available on haemocyte populations and their classification in adult mosquitoes are limited. Foley (1978) categorized the haemocytes of *Anopheles stephensi* adults as prohaemocytes, plasmatocytes, granular haemocytes, adipohaemocytes and spherule cells and Kaaya & Ratcliffe (1982) identified prohaemocytes, plasmatocytes, adipohaemocytes and oenocytoids in both adult *A. aegypti* and *Culex quinquefasciatus*. However, Drif & Brehélin (1983), from their work with *C. pipiens* and *A. aegypti* larvae, believe that mosquitoes have only two haemocyte types, plasmatoctyes and oenocytoids. They consider granular haemocytes to be plasmatocytes, and adipohaemocytes and spherule cells to be adipose tissue cells and not true haemocytes. Foley (1978) perfused mosquitoes with 100 μl of buffered salt solution or medium and reported mean total haemocyte counts of 10,220, 9,262 and 1,725 haemocytes/mosquito in newly-emerged, two-day-old and ten-day-old *A. stephensi*, respectively. These counts were obtained using a haemocytometer following dilution of the perfusate, and seem high compared to preliminary unpublished data we have obtained for *A. aegypti* LVP and *A. trivittatus* adults. At emergence and at 28 days of age, the average number of haemocytes per individual mosquito was 436 (N=19) and 460 (N=17) for *A. aegypti*, and 256 (N=22) and 294 (N=20) for *A. trivittatus*. Data for *Aedes* mosquitoes were obtained using the methods of Foley (1978) except that individual mosquitoes were perfused with 20 μl of *Aedes* saline and every cell contained in the perfusate was counted. These figures represent approximately 80% of the free haemocyte population perfused as determined by a [^{14}C] carboxyinulin dilution technique.

No data are available that correlate haemocyte population changes in adult mosquitoes with the melanization response against filarioid nematodes, and few studies have clarified the involvement of haemocytes in the response. Esslinger (1962) suggested the melanization of *B. pahangi* mff in *A. quadrimaculatus* was not directly associated with any haemocyte type. Bartlett (1984), working with *Dirofilaria scapiceps* and several species of *Aedes*, reported that the melanization response within the fat body was probably of humoral origin since haemocytes were never associated with these mff; however, haemocytes were present in close association with mff melanized in the haemocoel. Studies with *A. aegypti* and *Breinlia booliati* suggested that the initial melanization reaction was followed by haemocyte encapsulation (Ho *et al.* 1982).

Recent ultrastructural studies have verified that the haemocytes of adult *A. trivittatus* are involved in the melanization of intrathoracically inoculated mff of *D. immitis* (Forton, Christensen & Sutherland 1985). The melanization response, which was initiated almost immediately following exposure of the parasite to the haemolymph environment, was always accompanied by haemocyte lysis. Although intact haemocytes were never abundant, those present showed an active formation of membrane-bound vacuoles. These 'activated' haemocytes, which were often seen in different stages of lysis, were in close association but not in direct contact with mff. The classic haemocytic encapsulation generally described for insects was not evident, but observations suggested that the haemocytes functioned in immune recognition and the activation of melanin synthesis. Following the death of melanized mff, at approximately three days post-inoculation (PI), a single membrane began to develop around the reaction site. This eventually developed into a double membrane-like structure of 25–30 nm thickness that effectively enclosed and isolated the mff, melanin deposits and haemocyte remnants from the haemolymph components. This membrane served as an isolating boundary that prevented additional haemocyte involvement and thereby stopped the reaction. The membrane formed around melanized mff of *D. immitis* is likely what Ho *et al.* (1982) described as a 'translucent, gelatinous envelope around the melanized larva' formed by *A. aegypti*. Ultrastructural studies of the immune response (Christensen & Forton 1986) of *A. aegypti* LVP against *D. immitis* suggest a more extensive involvement of haemocytes than in *A. trivittatus* (Figs 1, 2), but the reaction is similar in the formation of an isolating membrane (Figs 2, 3). In this species, the initial reaction involves the lysis of haemocytes at or near the surface of the parasite prior to the synthesis of pigment. Subsequently, melanin formation occurs in the area of lysed cells and appears to be deposited onto the parasite surface (Christensen & Forton 1986). These studies also suggest that melanin may be synthesized within certain haemocytes and released by exocytosis or upon cell lysis. Haemocyte lysis may also precede melanization deposition in *A. trivittatus*, but because the immune reaction occurs so rapidly in *A. trivittatus* compared with *A. aegypti* it may not have been possible to recover and fix mff for EM studies rapidly enough from the *in vivo* system used by Forton *et al.* (1985).

Ultrastructural studies of the melanization response of *Anopheles quadrimaculatus* against *B. pahangi* suggest that a slightly different sequence of events might occur in this genus (Chen & Laurence 1985). These authors reported 'humoral' melanization followed by cellular encapsulation and did not observe any evidence of haemocyte lysis in any of their micrographs. Melanin deposition also occurs very rapidly in this mosquito and mff recovered as soon as 10 min following exposure were completely encased in pigment. The development of an *in vitro* system would greatly facilitate studies concerning the early events in the melanization reaction of mosquitoes against mff.

Although the melanization response has been noted in mosquito species that are highly susceptible to filarial worm development (Intermill 1973; Christensen 1981), the response generally has been associated with mosquitoes that will not support the development of a particular species of filarioid

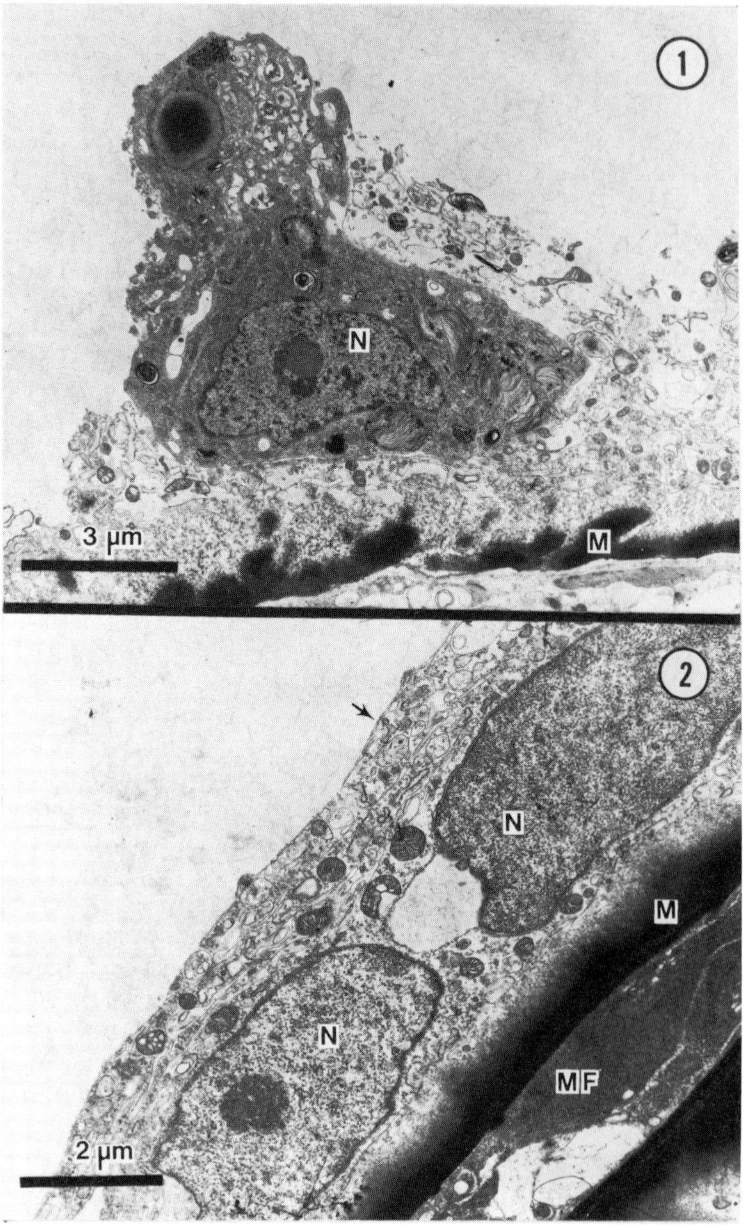

Figs 1, 2. Melanization response of *Aedes aegypti* against inoculated *Dirofilaria immitis* microfilariae. Fig. 1. Intact haemocyte beginning to lyse near the surface of melanized microfilaria. N, haemocyte nucleus; M, melanin deposits. Fig. 2. Two intact haemocyte nuclei and numerous haemocyte remnants on the surface of melanized microfilaria. Note membrane formation (arrow). MF, microfilaria.

nematode, e.g., *Anopheles labranchiae atroparvus* and *B. patei* (Oothuman *et al.* 1974). It recently has become evident, however, that both susceptible and insusceptible mosquitoes possess the ability to destroy mff by melanization. Christensen, Sutherland & Gleason (1984) studied two susceptible species, *A. aegypti* LVP and *A. trivittatus*, and a mosquito that does not support filarial worm development, *A. aegypti* Rockefeller strain (RKF).

Microfilariae of *B. pahangi* and *D. immitis* intrathoracically inoculated into *A. trivittatus* were melanized and killed within two days PI and the response of this species was significantly more rapid and effective than that seen in either the LVP or RKF strain of *A. aegypti*. Inoculation of large numbers of mff overloaded the immune capabilities of *A. aegypti*, but not those of *A. trivittatus*. Likewise, the melanization response of *A. aegypti* could be effectively reduced for up to four days PI, but for only one day PI in *A. trivittatus*, when mosquitoes were maintained on a 0.3 M sucrose diet containing from 0.1 to 1.0% phenylthiourea. Although *A. trivittatus* is a principal natural vector of *D. immitis* (Christensen & Andrews 1976; Christensen 1977; Pinger 1982), it has a much greater immunological capability against mff than either strain of *A. aegypti*.

The ability of either *A. aegypti* or *A. trivittatus* adults to melanize *D. immitis* mff after intrathoracic inoculation was easily explained by the abnormal environment in which mff were placed (haemocoel versus the normal Malphighian tubule environment), but the response observed against *B. pahangi* mff inoculated into the haemocoel of susceptible mosquito vectors was not expected. This filarial worm normally is exposed to the haemocoel environment during the migration from the midgut to the thoracic musculature where development occurs. It was postulated that perhaps the sheath present on inoculated mff stimulated a response in mosquitoes that would not occur if mff were exsheathed. Experiments with LVP strain *A. aegypti* inoculated with chemically exsheathed *B. pahangi* mff produced a significantly reduced melanization response as compared with sheathed mff, but still over 50% of the exsheathed mff were melanized by Day 5 PI (Sutherland, Christensen & Forton 1984). In unpublished studies involving hundreds of LVP mosquitoes naturally exposed to *B. pahangi* via an infected blood meal, we have rarely seen a melanization response against mff or developing larvae. To assess if artificial exsheathment damaged mff in some way and thereby caused an increased rate of melanization, Sutherland, Christensen & Forton (1984) allowed mff to penetrate LVP midguts *in vitro* before inoculation into intact *A. aegypti* LVP. Essentially 100% of the inoculated parasites recovered at five days PI had avoided the response and were developing as normal first-stage larvae. However, it also was determined that *B. pahangi* mff do not exsheath within the midgut prior to penetration. The great majority of mff penetrate the midgut and exsheath within the haemocoel of their vectors (Christensen & Sutherland

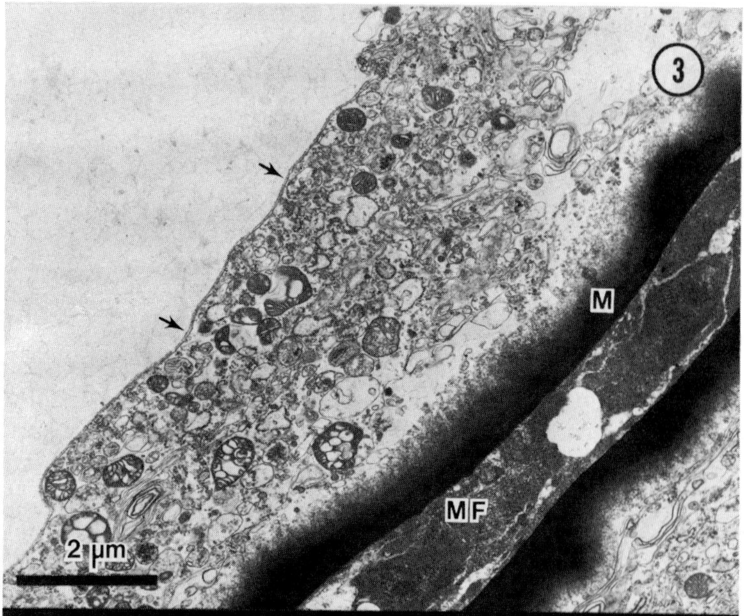

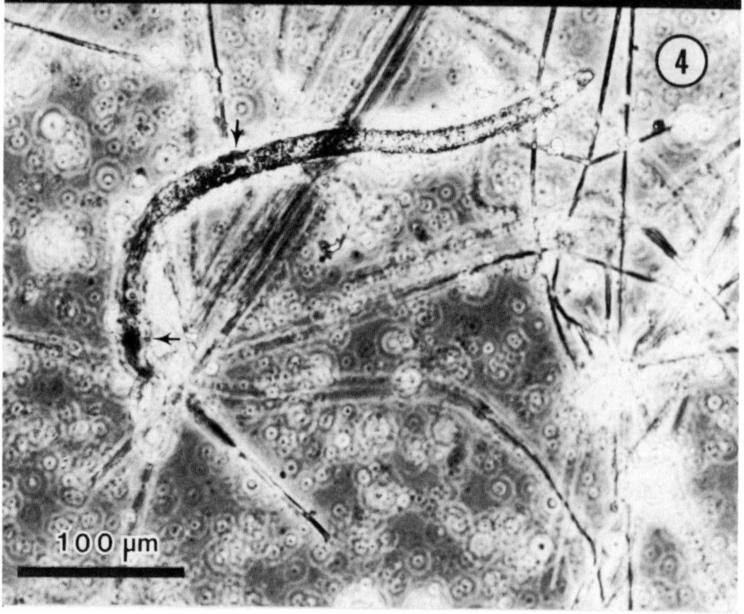

Fig. 3. Melanized microfilaria at Day 5 post-inoculation showing numerous haemocyte remnants contained within a double membrane-like boundary (arrows). M, melanin deposits; MF, microfilaria.

Fig. 4. *Dirofilaria immitis* microfilaria recovered from a male *Aedes aegypti* at five days post-inoculation. Note the very light melanin deposits (arrows).

1984). Studies by Agudelo-Silva & Spielman (1985) confirmed these findings. The presence or absence of the microfilarial sheath evidently plays a very small role in determining if a melanization response will or will not occur.

In addition to species and strain differences in the immune response of mosquitoes to filarial worms, there are significant differences in the immune capabilities that are associated with both age and sex of the host. Studies by Duxbury, Moon & Sadun (1961) and Desowitz & Chellapah (1962) have suggested that older mosquitoes are more effective vectors of filarial worms than are younger hosts, but no study has been published on the effects of host age on the immune capacity of adult insects. Christensen, LaFond & Christensen (1986) intrathoracically inoculated *D. immitis* mff into newly-emerged, 14-, 21-, and 28-day-old *A. aegypti* LVP and *A. trivittatus* and assessed the melanization response at one, three, and five days PI. The melanization response of newly-emerged mosquitoes was the same as that previously reported for four- to six-day-old mosquitoes by Christensen, Sutherland *et al.* (1984); however, the ability of 14- to 28-day-old *A. aegypti* to respond effectively against inoculated mff was significantly reduced compared with younger mosquitoes at one, three, and five days PI (Table 1). The melanization response of *A. trivittatus* also was delayed in older mosquitoes at one and three days PI, but by Day five PI essentially all mff recovered from any age group were completely melanized (Table 1). Although all mff in older *A. trivittatus* were melanized and killed by five days PI, melanin deposits on mff recovered from older mosquitoes were much thinner and lighter in colour. This reduced immune capability in older mosquitoes could be explained by a reduction in haemocyte numbers as reported by Foley (1978), a reduction in phenoloxidase activity or in the ability of its proenzyme to be activated, a change in levels of tyrosine or other melanin precursors, or an alteration in the ability of haemocytes or humoral factors to recognize and respond to non-self material.

Preliminary studies in our laboratory (Harris & Christensen in prep.) with male *A. aegypti* LVP and *A. trivittatus* inoculated with *D. immitis* mff have shown that males are virtually immune-incompetent. Even after five days PI, the majority of mff were alive and showed no signs of melanization. The few effective host responses (<5 %) that were noted occurred only in *A. trivittatus* and usually consisted of small, light patches of melanin deposited randomly on the cuticular surface (Fig. 4). Moreover, most of the mff that were melanized were still alive at five days PI. Initial studies of haemocyte populations indicate that males have fewer blood cells than females, but experimental data are needed to determine if this is the cause of the reduced immune competence in males.

It is apparent that both refractory and susceptible female mosquitoes of different species and strains possess the inherent ability to destroy mff by melanization (Christensen, Sutherland *et al.* 1984). The ability of the para-

Table 1. Effect of host age on the melanization response of adult *Aedes trivittatus* and *Aedes aegypti* against inoculated *Dirofilaria immitis* microfilariae (mff).

Mosq. species	Age (days)	% mff melanized ± S.E.		
		Day 1 PI[a]	Day 3 PI	Day 5 PI
A. aegypti	<1	34.4 ± 6.8	64.2 ± 7.0	91.9 ± 2.6
	14	3.9 ± 1.5	11.8 ± 3.6	20.5 ± 5.7
	21–28	3.3 ± 1.1	14.0 ± 4.2	25.3 ± 6.3
A. trivittatus	<1	100	100	100
	14	73.6 ± 11.8	96.7 ± 3.3	100
	21	76.3 ± 13.2	73.2 ± 11.6	100
	28	70.2 ± 14.6	62.4 ± 15.7	89.6 ± 10.4

[a] Days post-inoculation on which mosquitoes were examined.

site to evade immune recognition and/or suppress the immune response of the host are important factors in the development of compatible or incompatible host-parasite relations.

Immune evasion by parasites

The ability of parasites to avoid immune recognition or to inhibit the immune response is essential for their survival in hosts with internal defence mechanisms ostensibly designed to retain the integrity of self. Recent studies have provided data that suggest *B. pahangi* avoids immune recognition upon initial entry into a compatible mosquito vector and, after development begins, these parasites have the ability to actively suppress the immune response.

In preliminary studies, Sutherland, Christensen & Forton (1984) suggested that penetration of the midgut in *A. aegypti* LVP provided *B. pahangi* mff with protection from immune recognition. They proposed that mff might acquire host materials on their surface during midgut penetration and therefore not be recognized as foreign by the mosquito, or that upon exposure to the midgut environment the mff are modified or stimulated in such a way that they are able to actively inhibit the immune response in the mosquito. Subsequently, LaFond, Christensen & Lasee (1985) studied this phenomenon in more detail by examining the melanization response of *A. trivittatus* and the LVP and RKF strains of *A. aegypti* against intrathoracically inoculated *B. pahangi* mff that had previously penetrated LVP, RKF, or *A. trivittatus* midguts *in vitro*. Penetration of either LVP or RKF midguts not only provided mff with protection from melanization in both strains of *A. aegypti*, but also enabled 31–43% of the mff to survive in the highly immunocompetent *A. trivittatus*. In contrast, exposure to the midgut environment in *A. trivittatus* did not significantly affect the ability of mff to avoid the melanization response in any of the mosquitoes examined.

LaFond *et al.* (1985) also reported that the LVP and RKF strains of *A. aegypti* and *A. trivittatus* also lacked the ability to recognize either allogeneic or xenogeneic midgut tissue implants. They suggested that if substances were being accreted onto the surface of mff during their migration through the midgut wall they would prevent the hosts from recognizing the parasites as foreign.

In an attempt to provide direct evidence that antigen sharing might be the mechanism whereby mff avoid immune recognition following midgut penetration, Christensen, Forton, LaFond & Grieve (in press) utilized an indirect fluorescent antibody assay employing anti-midgut antibodies. Antibodies to both blood-fed (to account for the possible involvement of the peritrophic membrane) and non-blood-fed midguts were generated in rabbits following the methods of Vaitukaitus, Robbins, Nuschlag & Ross (1971). Antibody responses in rabbits were monitored using homologous antigen in an enzyme-linked immunosorbent assay (ELISA) modified from an assay described by Grieve, Mika-Johnson, Jacobsen & Cypess (1981). Under ELISA conditions of 0.75 μg midgut protein/microtitre well, 50 μl serum, and 1:1,000 horseradish peroxidase-conjugated anti-rabbit immunoglobulin G, titres of 1:32,728 were measured within four weeks after immunization. Titres of 1:2–1:8 were obtained from the sera of an adjuvant control rabbit and pre-immunization sera from each rabbit. Microfilariae isolated directly from blood and following LVP mosquito midgut penetration were incubated in immune sera and subsequently in fluorescein-conjugated goat anti-rabbit IgG. Microfilariae incubated with control sera served as controls. Fluorescence microscopy revealed no significant fluorescence on the surface of any mff examined. Although these results were negative, one cannot discount the possibility of antigen sharing as a viable mechanism. The assay employed could possibly lack the sensitivity required to detect small quantities of antigen and it is possible that the relevant material might be poorly antigenic in rabbits.

Recent transmission electron microscopy studies involving cationic and anionic colloidal iron incubation (after the methods of Gasic, Berwick & Sorrentino (1968) and Lumsden (1972)) show that *B. pahangi* mff derived directly from vertebrate host blood have a high electronegative charge on the entire outer surface of the sheath, but that this surface electronegativity on the sheath is lost following penetration of the *A. aegypti* midgut (Christensen, Forton *et al.* in press). Neither blood-isolated or midgut penetrated mff showed an affinity for anionic colloidal iron. This loss of electronegativity on the surface of sheathed mff following migration through the midgut could be caused by the chemical environment within the midgut itself, or this negative charge could be effectively negated by the acquisition of midgut materials during penetration.

Once *B. pahangi* mff reach the thoracic musculature and begin develop-

Table 2. Percentage of inoculated *Brugia pahangi* or *Dirofilaria immitis* microfilariae (mff) melanized in adult *Aedes aegypti* either infected with developing *B. pahangi* or not infected.

No. mosquitoes examined	Infection status	Inoculated with	% mff melanized ± S.E.
63	*B. pahangi*	*B. pahangi*	38.2 ± 6.3
68	Not infected	*B. pahangi*	64.5 ± 6.5
17	*B. pahangi*	*D. immitis*	51.2 ± 6.6
20	Not infected	*D. immitis*	71.3 ± 3.9

ment as first-stage larvae, melanization never seems to occur (Christensen, Sutherland *et al.* 1984). It is possible that when mff reach their site of development within the muscle fibres they are shielded from immune recognition. In addition, developing larvae could actively suppress the immune response of the mosquito, as has been suggested for *Trypanosoma cruzi*-infected *Triatoma infestans* by Bitkowska, Dzbenski, Szadziewska & Wegner (1982). To test the hypothesis that developing larvae are capable of suppressing the immune response in LVP strain *A. aegypti*, Christensen & LaFond (1986) inoculated mosquitoes with midgut-derived *B. pahangi* mff, maintained these mosquitoes for five to eight days to permit larvae to develop and then inoculated these with *D. immitis* or *B. pahangi* mff isolated from blood. Mosquitoes initially inoculated with saline only and then receiving blood-isolated mff served as controls. In all experiments, the melanization response was significantly reduced in mosquitoes harbouring developing larvae (Table 2). In other insect-parasite systems it has been suggested that parasites actively suppress the melanization response in the host by inhibiting phenoloxidase activity (Stoltz & Cook 1983), by affecting substrate levels (Sroka & Vinson 1978), or by interfering with normal haemocyte function (Rizki & Rizki 1984).

Knowledge of the mechanisms whereby filarial worms are able to circumvent the internal immune systems of their mosquito hosts would undoubtedly provide us with a better appreciation of those factors that contribute to a successful or unsuccessful association between vector and parasite. However, a better understanding of the basic immune processes in mosquitoes is essential before we can determine how these immune functions might be altered in the presence of parasites.

Acknowledgments

I sincerely thank Dr A. J. Nappi for critically reading this manuscript. These studies were supported in part by National Institutes of Health Grants AI 16472 and AI 19769.

References

Agudelo-Silva, F. & Spielman, A. (1985). Penetration of mosquito midgut wall by sheathed microfilariae. *J. Invert. Path.* **45**: 117–9.

Bartlett, C.M. (1984). Development of *Dirofilaria scapiceps* (Leidy, 1886) (Nematoda: Filarioidea) in *Aedes* spp. and *Mansonia perturbans* and responses of mosquitoes to infection. *Can. J. Zool.* **62**: 112–29.

Bitkowska, E., Dzbenski, T.H., Szadziewska, M. & Wegner, Z. (1982). Inhibition of xenograft rejection reaction in the bug *Triatoma infestans* during infection with a protozoan, *Trypanosoma cruzi. J. Invert. Path.* **40**: 186–9.

Bradley, T.J. & Nayar, J.K. (1985). Intracellular melanization of the larvae of *Dirofilaria immitis* in the Malpighian tubules of the mosquito, *Aedes sollicitans. J. Invert. Path.* **45**: 339–45.

Brug, S.L. (1932). Chitinisation of parasites in mosquitos. *Bull. ent. Res.* **23**: 229–31.

Bryan, J.H., Oothuman, P., Andrews, B.J. & McGreevy, P.B. (1974). Effects of pharyngeal armature of mosquitoes on microfilariae of *Brugia pahangi. Trans. R. Soc. trop. Med. Hyg.* **68**: 14.

Chen, C.C. & Laurence, B.R. (1985). An ultrastructural study on the encapsulation of microfilariae of *Brugia pahangi* in the haemocoel of *Anopheles quadrimaculatus. Int. J. Parasit.* **15**: 421–8.

Christensen, B.M. (1977). Laboratory studies on the development and transmission of *Dirofilaria immitis* by *Aedes trivittatus. Mosquito News* **37**: 367–72.

Christensen, B.M. (1978). *Dirofilaria immitis*: Effect on the longevity of *Aedes trivittatus. Expl Parasit.* **44**: 116–23.

Christensen, B.M. (1981). Observations on the immune response of *Aedes trivittatus* against *Dirofilaria immitis. Trans. R. Soc. trop. Med. Hyg.* **75**: 439–43.

Christensen, B.M. & Andrews, W.N. (1976). Natural infection of *Aedes trivittatus* (Coq.) with *Dirofilaria immitis* in central Iowa. *J. Parasit.* **62**: 276–80.

Christensen, B.M. & Forton, K.F. (1986). Hemocyte-mediated melanization of microfilariae in *Aedes aegypti. J. Parasit.* **72**: 220–5.

Christensen, B.M., Forton, K.F., LaFond, M.M. & Grieve, R.B. (In press). Surface changes on *Brugia pahangi* microfilariae and their association with immune evasion in *Aedes aegypti. J. Invert. Path.*

Christensen, B.M. & LaFond, M.M. (1986). Parasite induced suppression of the immune response in *Aedes aegypti* by *Brugia pahangi. J. Parasit,* **72**: 216–9.

Christensen, B.M., LaFond, M.M. & Christensen, L.A. (1986). Defense reactions of mosquitoes to filarial worms: Effect of host age on the immune response to *Dirofilaria immitis* microfilariae. *J. Parasit,* **72**: 212–5

Christensen, B.M. & Sutherland, D.R. (1984). *Brugia pahangi*: Exsheathment and midgut penetration in *Aedes aegypti. Trans. Am. microsc. Soc.* **103**: 423–33.

Christensen, B.M., Sutherland, D.R. & Gleason, L.N. (1984). Defense reactions of mosquitoes to filarial worms: Comparative studies on the response of three different mosquitoes to inoculated *Brugia pahangi* and *Dirofilaria immitis* microfilariae. *J. Invert. Path.* **44**: 267–74.

Coluzzi, M. & Trabucchi, R. (1968). Importanza dell'armatura bucco-faringea in *Anopheles* e *Culex* in relazione alle infezioni con *Dirofilaria. Parassitologia* **10**: 47–59.

Desowitz, R.S. & Chellapah, W.T. (1962). The transmission of *Brugia* sp. through *Culex pipiens fatigans*: The effect of age and prior non-infective blood meals on the infection rate. *Trans. R. Soc. trop. Med. Hyg.* **56**: 121–5.

Drif, L. & Brehélin, M. (1983). The circulating hemocytes of *Culex pipiens* and *Aedes aegypti*: Cytology, histochemistry, hemograms and functions. *Devl comp. Immunol.* **7**: 687–90.

Duxbury, R.E., Moon, A.P. & Sadun, E.H. (1961). Susceptibility and resistance of *Anopheles quadrimaculatus* to *Dirofilaria uniformis*. *J. Parasit.* **47**: 687.

Esslinger, J.H. (1962). Behavior of microfilariae of *Brugia pahangi* in *Anopheles quadrimaculatus*. *Am. J. trop. Med. Hyg.* **11**: 749–58.

Ewert, A. (1965). Comparative migration of microfilariae and development of *Brugia pahangi* in various mosquitoes. *Am. J. trop. Med. Hyg.* **14**: 254–9.

Foley, D.A. (1978). Innate cellular defense by mosquito hemocytes. *Comp. pathobiol.* **4**: 113–44.

Forton, K.F., Christensen, B.M. & Sutherland, D.R. (1985). Ultrastructure of the melanization response of *Aedes trivittatus* against inoculated *Dirofilaria immitis* microfilarie. *J. Parasit.* **71**: 331–41.

Gasic, G.J., Berwick, L. & Sorrentino, M. (1968). Positive and negative colloidal iron as cell surface electron stains. *Lab. Invest.* **18**: 63–71.

Grieve, R.B., Mika-Johnson, M., Jacobsen, R.H. & Cypess, R.H. (1981). Enzyme-linked immunosorbent assay for measurement of antibody responses to *Dirofilaria immitis* in experimentally infected dogs. *Am. J. vet. Res.* **42**: 66–9.

Harris, K.L. & Christensen, B.M. (In preparation). *Observations on the melanization response of adult male mosquitoes against inoculated microfilariae.*

Highby, P.R. (1943). Mosquito vectors and larval development of *Dipetalonema arbuta* Highby (Nematoda) from the porcupine, *Erethizon dorsatum*. *J. Parasit.* **29**: 243–52.

Ho, B.-C., Yap, E.-H. & Sin, M. (1982). Melanization and encapsulation in *Aedes aegypti* and *Aedes togoi* in response to parasitization by a filarial nematode (*Breinlia booliati*). *Parasitology* **85**: *567–75.*

Hu, S.M.K. (1939a). Observations on the development of filarial larvae during the winter season in Shanghai region. *Am. J. Hyg.* **29**: 67–74.

Hu, S.M.K. (1939b). Studies on the susceptibility of Shanghai mosquitoes to experimental infection with *Wuchereria bancrofti* Cobbold. VII: *Culex vorax* Edwards. *Peking nat. Hist. Bull.* **13**: 113–6.

Intermill, R.W. (1973). Development of *Dirofilaria immitis* in *Aedes triseriatus* Say. *Mosquito News* **33**: 176–81.

Kaaya, G.P. & Ratcliffe, N.A. (1982). Comparative study of hemocytes and associated cells of some medically important dipterans. *J. Morph.* **173**: 351–65.

Kartman, L. (1953a). Factors influencing infection of the mosquito with *Dirofilaria immitis* (Leidy, 1856). *Expl Parasit.* **2**: 27–78.

Kartman, L. (1953b). Effect of feeding mosquitoes upon dogs with differential microfilaremias. *J. Parasit.* **39**: 572.

Kershaw, W.E., Lavoipierre, M.M.J. & Chalmers, T.A. (1953). Studies on the intake of microfilariae by their insect vectors, their survival and their effect on the survival of their vectors. I.—*Dirofilaria immitis* and *Aedes aegypti*. *Ann. trop. Med. Parasit.* **47**: 207–21.

LaFond, M.M., Christensen, B.M. & Lasee, B.A. (1985). Defense reactions of mosquitoes to filarial worms: Potential mechanism for avoidance of the response of *Brugia pahangi* microfilariae. *J. Invert. Path.* **46**: 26–30.

Lavoipierre, M.M.J. (1958). Studies on the host-parasite relationships of filarial nematodes and their arthropod hosts II.—The arthropod as a host to the nematode: a brief appraisal of our present knowledge, based on a study of the more important literature from 1878 to 1957. *Ann. trop. Med. Parasit.* **52**: 326–45.

Lindemann, B.A. (1977). *Dirofilaria immitis* encapsulation in *Aedes aegypti*. *Mosquito News* **37**: 293–5.

Lumsden, R.D. (1972). Cytological studies on the absorptive surfaces of cestodes. VI. Cytochemical evaluation of electrostatic charge. *J. Parasit.* **58**: 229–34.

Mather, T.N., Hamill, B.J., Ribeiro, J.M.C., Rossignol, P.A. & Spielman, A. (Unpublished). *Accelerated microfilarial invasion of mosquitoes feeding on mosquito-immune hosts.* Abstract presented at the 33rd annual meeting of the American Society of Tropical Medicine and Hygiene, Baltimore, 1984.

Nayar, J.K. & Sauerman, D.M., Jr. (1975). Physiological basis of host susceptibility of Florida mosquitoes to *Dirofilaria immitis*. *J. Insect Physiol.* **21**: 1965–75.

Newton, W.L., Wright, W.H. & Pratt, I. (1945). Experiments to determine potential mosquito vectors of *Wuchereria bancrofti* in the continental United States. *Am. J. trop. Med.* **25**: 253–61.

Oothuman, P., Simpson, M.G. & Laurence, B.R. (1974). Abnormal development of a filarial worm, *Brugia patei* (Buckley, Nelson and Heisch), in a mosquito host, *Anopheles labranchiae atroparvus* van Thiel. *J. Helminth.* **48**: 161–5.

Orihel, T.C. (1975). The peritrophic membrane: Its role as a barrier to infection of the arthropod host. In *Invertebrate immunity* 65–73. (Eds Maramorosch, K. & Shope, R.E.). Academic Press, New York.

Pinger, R.R. (1982). Presumed *Dirofilaria immitis* infections in mosquitoes (Diptera: Culicidae) in Indiana, USA. *J. med. Ent. Honolulu* **19**: 553–5.

Poinar, G.O., Jr. (1974). Insect immunity to parasitic nematodes. In *Contemporary topics in immunobiology* **4**: 167–78. (Ed. Cooper, E.L.). Plenum Press, New York.

Ramachandran, C.P. (1966). Biological aspects in the transmission of *Brugia malayi* by *Aedes aegypti* in the laboratory. *J. med. Ent. Honolulu* **3**: 239–52.

Rizki, R.M. & Rizki, T.M. (1984). Selective destruction of a host blood cell type by a parasitoid wasp. *Proc. natn. Acad. Sci. USA* **81**: 6154–8.

Sambon, L.W. (1902). Remarks on the life-history of *Filaria bancrofti* and *Filaria immitis*. *Lancet* **163**: 422–6.

Schacher, J.F. (1973). Laboratory models in filariasis: A review of filarial life-cycle patterns. *S.E. Asia J. trop. Med. pub. Hlth* **4**: 336–49.

Sroka, P. & Vinson, S.B. (1978). Phenoloxidase activity in the hemolymph of parasitized and unparasitized *Heliothis virescens*. *Insect Biochem.* **8**: 399–402.

Stoltz, D.B. & Cook, D.I. (1983). Inhibition of host phenoloxidase activity by parasitoid Hymenoptera. *Experientia* **39**: 1022–4.

Sutherland, D.R., Christensen, B.M. & Forton, K.F. (1984). Defense reactions of mosquitoes to filarial worms: Role of the microfilarial sheath in the response of mosquitoes to inoculated *Brugia pahangi* microfilariae. *J. Invert. Path.* **44**: 275–81.

Sutherland, D.R., Christensen, B.M. & Lasee, B.A. (1986). Midgut barrier as a possible factor in filarial worm vector competency in *Aedes trivittatus*. *J. Invert. Path.* **47**: 1–7.

Vaitukaitus, J., Robbins, J.B., Nuschlag, E. & Ross, G.T. (1971). A method for producing specific antisera with small doses of immunogen. *J. clin. Endocr.* **33**: 988–91.

Wharton, R.H. (1957). Studies on filariasis in Malaya: Observations on the development of *Wuchereria malayi* in *Mansonia* (*Mansonioides*) *longipalpis*. *Ann trop. Med. Parasit.* **51**: 278–96.

Yamamoto, H., Kobayashi, M., Ogura, N., Tsuruoka, H. & Chigusa, Y. (1985). Studies on filariasis VI: The encapsulation of *Brugia malayi* and *B. pahangi* larvae in the mosquito, *Armigeres subalbatus*. *Jap. J. sanit. Zool.* **36**: 1–6.

Symp. zool. Soc. Lond. (1986) No. 56 161–178

Evasion of insect immunity by helminth larvae

Ann M. LACKIE

Department of Zoology,
The University,
Glasgow G12 8QQ

Synopsis

That 'blindspots' exist in the immunorecognition system of insects is indicated by the results of several different types of assay: allogeneic and, in some species-combinations, xenogeneic, recognition appear to be absent. Since the occurrence and extent of the haemocytic response is also determined by the rather general properties of the foreign surface such as charge and hydrophobicity it is possible that a parasite could exploit these weaknesses in the recognition system by entirely evading recognition and encapsulation, or by suppressing a weak response.

The question of whether or not helminth larvae survive by immunosuppression of their insect hosts has been examined by investigating the relationship between dose, number of exposures, and resultant intensity of infection in two host-parasite systems. There seems to be little effect on establishment of multiple homologous or heterologous superinfections of cysticercoids of *Hymenolepis diminuta* or cystacanths of *Moniliformis moniliformis*. Immunosuppression is a dangerous strategy for a parasite to adopt since the host might succumb to other diseases before transmission was effected, but it might, in the short term, allow a parasite time to develop other evasion mechanisms. Considerably more quantitative information is needed about the cellular and humoral responses of naive hosts before questions about the existence of immunosuppression can be answered.

The alternative possibility, evasion of recognition by inherent mimicry or selective requisitioning of host-derived molecules has been examined, with particular reference to *H. diminuta* and *Moniliformis*.

The acuity of recognition and the efficacy of the haemocytic response can be enhanced non-specifically and for a period of several days, by prior injection of biotic or abiotic particles. Repeated or chronic stimulation of the immune system might thus allow a host to overcome the evasion strategies of its parasites, a result which would have interesting epidemiological implications.

Introduction

The biological control of agricultural insect pests and the interruption of transmission of parasitic diseases by their insect intermediate hosts or vectors are both dependent on our ability to manipulate the insect host-parasite

ZOOLOGICAL SYMPOSIUM No. 56
ISBN 0-19-854004-3

relationship. To kill pests, we attempt to select for pathogenic strains of parasite that can evade the immune response and kill the insect host. To combat parasitic diseases, we might hope to select for strains of insect whose immune response can effectively combat and kill the parasite. Whether or not these strains would be at a selective advantage in regions in which the disease is endemic is a separate matter for investigation (Minchella & Loverde 1983; Curtis & Graves 1983).

In attempting to interfere with transmission by enhancing host immunity we are faced with the necessity of understanding why the immune response to that parasite species is normally ineffective. As pointed out by Salt in his review of the subject in 1963(a), the majority of species of helminth in their 'habitual' hosts are not encapsulated by haemocytes. Lack of a haemocytic response to a parasite—assuming the host was healthy and unstressed at the start of infection—can be due to two main causes. First, the parasite may be recognized by the immune system as being 'foreign' but may suppress or disrupt the effector arm of the immune response or, secondly, the parasite may not be recognized as foreign because it appears to the immune system to be like host 'self'. In the latter case, the mistaken identity of the parasite might be due either to its inherent ability to mimic the host or to its ability to requisition and coat itself with host-derived materials. In terms of cost-effectiveness, evasion of the recognition arm is perhaps the more difficult strategy to evolve but is potentially cheaper to maintain.

That the immune system should fail to recognize such a markedly foreign surface as that of a parasite seems an unlikely concept. But how unlikely is it? In order to answer this, we need to define the system's discriminatory ability.

What does the immune system recognize?

The immune system of insects recognizes and responds to a wide variety of soluble and particulate materials which enter the haemocoel. Particulate material is sequestered within phagocytes or within haemocytic aggregates such as nodules or, in the case of larger objects, capsules.

The foreign materials which stimulate a response are apparently rather disparate and there are, of course, notable exceptions amongst the haemocoelic parasites and pathogens. Confronted with this diversity it would seem impossible to discern any common properties amongst those surfaces which are recognized as foreign and those which are not. However, haemocytic encapsulation of foreign objects introduced into the haemocoel provides a useful measure of whether or not a cellular response, and thus immunorecognition, has occurred. A variety of assays have been devised whose aim is to define the discriminatory ability of the immunorecognition system. In general, the assays fall into two main categories, those investigating the extent of the encapsulation response to biotic material and those which measure the response to abiotic materials of known surface properties. The results of these types of experiment will be discussed here in so far as they help to answer the question of the likelihood of a parasite evading immunorecognition through the strategy of looking like its host. The assays and their interpretation are all discussed in a wider context and in much greater detail elsewhere (A. M. Lackie in press).

Haemocytic response to biotic transplants

Transplantation experiments have utilized (a) intrahaemocoelic implantation of pieces of intact tissue such as ovariole, testis or nerve-cord (Salt 1970; Scott 1971; Kambysellis 1970; A. M. Lackie 1976, 1979), (b) 'skin grafting' of pieces of cuticle with the epidermis and connective-tissue layer intact (Thomas & Ratcliffe 1982; Jones & Bell 1982; A. M. Lackie 1983a; George, Karp & Rheins 1984) or (c) intra-haemocoelic implantation of larval cuticle with intact epidermis (Wigglesworth 1954; Riddiford 1976; A. M. Lackie in press). Additionally (d), whole haemolymph has been transferred between insects either by parabiosis (Wigglesworth 1954) or by injection (A. M. Lackie 1986a). In all four assays, the haemocytic response of the recipient—either by the formation of multicellular capsules around the implants, adhesion of layers of haemocytes beneath a graft followed by its rejection or by the formation of haemocytic nodules within the haemolymph—can be measured.

It is unfortunate that many of the insect vectors of medically-important diseases are so small; to investigate transplantation immunology in adult mosquitoes and sandflies is obviously difficult. Our own work has been concerned predominantly with immunorecognition in the cockroach *Periplaneta americana* and the desert locust *Schistocerca gregaria*, because these are large, easily-manipulated animals, whose immunorecognition systems differ quite markedly. *Periplaneta* is also an intermediate host of two species of helminth capable of infecting man.

Results of the transplantation assays described above are summarized in Figs 1 and 2. It is immediately obvious that allogeneic (i.e. intraspecific) recognition is absent in the two species. This lack of allogeneic recognition in insects has been confirmed, by implantation or grafting, for a wide range of species from many different Orders and the abundant information on this subject (reviewed by Salt (1961) and A. M. Lackie (in press)) includes similar results for transplantation in the fruitfly *Drosophila melanogaster* (Rizki & Rizki 1980) and in the tsetse fly *Glossina* (J.East pers. comm.).

Another observation is that xenogeneic recognition is also lacking in some species combinations, notably that of *Periplaneta–Blatta*, and for the combinations with *Schistocerca* as recipient. The lack of recognition

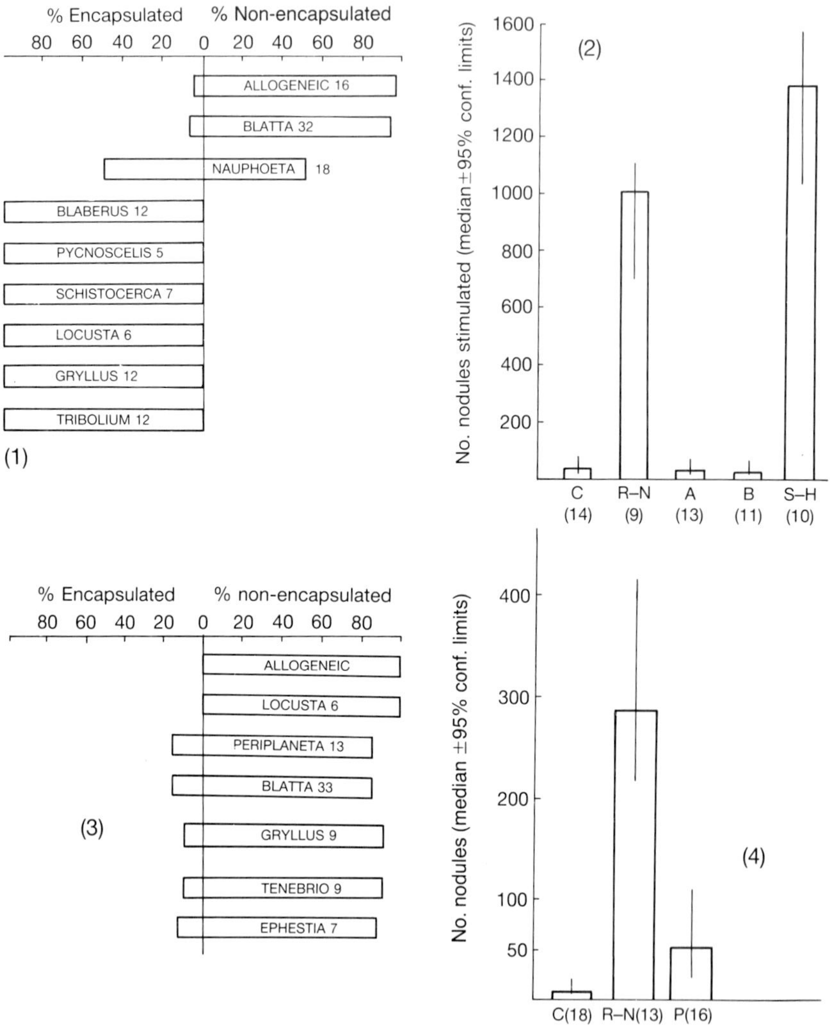

Fig. 1. Transplantation reactions in *Periplaneta* (1,2) and *Schistocerca* (3,4). Haemocytic encapsulation of tissue implants is shown in (1) and (3). *In vivo* haemocytic aggregation (nodule formation) is shown in (2) and (4); insects injected with 20 µl of saline (C); a suspension of rabbit neutrophils (R-N); allogeneic haemolymph (A); *Blatta* haemolymph (B); *Schistocerca* haemolymph (S-H); *Periplaneta* haemolymph (P).
Number of recipients indicated inside bars (in 1 and 3) and in brackets (in 2 and 4).
Modified from A. M. Lackie (1986a, in press).

between *Schistocerca* and *Periplaneta* is not reciprocal, however, a result which provides interesting information about the relative powers of discrimination of the two species. Xenogeneic recognition has been found to be absent between many pairs of closely-related species, including those of the

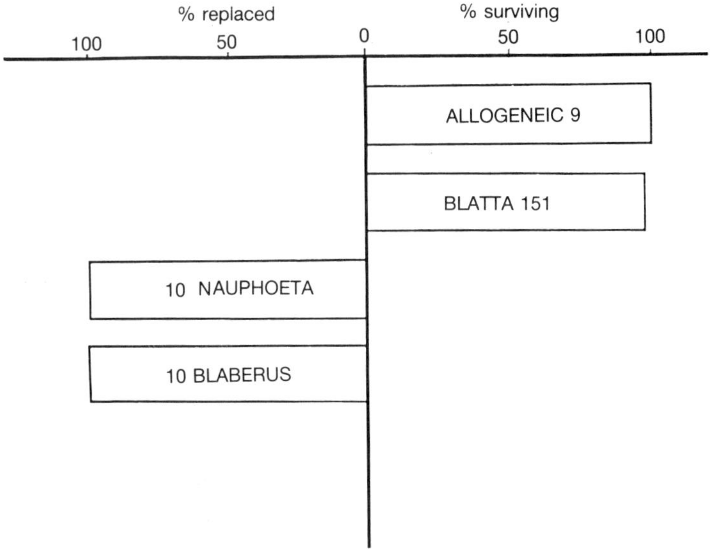

Fig. 2. Recognition of cuticular 'skingrafts' by *Periplaneta*. Numbers of recipients indicated inside bars. Data from A.M. Lackie (1983a); A.M. Lackie & B. Dularay (unpubl.).

Drosophila species-complex (Kambysellis 1970), the mosquitoes *Aedes aegypti* and *Culex molestus* (Larsen & Bodenstein 1959) and the reduviid bugs *Rhodnius* and *Triatoma* (Locke 1974).

Incidentally, although few transplantation experiments on xenogeneic combinations have been carried out on molluscs, mainly because of the technical difficulties involved, allogeneic recognition has been shown to be lacking in this other very important group of vectors (Sminia, Borghart-Reinders & van de Linde 1974).

Haemocytic response to abiotic implants

In order to define some of the properties of foreign surfaces that influence whether or not they are adhesive for haemocytes, abiotic surfaces with known physicochemical properties can be implanted and the extent of the haemocytic encapsulation response subsequently examined.

That the surface charge of a foreign object influences whether or not it will be encapsulated has been reported several times, mainly for the Lepidoptera and the orthopteroid complex. Negatively-charged beads are not encapsulated in the caterpillars of *Heliothis zea* (Vinson 1974), and negatively-charged Sepharose or polystyrene beads escape encapsulation in the locust *Schistocerca*, even after nine days *in vivo* (A. M. Lackie 1983b). The thickness of the capsule formed around objects that stimulate a response is very much dependent on the charge and wettability of the foreign surface. Thus, in *Periplaneta*, capsule thickness around Sepharose beads decreases with bead charge in the order positive > negative > neutral, whereas in *Schistocerca* neutral surfaces stimulate thicker capsules than positive (A. M. Lackie 1983b; A. M. Lackie, Takle & Tetley 1985). Differences in haemocytic adhesion *in vitro* to surfaces of different negativity parallel the differences in haemocytic response *in vivo* (A. M. Lackie 1986b). It has been recently shown, by cell electrophoresis and by measurement of the binding of cationized ferritin to haemocyte surfaces, that locust cells bear a much greater negative charge than cockroach cells, an observation which might help to explain why locust haemocytes adhere less readily to negatively charged substrata (Takle & Lackie 1985).

The plausibility of evading recognition

The general efficacy of the insect immune system in protecting the organism against disease has been described elsewhere in this volume, and this is well exemplified by an animal such as the cockroach which can survive for 12-18 months in a filthy, bacteria-rich environment (Cornwell 1968). In view of the obvious skills of the effector arm of the immune system it is perhaps the more surprising that the recognition arm is apparently defective. It might be expected that the biological consequences of the failure to recognize allogeneic transplants would mean that the insect would be incapable of recognizing altered 'self' in the form either of wounds or of tumours, but this appears not to be so: wounded tissues, and thus altered basal laminae or connective tissue layers, are rapidly encapsulated by haemocytes and repaired (Salt 1961; Lackie in press) and certain types of tumour, particularly those where the surrounding basal lamina is altered, provoke haemocytic attack (Rizki & Rizki 1980). Non-recognition of certain xenogeneic transplants implies a more serious defect in the discriminatory ability.

The implications for parasitism are clear—within each insect species there are discrete blindspots in the immunorecognition system which can be exploited.

An insect species whose haemocytes do not encapsulate negatively-charged surfaces might be expected to experience some problems in responding not only to tissue transplants but also to potential parasites, since most biotic surfaces are negatively-charged (Pethica 1980). A survey of the literature suggests that the immune system of *Schistocerca* is indeed less able to cope with helminth parasites than that of *Periplaneta* (Table 1).

Parasite survival

The host-parasite relationship must be considered as a balance, not merely at the level of the individual participants but also in terms of host and parasite populations (Anderson & May 1982), between the needs of the parasite to exploit the resources of the host and the necessity of the host to avoid

Table 1. Haemocoelic parasites of *Schistocerca* and *Periplaneta*.

	Reference
Periplaneta	
Moniliformis moniliformis	See text
Hymenolepis nana[a]	Cavier & Leger (1965)
Hymenolepis diminuta[a]	A.M. Lackie (1976)
Schistocerca	
Moniliformia moniliformis	A.M. Lackie & Lackie (1979)
Hymenolepis diminuta	A.M. Lackie (1976); Lethbridge (1971a)
Mermis nigrescens	
Ephestia kuehniella	Salt (1963b)

[a] Low proportion of injected oncospheres develop unencapsulated.

over-exploitation. Parasite pathogenicity (over-exploitation) inducing host mortality will not only eliminate part of the parasite population and reduce the chances of transmission but could also reduce the numbers of hosts available to be parasitized. Larvae of the tapeworm *Hymenolepis citelli* cause high levels of mortality in the population of beetle hosts and there is some evidence, from work using laboratory stocks of the beetle *Tribolium confusum* and the parasite, that the hosts exert strong selection pressures on the tapeworm to reduce the number of lines of parasite with high infectivity, thereby increasing host survivorship and, in the long term, ensuring maintenance of the parasite population (Schom, Novak & Evans 1981).

Survival of the individual parasite within the individual host will not only be affected by the physiological suitability of the host but also will be dependent on the parasite's circumventing the immune response.

In this section, the evidence for survival of helminths by immunosuppression of their insect hosts or by evading immunorecognition will be examined, with particular reference to larvae of the tapeworm *Hymenolepis diminuta* (Platyhelminthes: Cestoda) and of *Moniliformis moniliformis* (= *dubius*) (Acanthocephala) in natural and experimental hosts. The adults of both species are naturally parasitic in the small intestine of rats, but both worms are also known to infect humans (Sahba, Arfaa & Rastegaar 1970; Goldsmid, Smith & Fleming 1974).

The natural intermediate hosts of *H. diminuta* are the flourbeetles, *Tenebrio molitor* and *Tribolium confusum*, and the natural host of *Moniliformis* is the cockroach, *Periplaneta americana*. Both species of worm are easy to maintain, the developmental rates of the larvae at different temperatures are well documented (Voge & Turner 1956; J. M. Lackie 1972a; Robinson & Strickland 1969) and the eggs of both species can be hatched *in vitro* and clean viable larvae obtained for direct intrahaemocoelic injection (J. M. Lackie 1975; A. M. Lackie & Lackie 1979; A. M. Lackie 1976) or for *in vitro* culture (J. M. Lackie & Rotheram 1972; Voge & Green 1975).

Immunosuppression

The response to superimposed infections

The ability of an immunosuppressed host to protect itself against subsequent infection or to respond to immunogenic stimuli is likely to be compromised. There have been several investigations into the ability of insects to support multiple homologous or heterologous helminth infections and although these studies have not always been aimed at examining host immunity, the results provide much relevant information.

In an orally administered infection, the percentage of eggs which produce viable larvae is usually rather low and presumably dependent on mechanical and physiological stimuli in the gut (Holmes & Fairweather 1982), the success of the hatched larva in penetrating the gut wall (Lethbridge 1971b) and the ability of the larva to protect itself once it reaches the haemocoel. Only about 15% of the administered mature eggs of *Moniliformis* develop to cystacanths (J. M. Lackie 1973), and encapsulated and melanized acanthors are found attached to the haemocoelic surface of the midgut wall of cockroaches (Rotheram & Crompton 1972). If the developing parasites of a primary infection immunosuppress the host, it might be expected that a larger proportion of subsequent doses might be able to survive. However, the results of double or triple oral infections of *Moniliformis* indicated that there was neither diminished (immunized) nor enhanced (immunosuppressed) susceptibility to challenge infections (J. M. Lackie 1972b). Similarly, Keymer (1980) and Gordon & Whitfield (1985), working with *H. diminuta* in *Tribolium*, found that there was a linear relationship between the number of cysticercoids devleoping in each host and the number of times eggs had been administered.

Cysticercoids of the fowl tapeworm, *Raillietina cesticillus*, also use *Tribolium* as intermediate host. In an investigation into the possible effects of heterologous infections on transmission dynamics, Gordon & Whitfield (1985) counted the numbers of cysticercoids developing in beetles infected with both *Raillietina* and *H. diminuta*. If the primary infection was of *Hymenolepis*, challenge doses of *Raillietina* developed as though in a homologous infection. If *Raillietina* was given first, fewer of the *Hymenolepis* challenge developed. The reason for this anomaly was not investigated, but the authors suggested that *Raillietina* might have activated the beetle's immune system in some manner which was deleterious to the incoming *Hymenolepis* but which had no effect on *Raillietina* itself. Whatever the mechanism, it is quite apparent that there was no increased host susceptibility as a result of the primary infection.

Moniliformis, in its natural host, *Periplaneta*, overcomes any haemocytic response at an early stage (see p. 171); whether its survival, unmolested by haemocytes, is due to immunosuppression can be tested by submitting the host to heterologous infection with *H. diminuta*. The latter species is not infective, *per os*, to *Periplaneta*, but if the eggs are hatched *in vitro* and the oncosphere larvae injected into the cockroach haemocoel, a very small proportion of the larvae survive, unencapsulated by haemocytes, and develop to the cysticercoid stage (A. M. Lackie 1976). Survival depends partly on the infectivity of the individual parasite and partly on the susceptibility of the individual host; for example, if 1000 oncospheres are injected into each of 20 cockroaches, two to four of the cockroaches might be infected and contain from two to 100 parasites. The remainder of the injected dose will be visible as small melanized and encapsulated objects in the haemocoel (A. M. Lackie 1976, and unpubl). The physiological environment would thus seem to be adequate—*H. diminuta* can be cultured to cysticercoids in media developed originally for *Periplaneta* haemocyte culture (Voge & Green 1975)—but the parasites are on the threshold of evading the immune response. Theoretically, then, any diminution in the effectiveness of the immune system should allow a greater proportion of larvae to develop in a greater proportion of the cockroaches. However, cockroaches carrying *Moniliformis* infections of known age, whether infected orally—and thus containing variable numbers of parasites—or injected with known doses of hatched acanthors, were no more susceptible to infection with *H. diminuta* than were uninfected insects (Fig. 3 and unpublished).

Preliminary experiments have suggested that a strain of *H. diminuta* infective for *Periplaneta* can be selected; whether the presence of these 'cockroach-adapted' larvae will permit development of the normal, unadapted strain, is worth exploring.

The response to abiotic particles

One problem with using the survival of challenge infections of living parasites as an assay for immunosuppression is that we are dealing with undefined dynamic surfaces—the 'test particles' might themselves respond to the immune response. This complication can be avoided by using abiotic test particles such as beads of known charge or wettability, as described in an earlier section. Ion-exchange beads or nylon thread have been implanted into insects parasitized by wasps by Vinson (1974), Salt (1980) and Guzo & Stoltz (1985) with variable results depending on the species of parasitoid wasp and the time of implantation post-infection.

The time course of encapsulation and the thickness of the 24-h haemocytic capsule formed around Sepharose beads in naive (unparasitized) cockroaches are known (A. M. Lackie 1983b; G. B. Takle unpubl.) and can thus be compared with the haemocytic response to these particles in the parasitized host. There is no significant difference in thickness of the haemocytic capsules around neutral Sepharose beads implanted into *Periplaneta*, either unparasitized or parasitized with stage I acanthellae of *Moniliformis* (A. M.

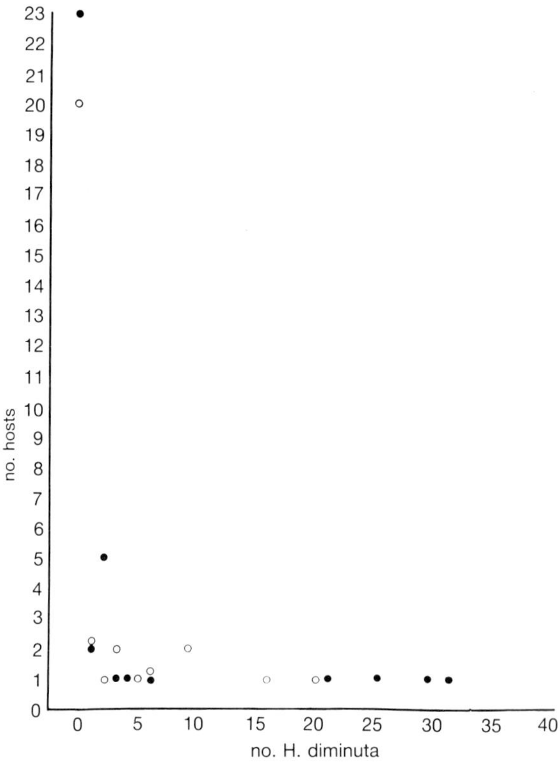

Fig. 3. Effect of prior infection of *Periplaneta* with *Moniliformis moniliformis* on establishment of *Hymenolepis diminuta*.
Moniliformis (orally-derived infection) stage I acanthellae at time of superinfection with *H. diminuta* (10^3 oncospheres, injected intrahaemocoelically).
○ — naive insects injected with *H. diminuta*
● — *Moniliformis*-infected insects, injected with *H. diminuta*.

Lackie unpubl.); this is the stage of parasite development at which the larva is starting to evade the haemocytic response (see p. 172).

General considerations

To survive by compromising the immunological defences of its host is a dangerous strategy for a parasite to adopt: the host might succumb to other diseases before the parasite developed to infectivity. The risks might be worthwhile if the parasite's developmental period is short and the chances of transmission are consistently high, but this is rarely the case. Selective immunosuppression is, however, at least theoretically possible. That the extent of the haemocytic response is graded, depending on the nature of the foreign surface, is well known (A. M. Lackie 1983b; Jones & Bell 1982) and it has been postulated that recognition and response occur only if a critical threshold of difference from 'self' is exceeded (J. M. Lackie 1975; A. M.

Lackie in press). Activation of the immune system, having the effect of lowering the threshold so that previously unrecognized surfaces will provoke a response, is described in a later section; perhaps the converse is also possible and the threshold can be raised to include a small range of hitherto recognized surfaces, thereby depressing the acuity of recognition. It is clear that the possibility of immunosuppression, whether total or selective, needs detailed investigation and the answer is dependent upon possession of comparative data on the nature of the cellular and humoral response of the naive host.

There are a variety of assays available for assessing immunocompetence. For haemocytes, these include measurements of the *in vivo* phagocytic index; of the number of haemocytic aggregates (nodules) produced in response to particulate (Ratcliffe & Walters 1983) or soluble (Gunnarsson & Lackie 1985) material; of capsule thickness around abiotic particles (A. M. Lackie 1983b); of the spreading, adhesion and locomotory rates of haemocytes *in vitro* (Takle 1985; Huxham, Takle & Lackie in prep.) and of changes in differential or total haemocyte counts measured either directly or after separation by gradient centrifugation (Huxham & Lackie in prep). Changes in the agglutination titre of serum or in the combining specificity of serum lectins may also provide some relevant information.

Finally, there is the question of timing—immunosuppression may only be needed at a particular stage in the parasite's development, such as when a nematode larva moults or when an oncosphere or acanthor larva struggles through the gut wall into the haemocoel, still relying on its endogenous glycogen reserves for energy. The responsiveness of the host need only be reduced until the parasite can initiate or regain its own localized defences which might enable it to escape recognition.

Evasion of recognition

Larvae of *H. diminuta* develop, unencapsulated, in their beetle hosts, from the time of haemocoelic penetration as an oncosphere to the infective cysticercoid stage (Heyneman & Voge 1971), the tegumental surface being extended by microvillar outgrowths (Ubelaker, Cooper & Allison 1970; Richards & Arme 1985). The pathological effects of the larvae are apparently fairly minor, although they may lower the fecundity of the female beetles (Keymer 1980), perhaps as a result of imbalance in vitellogenin metabolism (Hurd & Arme 1984). Conversely, the hatched acanthor of *Moniliformis* becomes encapsulated as soon as it reaches the haemocoel; the capsule, however, comprises only a few layers of haemocytes and is about 15 μm thick (Rotheram & Crompton 1972).

Within about seven to ten days (at 28°C), during which time the acanthor has been growing and taking up nutrients through its tegument, the number of adherent haemocytes decreases and the cells finally disappear altogether so that the surface of the larva is completely free. The decline in thickness of the capsule coincides with production of a 'membranous coat', produced as a microvillous proliferation of the tegumental membrane (Rotheram & Crompton 1972). By the mid-acanthella stage the membranous coat has become elevated from the tegument, separated from it by viscous material and compacted to form a transparent envelope. The envelopes surrounding acanthocephalan larvae seem to be associated with protection from haemocytic attack: if the envelopes are carefully removed from cystacanths of *Moniliformis* or *Polymorphus minutus* and the larvae are injected into uninfected cockroaches or *Gammarus pulex*, respectively, the parasites are immediately encapsulated (Robinson & Strickland 1969; Crompton 1964).

It is thus possible that these two species of helminth exploit the deficiencies in their hosts' immunorecognition systems and that the tegumentary surface of *H. diminuta* and the envelope surface of *Moniliformis* appear indistinguishable from host 'self'.

When larvae of *H. diminuta* and intact tissues from *Tribolium* or *Tenebrio* were transplanted to xenogeneic recipients, and the extent of the haemocytic encapsulation of the implants assessed, it was found that the degree of encapsulation of the parasite closely followed that of transplanted host tissue and this was interpreted as indicating some similarity between the two types of surface (A. M. Lackie 1976). Amongst other species, the locusts *Schistocerca* and *Locusta* were used as recipients and failed to encapsulate either beetle tissues or parasites—indeed, hatched injected oncospheres will develop normally within *Schistocerca* (Lethbridge 1971a; A. M. Lackie 1976). Since then, however, it has become clear that immunorecognition in *Schistocerca* is of particularly low acuity and the haemocytic response of this species does not provide a good indicator of surface similarity since such a wide range of surfaces fail to provoke a response. Of possible interest in this context is the observation that the serum of the locust does not agglutinate oncospheres *in vitro*, whereas the serum of *Periplaneta*, whose haemocytes encapsulate the larvae *in vivo*, does (A. M. Lackie 1981). Whether or not *H. diminuta* evades the immune response by evading recognition must thus remain an open question; whatever the mechanism, larvae grown entirely *in vitro* do not stimulate a response when injected into *Tenebrio* (A. M. Lackie 1976).

Transplantation experiments have also been carried out with *Moniliformis* (Table 2). That the disguise as 'host' is inexact is exemplified by the response of the oxyhaloinid cockroach *Leucophaea*, which strongly encapsulates *Periplaneta* tissue but not the envelope. Although nerve cord transplanted between *Periplaneta* and *Nauphoeta*, in either direction, provokes an equivocal response (Fig. 1.1; Scott 1971; J. M. Lackie 1975; A. M. Lackie 1979) 'skin grafts' and implants of *Nauphoeta* cuticle into *Peripla-*

Table 2. Xenogeneic transplantation of enveloped *Moniliformis* and tissue from *Periplaneta* (1–4) and from *Schistocerca* (5).

Recipient species	*Transplant*	*Encapsulated*
1. *Blatta* (Blattidae)	nerve cord	−
	acanthellae	−
2. *Nauphoeta* (Oxyhaloinidae)	nerve cord	±[a]
	acanthellae	−
3. *Leucophaea* (Oxyhaloinidae)	nerve cord	+
	acanthellae	−
4. *Blaberus* (Blaberoidea)	nerve cord	+
	acanthellae	+
5. *Periplaneta*	*Schistocerca*	
	nerve cord	+
	acanthellae	−

[a] See Fig. 1.
Table compiled from data from J.M. Lackie (1975); A.M. Lackie & Lackie (1979); and A.M. Lackie (unpubl.)

neta are clearly rejected (Fig. 2; A. M. Lackie in press); despite this *Periplaneta*-derived parasites are not encapsulated. The possibility that the envelope is a completely non-adhesive surface can be discounted since *Blaberus* haemocytes are able to adhere. The results of transplantation experiments led J. M. Lackie (1975) to propose that the envelope might be regarded as a surface 'having properties in common with, but not identical to, those of the surface of host tissues'.

In the case of tissue transplants, it is the properties of the ensheathing basal lamina that decide whether or not the transplant is treated as foreign: whether the overall charge or more specific molecular configurations such as the carbohydrate moieties of the glycosaminoglycans (GAGs) are the deciding factor is not yet clear. However, the envelope of *Moniliformis*, like the subepidermal connective tissue of its host, stains strongly with the dye Alcian blue (A. M. Lackie & L. Tetley, unpubl.) suggesting that 'acid mucopolysaccharides' (GAGs) are present in both tissues.

The question arises whether this mimicry, if it really is such, of *Periplaneta* tissue surfaces is inherent or derives from incorporation of the relevant host molecules into the membranous coat and envelope. Evidence which could be interpreted in favour of inherent mimicry comes from experiments in which hatched acanthors were injected directly into *Schistocerca*, where they grew and developed normally. If the envelope had incorporated host molecules as disguise, then, like its host's tissues (Fig. 1.3), it should have been encapsulated when transplanted into *Periplaneta*. However, this did not happen—*Periplaneta* still treated the enveloped parasite as though it were its own, not the locust's, tissue (A. M. Lackie & Lackie 1979).

An alternative interpretation is that the parasite continuously takes up components of the host plasma, thereby masking its own foreign-ness; rapid turnover of these components is a necessary postulate to explain the results of transplanting locust-derived parasites into *Periplaneta*—the 'locust' disguise must be rapidly lost and replaced by the mask of 'cockroach'. In *Blaberus* it must be presumed that successful exchange cannot occur.

Recent evidence suggests that the envelope acts like a 'sponge', taking up a variety of exogenous proteins and some proteins with the same electrophoretic mobility as those found in serum (V. O'Brien, J. Kusel & A. M. Lackie unpubl.). Whether, as with sporocysts of *Schistosoma mansoni* in snails (see chapter by Loker & Bayne, this volume), some selectivity is also exerted, remains to be seen; parasite survival could be due to the ability to take up the relevant proteins and degrade them rapidly (localized immunosuppression?) or to use them to mask surface characteristics otherwise recognized by the host as 'not-self'.

Alterations in the acuity of recognition

The adhesion of phagocytes to bacteria is partly dependent on the charge (Pethica 1980) and the relative hydrophobicity of the two surfaces; modification of the bacterial surface by removal of the polysaccharide capsule, or opsonization with immunoglobulin (Edebo, Kihlstrom, Magnusson & Stendahl 1980) may more readily permit cell-bacterium adhesion and phagocytosis. The bacterium *Bacillus thuringiensis subtoxicus* cannot normally be phagocytosed by haemocytes of *Galleria* (Mohrig, Schittek & Hanschke 1979), presumably because its surface properties do not allow haemocyte adhesion. However, prior injection of *Galleria* larvae with phagocytosable latex beads activates the haemocytes so that phagocytosis of the bacilli can be achieved (Mohrig *et al.* 1979). There has therefore been an alteration in the responsiveness of the immune system.

The immune system of *Periplaneta* does not normally recognize tissues from *Blatta* as foreign (Figs 1 and 2) and attempts to 'immunize' the recipient by prior implantation of *Blatta* tissue were unsuccessful (A. M. Lackie 1983a, in press). However, if the recipients were given two injections of *Blaberus* tissue—which provokes a haemocytic response (Fig. 1.1)—and then subsequently grafted with *Blatta* cuticle, short-term immunization (up to four days) was achieved. Prior injections with Sepharose beads, which are also encapsulated, had the same effect (Dularay in prep.). In other words, the system had been persuaded to recognise *Blatta* tissue as foreign, and the acuity of recognition was temporarily increased. In addition, haemocyte behaviour was altered in such a way that their *in vitro* adhesion was reduced and their *in vivo* phagocytic ability increased (B. Dularay & A. M. Lackie unpubl.). Whether or not this heightened sensitivity can be maintained for longer periods by repeated stimulation of the recipient is as yet unknown, but such changes in both the recognition and the effector arms of the immune system could well have profound effects on the evasion strategies of helminth larvae, irrespective of whether this involved evasion of recognition or short-term immunosuppression. This would have interesting implications for the 'vectorial capacity' (White 1982) of the host population.

Acknowledgements

My grateful thanks to Bubbly Dularay, Garry Takle, Max Huxham, Vincent O'Brien and John Kusel for help, collaboration and useful discussions. Some of the work referred to has been supported by SERC grants GRCO1351 and GRC87959 and AFRC grant AG17/160.

References

Anderson, R.M. & May, R.M. (1982). Coevolution of hosts and parasites. *Parasitology* **85**: 411–26.

Cavier, R. & Leger, N. (1965). A propos de l'évolution d'*Hymenolepis nana* var. *fraterna* chez des hôtes intermédiares inhabituels. *Annls parasit. hum. comp.* **40**: 651–8.

Cornwell, P.B. (1968). *The cockroach* **1**. Hutchinson, London.

Crompton, D.W.T. (1964). The envelope surrounding *Polymorphus minutus* (Goeze, 1782) (Acanthocephala) during its development in the intermediate host, *Gammarus pulex*. *Parasitology* **54**: 721–35.

Curtis, C.G. & Graves, P.M. (1983). Genetic variation in the ability of insects to transmit filariae, trypanosomes and malarial parasites. In *Current topics in vector research* **1**: 31–62. (Ed. Harris, K.F.). Praeger Press, New York.

Dularay, B. (In prep.). *The effect of biotic and abiotic implants on the recognition of* Blatta orientalis *cuticular transplants by* Periplaneta americana.

Edebo, L., Kihlstrom, E., Magnusson, K.E. & Stendahl, O. (1980). The hydrophobic effect and charge effects in the adhesion of enterobacteria to animal cell surfaces and the influence of antibodies of different immunoglobulin classes. In *Cell adhesion and motility* 65–101. (*British Society for Cell Biology Symposium* **3**.) (Eds Curtis, A.S.G. & Pitts, J.D.). Cambridge University Press, Cambridge.

George, J.F., Karp, R.D. & Rheins, L.A. (1984). Primary integumentary xenograft reactivity in the American cockroach, *Periplaneta americana*. *Transplantation (Baltimore)* **37**: 478–84.

Goldsmid, J.M., Smith, M.E. & Fleming, F. (1974). Human infection with *Moniliformis* sp. in Rhodesia. *Ann. trop. Med. Parasit.* **68**: 363–4.

Gordon, D.M. & Whitfield, P.J. (1985). Interactions of the cysticercoids of *Hymenolepis diminuta* and *Raillietina cesticillus* in their intermediate host, *Tribolium confusum*. *Parasitology* **90**: 421–32.

Gunnarsson, S.G.S. & Lackie, A.M. (1985). Haemocytic aggregation in *Schistocerca gregaria* and *Periplaneta americana* as a response to injected substances of microbial origin. *J. Invert. Path.* **46**: 312–9.

Guzo, D. & Stoltz, D.B. (1985). Obligatory multiparasitism in the tussock moth, *Orgyia leucostigma*. *Parasitology* **90**: 1–10.

Heyneman, D. & Voge, M. (1971). Host-response of the flour-beetle, *Tribolium confusum*, to infections with *Hymenolepis diminuta*, *H. microstoma* and *H. citelii* (Cestoda: Hymenolepididae). *J. Parasit.* 57: 881–6.

Holmes, S.D. & Fairweather, I (1982). *Hymenolepis diminuta*: the mechanism of egg-hatching. *Parasitology* **85**: 237–50.

Hurd, H. & Arme, C. (1984). Pathophysiology of *Hymenolepis diminuta* infections in *Tenebrio molitor*: effect of parasitism on haemolymph proteins. *Parasitology* **89**: 253–62.

Jones, S.E. & Bell, W.J. (1982). Cell-mediated immune-type response of the American cockroach. *Devl comp. Immunol.* **6**: 35–42.

Kambysellis, M.P. (1970). Compatibility in insect tissue transplantation. *J. exp. Zool.* **175**: 169–80.

Keymer, A.E. (1980). The influence of *Hymenolepis diminuta* on the survival and fecundity of the intermediate host, *Tribolium confusum*. *Parasitology* **81**: 405–21.

Lackie, A.M. (1976). Evasion of the haemocytic defence reaction of certain insects by larvae of *Hymenolepis diminuta* (Cestoda). *Parasitology* **73**: 97–107.

Lackie, A.M. (1979). Cellular recognition of foreign-ness in two insect species, the American cockroach and the desert locust. *Immunology* **36**: 909–14.

Lackie, A.M. (1981). Humoral mechanisms in the immune response of insects to larvae of *Hymenolepis diminuta*. *Parasite Immunol.* **3**: 201–8.

Lackie, A.M. (1983a). Immunological recognition of cuticular transplants in insects. *Devl comp. Immunol.* **7**: 41–50.

Lackie, A.M. (1983b). Effect of substratum wettability and charge on adhesion *in vitro* and encapsulation *in vivo* by insect haemocytes. *J. Cell Sci.* **63**: 181–90.

Lackie, A.M. (1986a). Haemolymph transfer as an assay for immunorecognition in insects. *Transplantation (Baltimore)* **41**: 360–3.

Lackie, A.M. (1986b). The role of substratum surface charge in adhesion and encapsulation by locust haemocytes *in vivo*. *J. Invert. Path.* **47**: 377–8.

Lackie, A.M. (In press). Transplantation: the limits of recognition. In *Humoral and cellular defence in arthropods*. (Ed. Gupta, A.P.). John Wiley, New York.

Lackie, A.M. & Lackie, J.M. (1979). Evasion of the insect immune response by *Moniliformis dubius* (Acanthocephala): further observations on the origin of the envelope. *Parasitology* **79**: 297–301.

Lackie, A.M., Takle, G.B. & Tetley, L. (1985). Haemocytic encapsulation in the locust *Schistocerca gregaria* (Orthoptera) and in the cockroach *Periplaneta americana* (Dictyoptera). *Cell Tissue Res.* **240**: 343–51.

Lackie, J.M. (1972a). The effect of temperature on the development of *Moniliformis dubius* (Acanthocephala) in the intermediate host, *Periplaneta americana*. *Parasitology* **65**: 371–77.

Lackie, J.M. (1972b). The course of infection and growth of *Moniliformis dubius* (Acanthocephala) in the intermediate host *Periplaneta americana*. *Parasitology* **64**: 95–106.

Lackie, J.M. (1973). *Studies on interactions between* Moniliformis *(Acanthocephala) and its intermediate hosts*. Ph.D. thesis: University of Cambridge.

Lackie, J.M. (1975). The host specificity of *Moniliformis dubius* (Acanthocephala), a parasite of cockroaches. *Int. J. Parasit.* **5**: 301–7.

Lackie, J.M. & Rotheram, S. (1972). Observations on the envelope surrounding *Moniliformis dubius* (Acanthocephala) in the intermediate host, *Periplaneta americana*. *Parasitology* **65**: 303–8.

Larsen, J.R. & Bodenstein, D. (1959). The humoral control of egg maturation in the mosquito. *J. exp. Zool.* **140**: 343–81.

Lethbridge, R.C. (1971a). The locust as an intermediate host for *Hymenolepis diminuta*. *J. Parasit.* **57**: 445–6.

Lethbridge, R.C. (1971b). The hatching of *Hymenolepis diminuta* eggs and penetration of the hexacanths in *Tenebrio molitor* beetles. *Parasitology* **62**: 445–56.

Locke, M. (1974). The structure and formation of the integument in insects. In *The physiology of Insecta* **6**: 123–213 (2nd edn.). (Ed. Rockstein, M.). Academic Press, New York & London.

Minchella, D.J. & Loverde, P.T. (1983). Laboratory comparison of the relative success of *Biomphalaria glabrata* stocks which are susceptible and insusceptible to infection with *Schistosoma mansoni*. *Parasitology* **86**: 335-44.

Mohrig, W., Schittek, D. & Hanschke, R. (1979). Immunological activation of phagocytic cells in *Galleria mellonella*. *J. Invert. Path.* **34**: 84–7.

Pethica, B.A. (1980). Microbial and cell adhesion. In *Microbial adhesion to surfaces*: 19-45. (Eds Berkeley, R.C.W., Lynch, J.M., Melling, J., Rutter, P.R. & Vincent, B.). Ellis Horwood, Chichester.

Ratcliffe, N.A. & Walters, J.B. (1983). Studies on the *in vivo* cellular reactions of insects: clearance of pathogenic and non-pathogenic bacteria in *Galleria mellonella* larvae. *J. Insect Physiol.* **29**: 407–15.

Richards, K.S. & Arme, C. (1985). Phagocytosis of microvilli of the metacestode of *Hymenolepis diminuta* by *Tenebrio molitor* haemocytes. *Parasitology* **90**: 365–74.

Riddiford, L.M. (1976). Hormonal control of insect epidermal cell commitment *in vitro*. *Nature, Lond.* **259**: 115–17.

Rizki, R.M. & Rizki, T.M. (1980). Hemocyte responses to implanted tissues in *Drosophila melanogaster* larvae. *Wilhelm Roux Arch. dev. Biol.* **189**: 207–13.

Robinson, E.S. & Strickland, B.C. (1969). Cellular responses of *Periplaneta americana* to acanthocephalan larvae. *Expl Parasit.* **26**: 384–92.

Rotheram, S. & Crompton, D.W.T. (1972). Observations on the early relationship between *Moniliformis dubius* (Acanthocephala) and the haemocytes of the intermediate host, *Periplaneta americana*. *Parasitology* **64**: 15–21.

Sahba, G.H., Arfaa, F. & Rastegaar, M. (1970). Human infection with *Moniliformis dubius* (Acanthocephala) (Meyer 1932) (Syn. *M. moniliformis* (Bremser 1811)) in Iran. *Trans. R. Soc. trop. Med. Hyg.* **64**: 284–8.

Salt, G. (1961). The haemocytic reaction of insects to foreign bodies. In *The cell and the organism*: 175–92. (Eds Ramsay, J.A. & Wigglesworth, V.B.). Cambridge University Press, Cambridge.

Salt, G. (1963a). The defence reactions of insects to metazoan parasites. *Parasitology* **53**: 527–642.

Salt, G. (1963b). Experimental studies in insect parasitism XII. The reactions of six exopterygote insects to an alien parasite. *J. Insect Physiol.* **9**: 647–69.

Salt, G. (1970). *The cellular defence reactions of insects.* Cambridge University Press, Cambridge.

Salt, G. (1980). A note on the resistance of two parasitoids to the defence reactions of their insect hosts. *Proc. R. Soc. Lond.* (B) **207**: 351–3.

Schom, C., Novak, M. & Evans, W.S. (1981). Evolutionary implications of *Tribolium confusum/Hymenolepis citelli* interactions. *Parasitology* **83**: 77-90.

Scott, M.T. (1971). Recognition of foreign-ness in invertebrates: transplantation studies using the American cockroach, *Periplaneta americana. Transplantation (Baltimore)* **11**: 78–85.

Sminia, T., Borghart-Reinders, E. & van de Linde, A.W. (1974). Encapsulation of foreign materials experimentally introduced into the freshwater snail *Lymnaea stagnalis*: an electron-microscopic and autoradiographic study. *Cell Tissue Res.* **153**: 307–26.

Takle, G.B. & Lackie, A.M. (1985). Surface charge of insect haemocytes, examined using cell electrophoresis and cationized ferritin-binding. *J. Cell Sci.* **75**: 207–14.

Takle, G. (1985). The effects of components of the prophenoloxidase-activation pathway on the chemokinetic locomotion of insect haemocytes *in vitro. Devl comp. Immunol.* **9**: 164.

Thomas, I.G. & Ratcliffe, N.A. (1982). Integumental grafting and immunorecognition in insects. *Devl comp. Immunol.* **6**: 643–54.

Ubelaker, J.E., Cooper, N.B. & Allison, V. (1970). Possible defensive mechanism of *Hymenolepis diminuta* cysticercoids to haemocytes of the beetle *Tribolium confusum. J. Invert. Path.* **16**: 310–12.

Vinson, S.B. (1974). The role of the foreign surface and female parasitoid secretions on the immune response of an insect. *Parasitology* **79**: 297–306.

Voge, M. & Green, J. (1975). Axenic growth of oncospheres of *Hymenolepis citelli* (Cestoda) to fully developed cysticercoids. *J. Parasit.* **61**: 291–7.

Voge, M. & Turner, J.A. (1956). Effect of temperature on larval development of the cestode, *Hymenolepis diminuta. Expl Parasit.* **5**: 580–6.

White, G.B. (1982). Malaria vector ecology and genetics. *Br. med. Bull.* **38**: 207–12.

Wigglesworth, V.B. (1954). *The physiology of insect metamorphosis.* Cambridge University Press, Cambridge.

Symp. zool. Soc. Lond. (1986) No. 56 179–198

Interaction between the immune system of lymnaeid snails and trematode parasites

W.P.W. van der KNAAP
and Elisabeth A. MEULEMAN

*Laboratory of Medical Parasitology,
Faculty of Medicine,
Free University, P.O. Box 7161,
1007 MC Amsterdam, The Netherlands*

Synopsis

Trematode parasites can develop in their molluscan host because they can circumvent adverse activities of the host's internal defence system. The haemocytes and other defence cells, the humoral defence factors, and the way in which they co-operatively combat foreign materials are described for one lymnaeid species. If a trematode invades a lymnaeid and is confronted with an incompatible defence system, it attracts host haemocytes. These encapsulate, kill, invade and destroy the parasite. Only circumstantial evidence exists of humoral anti-trematode defence in lymnaeids.

Properties of both the trematode and the lymnaeid host's defence system determine whether there is immunological compatibility. It is likely that repulsive physicochemical forces are involved. The developmental stages of the parasite that are in contact with the lymnaeid's defence system for only a very short period, the miracidia and cercariae, seem to rely principally on molecular disguise, thus evading immunorecognition. The true parasitic stages that encounter the potentially harmful defence system over prolonged periods of time, the sporocysts and rediae, are probably able to suppress anti-trematode defence activities. During the period in which these stages are present in the snail, alterations occur in the defence system which suggest that the system is interfered with. Using the above strategies, a trematode has more chance of successfully establishing itself in a juvenile snail with a still poorly developed defence system, than in an adult host with a more active defence system.

Introduction

Digenetic trematodes, members of the phylum Platyhelminthes, have a complicated life cycle. They reproduce sexually in a vertebrate host. When an egg, released from the definitive host, hatches, a ciliated miracidium larva is produced. A miracidium which succeeds in penetrating a suitable molluscan host, usually a specific species of freshwater gastropod, transforms into a sporocyst. Each sporocyst may give rise to one or more generations of either

ZOOLOGICAL SYMPOSIUM No. 56
ISBN 0-19-854004-3

rediae or daughter sporocysts, depending on the trematode species. Eventually many cercariae are produced in the daughter sporocysts or in the rediae. Once emerged from the intermediate host, the free-swimming cercariae may, directly or after encystment on foliage or in a second intermediate host, again depending on the trematode species, infect a definitive vertebrate host and mature.

A digenetic trematode can establish itself, develop and reproduce asexually in a snail if it encounters both a physiologically suitable environment and an immune system which fails to eliminate the parasite. The present paper reviews matters pertaining to immunological compatibility and incompatibility between lymnaeid snails and trematode parasites. A description will be given of the internal defence system (immune system) of the snails. Then it will be discussed how this defence system responds to trematodes when compatibility is lacking. Finally, an attempt will be made to explain why these responses fail to appear in case of susceptibility of the snail to the parasite.

The internal defence system of lymnaeid snails

Introductory remarks

Our knowledge of the internal defence system of snails has mainly been gathered from studies of the commercially important vineyard snail *Helix pomatia* and of two freshwater pulmonate species. Of the latter two, *Biomphalaria glabrata* has received much attention because it is an intermediate host of the human blood fluke *Schistosoma mansoni* (see chapter by Loker & Bayne, this volume); the common pond snail *Lymnaea stagnalis* is the only lymnaeid of which the immune system *per se* has been studied extensively. The results of the studies performed with these three not very closely related species show us three remarkably similar pictures of defence systems. It is thus very likely that the description we give here of the defence system of *L. stagnalis* is applicable to other lymnaeids as well.

Defence cells

A number of different cell types are involved in internal defence in *L. stagnalis*. The haemocyanin-producing pore cells, present in the connective tissue, ingest especially foreign proteins, which are then slowly degraded by lysosomal enzymes (Sminia 1981). The fixed phagocyte or reticulum cell, which lies in close association with connective tissue fibrils, is less selective; it endocytoses not only foreign proteins but also larger abiotic particles, foreign cells and micro-organisms. The cells have an extensive lysosomal system which contains the enzymes non-specific esterase and acid phosphatase; however, the enzyme peroxidase is lacking (Sminia, Van der Knaap &

Kroese 1979; Sminia 1981). Thus it is unlikely that the cells are able to kill bacteria through toxic oxygen metabolites, generated in a peroxidase-dependent system (Klebanoff 1968). A third type of non-motile cell that may be involved in defence is the endothelial cell. Renwrantz, Schäncke, Harm, Erl, Liebsch & Gercken (1981) found that in *H. pomatia*, endothelial cells trap certain types of foreign cells before these are processed by motile defence cells. In *L. stagnalis*, a close association of injected bacteria with vascular walls has been observed, reminiscent of this trapping (Van der Knaap, Sminia, Kroese & Dikkeboom 1981), but never to the extent described in *H. pomatia*.

The best-studied defence cell is the motile haemocyte, also called amoebocyte, occurring both in the open circulatory system and in the connective tissue. The circulatory cells can move into the tissues and the sedentary cells can move into the circulation (Sminia 1972). As regards their morphology and functioning, they can best be compared with monocytes or macrophages of vertebrates. The cells have a cytoskeleton which enables them to perform amoeboid movements, to extend pseudopodia and to engulf foreign particles. In the cytoplasm mitochondria, glycogen, free ribosomes, rough endoplasmic reticulum, Golgi bodies, primary and secondary lysosomes and multivesicular bodies occur. The cells synthesize the enzymes acid and alkaline phosphatase, non-specific esterase and peroxidase (Sminia & Barendsen 1980; Dikkeboom, Van der Knaap, Meuleman & Sminia 1984). The haemocyte population comprises small spherical cells with relatively little cytoplasm which contains only few organelles, large cells which are more irregular in outline and which contain relatively much cytoplasm loaded with organelles, and intermediate forms. In the circulation of young snails, fewer cells are present and most of these are the small round cells which have a relatively low immunological capability. These cells show a rather high mitotic activity and as a result, adult snails have more haemocytes per μl haemolymph, the majority of which are large cells with many cytoplasmic inclusions, a high lysosomal enzyme content and a high immunocompetence (Dikkeboom, Van der Knaap, Meuleman & Sminia 1984, 1985). All this suggests that only one cell type exists which goes through a series of differentiation steps, with a rather heterogeneous blood picture as a result. Recent studies using monoclonal antibodies directed against antigenically distinct sub-populations of haemocytes (Dikkeboom, Van der Knaap, Maaskant & De Jonge 1985) left this concept untouched but more detailed studies of the ultrastructure, functioning, proliferation and differentiation of the different sub-populations of haemocytes, both in the tissues and in the circulation, are needed. As to the origin of the haemocytes, it is well-established that the haemocytes in the tissues and those in the haemolymph are equally well able to divide under normal (unstimulated) conditions (Sminia 1974); thus the snail does not depend on a haematopoietic organ for its regular production and replenishment of haemocytes. The haemocytes discriminate between self and not-self, including damaged and effete self. Not-self materials are phagocytosed and, if the nature of the material allows it, digested by the lysosomal system (Sminia 1972; Van der Knaap, Sminia, Kroese *et al.* 1981). Objects too big to be phagocytosed as a whole are encapsulated, usually by a multi-layered capsule of haemocytes, and subsequently phagocytosed. In older capsules haemocytes flatten and many transform into fibroblasts, which eventually convert the capsule into collagenous connective tissue (Sminia, Borghart-Reinders & Van de Linde 1974).

Contact between haemocytes and foreign material can come about by incidental collision. The chance of collision might be increased if the presence of foreign material alters the locomotory behaviour of the haemocytes. There are indications that the locomotion of haemocytes is influenced by environmental stimuli. After having phagocytosed bacteria in the circulation, the haemocytes migrate into the tissues; later, new cells are recruited from the tissues into the circulation (Van der Knaap, Sminia, Kroese *et al.* 1981). When an abiotic particle, which does not emit soluble substances, is encapsulated, the first layer of haemocytes which covers it must make contact by chance. In snails fixed soon after injection of Sepharose beads, accumulations of haemocytes can be observed in the immediate surroundings of beads which have already been covered by haemocytes. This observation suggests not only that locomotion of haemocytes is influenced by environmental stimuli, but also that haemocytes can emit substances which recruit more haemocytes.

Humoral defence factors

In a very low percentage of *L. stagnalis* specimens, the cell-free haemolymph has the capacity to lyse erythrocytes of various vertebrate species, but the mechanism of haemolysis has not been thoroughly investigated. Bactericidal substances could not be demonstrated, but only two bacterial species were tried as targets (Van der Knaap, Sminia, Kroese *et al.* 1981). The plasma contains a bacteriostatic factor, which inhibits the multiplication of the Gram-positive bacterium *Staphylococcus saprophyticus* strongly but that of the Gram-negative *Escherichia coli* only marginally (W.P.W. van der Knaap & R. Dikkeboom unpublished). Moreover, the plasma contains a factor which has agglutinating properties (Van der Knaap, Doderer, Boerrigter-Barendsen & Sminia 1982). It is a proteinaceous molecule of about 60,000 daltons which agglutinates micro-organisms and vertebrate red blood cells. This lectin has a broad spectrum specificity as a number of carbohydrates inhibit the agglutination of erythrocytes. Two varieties of the lectin occur. The haemolymph of a small percentage of the snails, designated type I, agglutinates erythrocytes and micro-organisms with very high titres; the majority of the snails, named type II, have a lectin which agglutinates the same micro-organisms as does type I agglutinin, but of the erythrocytes only rabbit red cells are agglutinated and at very low titres. Whether a snail has the highly reactive type I or the poorly reactive type II agglutinin is genetically determined. Moreover, there is a plasma factor which binds to foreign cells, thereby rendering these more suitable for phagocytosis by haemocytes (Sminia, Van der Knaap & Edelenbosch 1979). This opsonin and the agglutinin appear to be one and the same substance (Van der Knaap, Sminia, Schutte & Boerrigter-Barendsen 1983). In plasma of juvenile snails, opsonizing and agglutinating activities are significantly lower than in plasma of adult snails (Dikkeboom, Van der Knaap, Meuleman *et al.* 1985). In an immunocytochemical study (Van der Knaap, Boerrigter-Barendsen, Van den Hoeven & Sminia 1981) it was demonstrated that the opsonin/agglutinin is synthesized by the haemocytes (see Fig. 1).

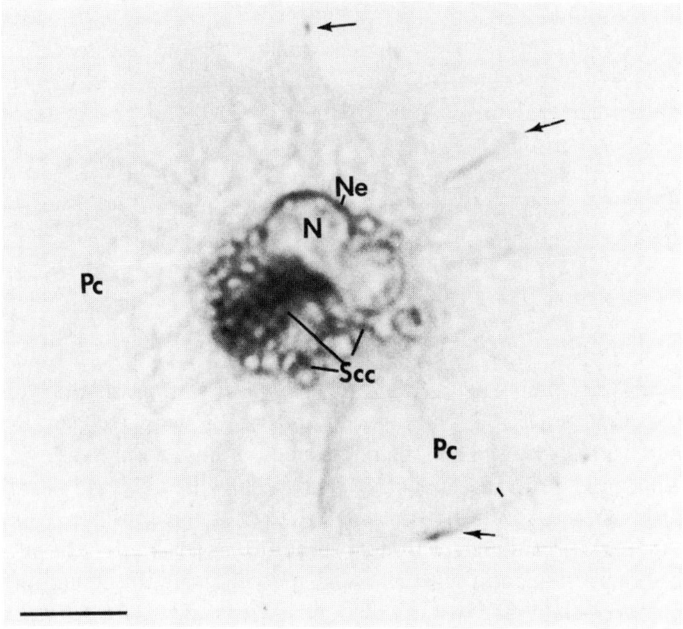

Fig. 1. Light micrograph of a haemocyte of *Lymnaea stagnalis*, settled on a glass slide and fixed with methanol. The cell was subjected to an indirect immunoperoxidase staining procedure with an antiserum raised against *L. stagnalis* opsonin/agglutinin. Nucleus (N) and peripheral cytoplasm (Pc) unstained. Staining product on the nuclear envelope (Ne) and on structures in the central cytoplasm (Scc) suggests that the rough endoplasmic reticulum contains opsonin/agglutinin. Staining of the cell's periphery (arrows) indicates that opsonin/agglutinin is present on the plasma membrane. Scale bar 5 μm.

Immunorecognition

In vivo, abiotic particles are recognized as foreign and phagocytosed (Sminia 1972) or encapsulated (Sminia, Borghart-Reinders *et al.* 1974). Particles with different surface characteristics evoke different reactions; whereas Sepharose beads provoke multilayered haemocytic capsules, latex beads of the same size get covered by only one layer of extremely flattened haemocytes. Also, haemocytes adhere *in vitro* to abiotic surfaces. On certain plastics the cells spread more eagerly than on glass. This indicates that physicochemical properties of the surface of both the haemocyte and the foreign object (e.g. wettability, net charge) can determine whether or not adhesion and engulfment take place. The haemocytes can discriminate between tissues of different nature. Implanted autografts and allografts are accepted, xenografts are encapsulated and destroyed. The severity of the reaction increases as the donor of the tissue is phylogenetically less related to *L. stagnalis* (Sminia, Borghart-Reinders *et al.* 1974). This cannot be explained merely by physicochemical properties, and a more subtle recognition system must be involved. Probably the opsonin plays a role here, since this substance binds to foreign cells and stimulates haemocytes to interact with them (Sminia, Van der Knaap & Edelenbosch 1979). Moreover, opsonin/agglutinin molecules are present on the outer membrane of haemocytes where they function as cytophilic receptors for foreignness (Van der Knaap, Sminia, Schutte *et al.* 1983). Since this recognition factor, as far as we know, recognizes only foreign carbohydrate moieties, the question remains whether additional systems exist, recognizing, for example, foreign proteins.

Acquired immunity

The response of the defence system of *L. stagnalis* can be enhanced through previous contact with foreign materials (Van der Knaap, Tensen, Kroese & Boerrigter-Barendsen 1982; Van der Knaap, Boots, Van Asselt & Sminia 1983). In snails previously injected with heat-killed bacteria (*E. coli* or *S. saprophyticus*) secondarily injected doses of live bacteria are eliminated from the circulation significantly faster than in naive snails. The reaction is non-specific: pre-injection with *E. coli* results in an enhanced elimination of *S. saprophyticus* and vice versa. It is the haemocytes that are activated. After injection of bacteria the number of circulating haemocytes drops drastically. The replenishment with new cells from the tissues is faster in immunized than in naive snails; these cells have an immensely increased surface area, are more active in phagocytosis and show a higher mitotic activity than do cells of unimmunized snails. Bacteria are still eliminated at an enhanced rate when administered 32 days after a primary injection and the faster replenishment with new haemocytes still occurs when secondary injections are given 64 days after the priming dose.

Immune responses to trematodes in lymnaeids

Introductory remarks

Defence responses to trematodes depend on whether physiological and/or immunological incompatibility exists. In a physiologically unsuitable host the trematode dies and may subsequently be attacked by the defence system; the defence system and the trematode may be compatible initially but owing to alterations in the defence system this compatibility may be lost (McReath, Reader & Southgate 1982); there may be a moderate or high degree of immunological incompatibility the moment that the trematode invades (Klühspies 1983). Moreover, even in highly susceptible hosts individual parasites may become attacked, for example cercariae which have been trapped in the dense tissues of the foot (Loker 1979).

Cellular defence reactions to trematodes

Most observations on cellular anti-trematode defence deal with encapsulation reactions. Although the course of events in such reactions may differ from one specific situation to another, they are generally made up of certain successive steps. Initially, haemocytes are induced to accumulate in the vicinity of the parasite. For example, accumulations of haemocytes have been observed around developmental stages of *Schistosomatium douthitti* in diverse locations in *Lymnaea catascopium* (Loker 1978, 1979), around mother sporocysts of *Trichobilharzia ocellata* in the head-foot of *L. stagnalis* (J.J. Mellink personal communication), and around sporocysts of *Hypoderaeum dingeri* and *Echinostoma audyi* in the heart of *Lymnaea rubiginosa* (Dondero, Ow-Yang & Lie 1977).

The next step to occur is that haemocytes attach to the parasite and cover it. Initially, capsules are loosely arranged and consist of a few layers of haemocytes which retain their original round shape. Such loosely arranged capsules have been observed around apparently vital sporocysts of *Isthmiophora melis* (Klühspies 1983) and around cercariae of *T. ocellata* (E.A. Meuleman & W.P.W. van der Knaap unpublished) in *L. stagnalis*, around sporocysts of *H. dingeri* and *E. audyi* in *L. rubiginosa* (Dondero *et al.* 1977) and around different developmental stages of *S. douthitti* in *L. catascopium* (Loker 1979). From ultrastructural observations of early encapsulation reactions around *Fasciola hepatica* sporocysts in the abnormal host *Lymnaea palustris*, McReath (1979) drew the conclusion that two types of cell might be involved: granulocytes and amoebocytes. The granulocytes have an electron-dense cytoplasm which contains, apart from very electron-dense granules and some mitochondria, only few inclusions. The cytoplasm of the amoebocytes is less electron-dense and contains more inclusions, such as mitochondria, aggregations of glycogen, dense granules and a few cisternae of endoplasmic reticulum. Our own ultrastructural observations of early encapsulations around injected *S. mansoni* miracidia in *L. stagnalis* (an incompatible parasite-snail combination) showed only haemocytes (amoebocytes) to be involved. The cells differ from the amoebocytes described by McReath (1979) in being more electron-lucent and having fewer organelles, especially where they lie closely apposed to the parasite. They actively phagocytose constituents of the parasite's body wall (see Fig. 2).

Several authors (Dondero *et al.* 1977; Loker 1978, 1979; Klühspies 1983) regard the following step of the encapsulation as the crucial one: cells in the capsule flatten and the capsule re-organizes into a structure of thin concentric layers of haemocytes. Ultrastructural observations indicate that the haemocytes which initially formed the capsule disappear and are replaced by others; the latter cells flatten and form the compact capsule. These cells are characterized, both in *L. stagnalis* and in *L. palustris*, by a moderately electron-dense cytoplasm with many mitochondria, an extensive rough endoplasmic reticulum, various lysosomal structures and glycogen deposits (see Fig. 3). Eventually, a number of these cells transform into fibroblasts (McReath 1979) which convert the capsule into collagenous connective tissue. The cells also infiltrate the encapsulated parasite, now degenerating, and phagocytose the contents.

A capsule is not a static structure; cells that initially form it disappear and others take their place. The function of the first cells seems to be to remove the parasite's body wall, that of the later arrivals to shut the parasite off from the surroundings and destroy it further. Since the two types of cell differ also in morphology, the question arises whether they are actually separate cell types, or different developmental stages of one type of cell. To answer this question, further ultrastructural studies are needed to characterize the cells involved in all phases of the encapsulation in more detail. Characterization at a molecular level may be obtained with monoclonal antibodies against different sub-populations of haemocytes; we have recently begun investigations in this respect.

Haemocyte adhesion and encapsulation does not necessarily mean that the parasite will be destroyed; Klühspies (1983) reported that *I. melis* sporocysts in *L. stagnalis* gave rise to rediae even after having been encapsulated. However, encapsulated trematodes are usually killed and destroyed. As to the mechanism used by the cells in the capsule to kill the parasite, only speculations can be made. The cause of death cannot be that the capsule inhibits exchange of molecules between the parasite and its surroundings since parasites also succumb which have been incompletely encapsulated (Loker

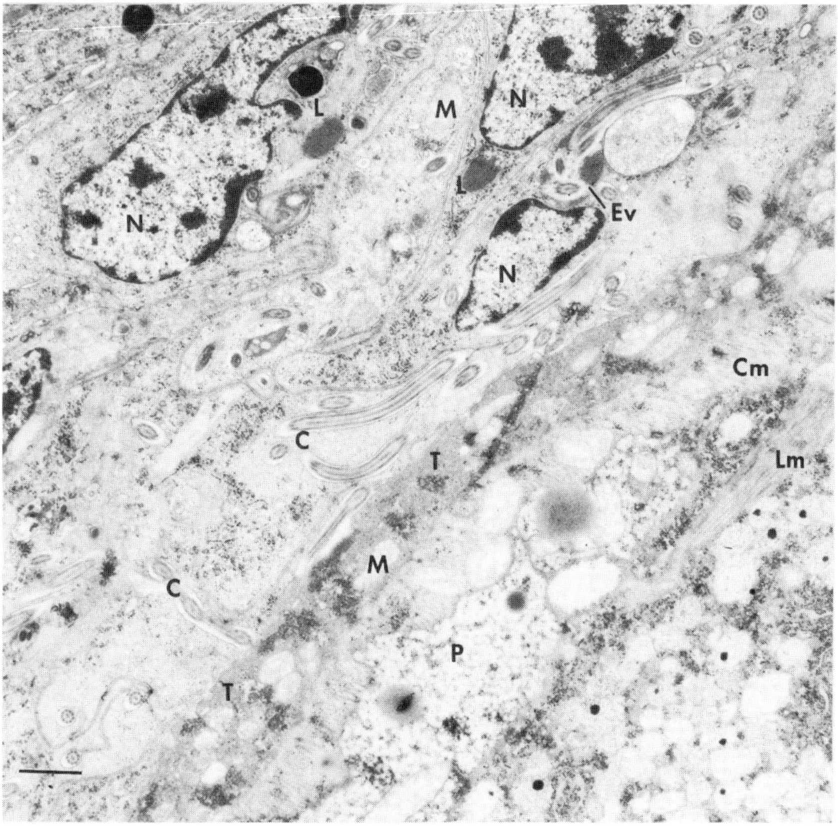

Fig. 2. Electron micrograph showing encapsulation of *Schistosoma mansoni* 3 h after injection of the miracidium into *Lymnaea stagnalis* (insusceptible snail). Haemocytes have made contact with the parasite, but have not yet flattened. Note that — especially where it is closely apposed to the parasite — the cytoplasm is electron-lucent and contains few organelles. The cilia (C) of the epidermal plates (not shown) are being phagocytosed by the haemocytes but the tegumental surface layer (T) is still intact. Cm, circular muscle; Ev, endocytotic vesicle with engulfed cilia; L, lysosome; Lm, longitudinal muscle; M, mitochondrion; N, nucleus of haemocyte; P, parenchyma. Scale bar 1 μm.

1979; McReath *et al.* 1982). Exocytosis of lytic enzymes from the innermost cells of the capsule has not been reported but the question of whether this phenomenon occurs deserves further study. Mammalian leucocytes are able to kill helminths through a peroxidase-dependent mechanism in which toxic forms of oxygen play an essential role (Buys, Wever, Van Stigt & Ruitenberg 1981; Jong, Mahmoud & Klebanoff 1981). Haemocytes of *L. stagnalis* generate these oxygen species (superoxide, hydrogen peroxide and possibly the hydroxyl radical and singlet oxygen) when stimulated by phagocytosis or by the surface-active agent phorbolmyristate acetate (Dikkeboom,

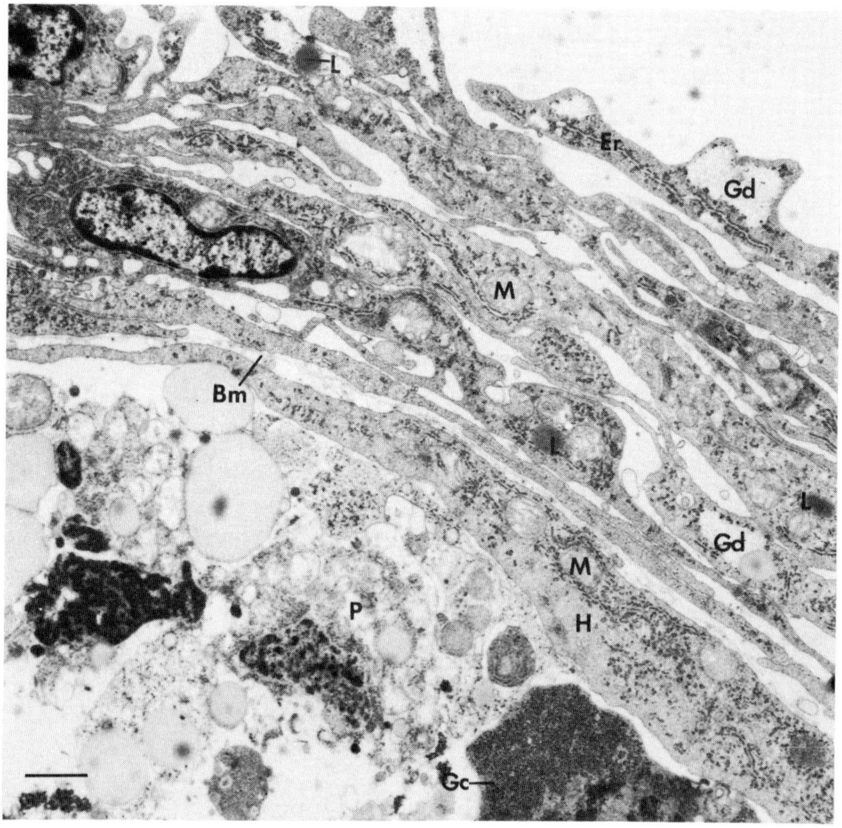

Fig. 3. Encapsulation of *Schistosoma mansoni* 17 h after injection of miracidium into *Lymnaea stagnalis*. The haemocytes in the capsule have flattened and contain glycogen deposits (Gd), endoplasmic reticulum (Er), mitochondria (M) and lysosomes (L). There is no sign of secondary lysosomes containing remnants of endocytosed cilia. One haemocyte (H) has invaded the parasite and lies beneath the basement membrane (Bm); the tegumental surface layer has disappeared. Gc, germinal cell; P, parenchyma. Scale bar 1 μm.

Mulder & Tijnagel in press). It seems worthwhile to investigate whether mollusc haemocytes are able to kill trematode parasites through a mechanism similar to that exerted by mammalian leucocytes.

It is unknown what governs the different events in the encapsulation reaction. Obviously, as yet unknown properties of each individual parasite are decisive for at least the initiation of the reaction; for example, in a susceptible host, one sporocyst may become encapsulated next to one which is left untouched (Loker 1978). The histological observations do not permit us to conclude what roles are played by surface receptors on host haemocytes and by humoral recognition factors. To our knowledge, no experimental work has been done with lymnaeids to clarify this problem.

Humoral defence reactions against trematodes

Sporocysts in abnormal hosts have been observed to be degenerating without having been encapsulated first, e.g. *F. hepatica* in *Lymnaea peregra* (McReath 1979), *Trichobilharzia elvae* in *Bulimnea megasoma* and in *Fossaria abrussa* (Sudds 1960). Mother sporocysts of *T. ocellata* in adult *L. stagnalis*, which have a low susceptibility, were retarded in their growth if compared to sporocysts developing in highly susceptible juvenile snails, but they were not encapsulated (Meuleman & Te Velde 1984). Neither were they killed immediately, which is consistent with the observations by Mellink & Van den Bovenkamp (1985) who demonstrated that *T. ocellata* mother sporocysts, transformed from miracidia *in vitro*, survived for two weeks when incubated in cell-free haemolymph of adult *L. stagnalis*. Production of *S. douthitti* cercariae was delayed in some specimens of *L. catascopium* which had previously been sensitized with irradiated miracidia of the same species (Loker 1978); since conspicuous encapsulation reactions were not observed around the sporocysts of the challenge dose, a humoral mechanism may be responsible for the delayed parasite development. Production of *T. ocellata* cercariae was delayed if the host snails were of the type I subpopulation of *L. stagnalis* with very reactive agglutinin instead of the type II with less reactive lectin (W.P.W. van der Knaap unpublished); this strongly suggests that the reactive agglutinin had hampered the parasite's development, but other mechanisms may also have been responsible, such as increased cellular reactivity, potentiated by a more efficient recognition system (cf. Van der Knaap, Sminia, Schutte *et al.* 1983). Miracidia of *I. melis* die after a few hours of *in vitro* incubation in water-diluted whole haemolymph of the moderately susceptible *L. stagnalis*; however, the incompatible *S. mansoni* was not affected in this system (Klühspies 1983). This seeming discrepancy may find its explanation in that *S. mansoni* is rather insensitive to humoral killing mechanisms (Loker & Bayne 1982). The above observations do not permit us to conclude whether humoral defence against trematodes takes place in lymnaeids, but they indicate the likelihood of its occurrence.

Acquired resistance to trematodes

Can lymnaeid hosts acquire resistance against trematodes? We have seen that humoral defence was probably induced by previous infection in *L. catascopium* (Loker 1978). Klühspies (1983) induced a rather non-specific activation of anti-trematode defence in *L. stagnalis*: exposure of the snails to several echinostome species and injection of various foreign substances resulted in an increase in both humoral (growth retardation) and cellular (encapsulation) defence against *I. melis* sporocysts. However, previous exposure of *L. catascopium* to irradiated miracidia of *S. douthitti* (Loker

1978) and exposure of *L. rubiginosa* to irradiated miracidia of *E. audyi* or *H. dingeri* (Dondero *et al.* 1977) did not lead to induction of increased cellular defence against sporocysts which developed from normal miracidia of the same species.

What causes immunological compatibility?

Introductory remarks

Immunological compatibility depends on properties of both the parasite and the host. The present section deals with the question of which properties, both of the trematode and of the lymnaeid's defence system, enable the parasite to circumvent adverse immunological activities in the host. In successive order attention will be paid to physicochemical properties of the surfaces of the parasite and of the host defence cell which may prevent cell adhesion, to the parasite's capacity to behave as 'self' in the snail, to the parasite's capacity to interfere actively with host defence activities, and to alterations in the snail's defence system which may alter the chance of success of the above three evasion strategies.

Physicochemical properties of haemocyte and parasite surfaces

Haemocytes adhere to abiotic surfaces and engulf and encapsulate abiotic materials. Responses to foreign material differ, depending on the type of material and the composition of the surrounding medium. There are no published reports on whether such phenomena play a role at the host-parasite interface in lymnaeid-trematode combinations, but the feasibility should be considered. Experiments can be thought of where, in compatible snail-parasite combinations, the composition or physical properties of the surface of either the haemocytes or the parasite are experimentally altered in order to abolish the compatibility. We have made a first attempt to study this by exposing *L. stagnalis* to *T. ocellata* miracidia which had been coated in specific antibodies. We found that of the seven snails that were examined, only one had developed a patent infection, compared with seven out of eight controls (snails exposed to miracidia which had been incubated in irrelevant antibodies). In the near future, monoclonal antibodies, to be used as a covering or as a vehicle for molecules with chosen physicochemical properties, may prove to be valuable tools in this area of research.

Molecular disguise

One of the most obvious strategies with which the parasite could circumvent host defences is to avoid being recognized as not-self. Two forms of molecular disguise can be thought of: the parasite synthesizes host-like molecules and presents these at its surface (molecular mimicry; see Damian

1979), or the parasite adopts molecules from the host and uses these as a disguising cover (molecular masking).

We have searched the surfaces of the different intra-snail developmental stages of *T. ocellata* for the presence of *L. stagnalis* (-like) molecules using polyvalent rabbit antisera, raised against cell-free haemolymph or tissue cells of *L. stagnalis*, in an indirect immunoperoxidase technique (Van der Knaap, Boots, Meuleman & Sminia 1985; Van der Knaap, Meuleman & Mellink 1985). Miracidia which have not been in contact with snail material are stained, especially at their surface, indicating that they synthesize snail-like molecules and express these at their surface. After the miracidia have penetrated the snail and have shed their ciliated plates, only a very thin film of host-like material is demonstrated at the surface of the mother sporocysts. On older mother sporocysts (see Fig. 4) and on daughter sporocysts (see Fig. 5) host-like or host-derived molecules could not be demonstrated. The surface of the cercariae appears to be devoid of host-like material when still within the daughter sporocyst, but becomes covered with host molecules once the cercariae have emerged from the sporocyst (see Fig. 5). The latter observation corroborates the results of Roder, Bourns & Singhal (1977) who demonstrated that antisera raised against cell-free haemolymph or tissue extract of uninfected *L. stagnalis* reacted with antigens in a homogenate of *T. ocellata* cercariae. Moreover, antisera prepared against intact miracidia and *in vitro*-cultured mother sporocysts of *T. ocellata* cross-reacted with haemocytes of uninfected *L. stagnalis* (J.J. Mellink personal communication). To our knowledge, trematodes other than *T. ocellata* have not been searched for cross-reactivity with antisera to their lymnaeid hosts, or vice versa. However, in all cases where lymnaeid-like molecules are demonstrated on intra-snail developmental stages of trematodes, it remains to be proven that they indeed provide protection against immunorecognition in the specific lymnaeid host (see also chapter by Yoshino & Boswell, this volume).

Interference with host defence

In field-collected *L. stagnalis appressa*, Bourns (1963) found significantly more snails shedding cercariae of two or three different trematode species than expected on the basis of their percentile incidences in single infections. Lie, Lim & Ow-Yang (1973) exposed *L. rubiginosa* to miracidia of *Echinostoma hystricosum*, and 16% of the snails developed a patent infection; however, when the snails had first been experimentally infected with *Trichobilharzia brevis*, exposure to *E. hystricosum* led to patent infections in 53% of the snails. In both cases, an explanation may be that the first trematode that established an infection increased the susceptibility of the snail to the next species. Proof that trematodes suppress anti-trematode

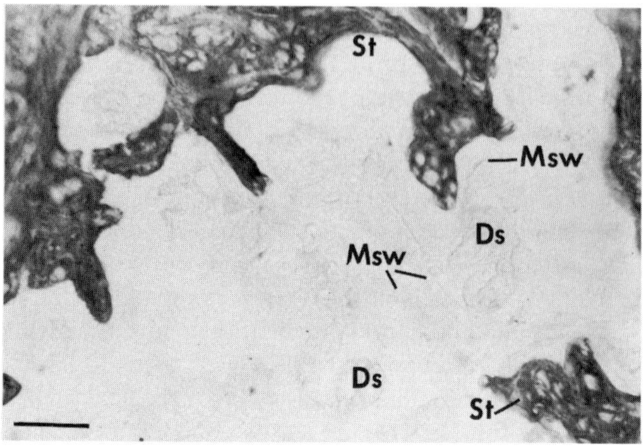

Fig. 4. Paraffin section of head-foot of *Lymnaea stagnalis*, fixed 25 days after exposure to miracidia of *Trichobilharzia ocellata*. The section was subjected to an indirect immunoperoxidase staining procedure with an antiserum raised against *L. stagnalis* tissue cells. Snail tissue (St) is intensely stained, parasite tissue remains unstained. Ds, daughter sporocyst, developing inside mother sporocyst; Msw, mother sporocyst wall. Scale bar 25 μm.

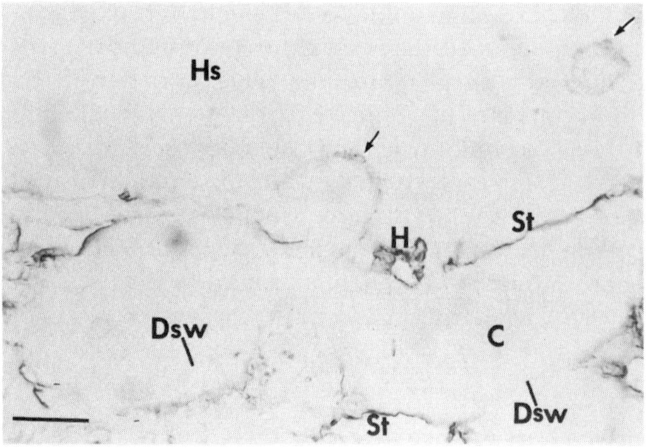

Fig. 5. Paraffin section of digestive gland area of *Lymnaea stagnalis*, fixed 60 days after exposure to miracidia of *Trichobilharzia ocellata*. The section was subjected to an indirect immunoperoxidase staining procedure with an antiserum, raised against *L. stagnalis* tissue cells. Haemocytes (H) which lie in a haemolymph space (Hs) and snail tissue (St) are stained. Daughter sporocyst wall (Dsw) and cercaria (C) inside daughter sporocyst are unstained. Cross-sections through tail of cercaria in host haemolymph covered with staining product (arrows). Scale bar 10 μm.

activities of the internal defence system of lymnaeids is still lacking, however. In *B. glabrata* the phenomenon has been demonstrated extensively. Much of the early work in this area was done by Lie and co-workers and has been reviewed by Lie (1982); more recent work is reviewed by Loker & Bayne in this volume.

The mechanism through which interference is thought to be exerted involves the secretion by the parasite of substances which would influence the host's defence system. In an immunocytochemical study, Mellink, Van den Bovenkamp & Van der Knaap (1986), using antisera prepared against products released by *T. ocellata* mother sporocysts in *in vitro* cultures, demonstrated that from two weeks of infection onwards *T. ocellata* in *L. stagnalis* releases substances which bind to host haemocytes. This is suggestive evidence for the concept that these substances would be the messengers and that the haemocytes would be the targets of interference. Indeed, in the period when increasing amounts of parasite products are found associated with the haemocytes, alterations occur in the haemocytes which may be reflections of interference (Van der Knaap, Meuleman & Sminia 1985). From five weeks of infection and onwards, the number of circulating haemocytes is much higher in infected snails than in controls. The newly arrived cells are larger, have more cytoplasmic inclusions and more and longer spike-like pseudopods than do cells of control snails; concurrently the capacity of these haemocytes to phagocytose sheep red blood cells *in vitro* is decreased. Moreover, titres of the agglutinin, which is produced by the haemocytes, are four times lower than in control snails. Haemocytosis such as observed in *L. stagnalis* has also been noted in *S. douthitti*-infected *L. catascopium* (Loker 1979) and in *F. hepatica*-infected *Lymnaea truncatula* (Rondelaud & Barthe 1981, 1982). In the latter species, the occurrence of haemocytosis appears to be related to an activation of the amoebocyte-producing organ; this tissue area in the reno-pericardial region becomes loaded with haemocytes and many cells can be seen in mitosis during the infection (Rondelaud & Barthe 1982). Massive accumulations of haemocytes occur in the lung roof, near the pericardium, of infected *L. catascopium* (Loker 1979) and in the reno-pericardial region of infected *L. stagnalis*; in these cases, mitotic figures are not abundant, however. The parasite-induced alterations become manifest when the snails harbour both mother sporocysts and daughter sporocysts (rediae in *L. truncatula*), but well before production of cercariae has commenced. It is noteworthy that in *B. glabrata*, the snail species in which interference takes place beyond doubt, an amoebocyte-producing organ occurs, that this organ is stimulated by trematode infection (Jeong, Lie & Heyneman 1983; Sullivan, Cheng & Howland 1984; Joky, Matricon-Gondran & Benex 1985) and that in this species infection may lead to haemocytosis (Jeong, Lie & Heyneman 1980) as is the case in trematode-infected lymnaeids. Since there are many signs indicative of trematodes exerting interference with host defence in lymnaeids, it can be anticipated that this phenomenon will also be demonstrated in this family.

Shifts in chance of success of evasive strategies

If a trematode succeeds in maintaining itself in its snail host, there is an equilibrium between the parasite's efforts to circumvent host defences and the efforts of the defence system to overpower the parasite. This equilibrium will not be established if the same parasite invades a host wherein it is confronted with a more powerful defence system; in the case of an established infection, the equilibrium may be lost if the defence system gains in power. The level of activity of the defence system of *L. stagnalis* can be raised by previous contact with foreign material. In snails where this has happened, fewer sporocysts of *I. melis* can evade assaults by the defence system than in naive snails (Klühspies 1983).

In *L. stagnalis*, the defence system gains in power with increasing age of the snail (Dikkeboom, Van der Knaap, Meuleman *et al.* 1984, 1985). This immunological maturation may partly explain why in many cases the susceptibility of the host decreases with increasing age (Boray 1969). Miracidia of *T. ocellata* invade both juvenile and adult *L. stagnalis* and transform into mother sporocysts; in juvenile snails they show further development but in adults their growth is inhibited, possibly by a humoral defence reaction (Meuleman & Te Velde 1984). The period of time during which sporocysts of *F. hepatica*, in the abnormal host *L. palustris*, can successfully withdraw from fatal encapsulation reactions decreases with increasing age of the host at the moment of infection (McReath *et al.* 1982). Boray (1966) transferred rediae of *F. hepatica* from young *L. stagnalis* to adult snails where they evoked defence reactions and died. These three examples indicate that a parasite's potential to circumvent host defences can be adequate in young snails, but falls short when confronted with the more powerful defence system of adult snails. After establishment in a juvenile *L. stagnalis*, *T. ocellata* and even *F. hepatica* (Boray 1969) can develop further, even when the host has reached an age at which a naive snail cannot be infected for the first time. This suggests that the parasite establishes itself in a young snail where it can cope with the defence system and that during the intra-snail development, its potential to exert evasive strategies co-develops with the defence potential of the host.

Concluding remarks

When immunological compatibility is lacking, the defence system of a lymnaeid host attacks and destroys a trematode. More information is needed on the role played by humoral defence factors in this process. What we know of cellular defence is that haemocytes encapsulate, kill and destroy the trematode. Much research must still be done before we will start to understand what triggers the cells to adhere and to kill, and of the nature of the killing mechanism itself.

The search for those properties of trematodes which enable them to evade defence activities of their lymnaeid host has hitherto yielded indications that molecular disguise and interference with host defences play a role. Definitive proof that these strategies of evasion are successfully employed must still be given. Finally, attention will also have to be paid to whether the outer surfaces of the different developmental stages of trematodes have physicochemical properties which discourage haemocytes of their specific host from adhering.

Acknowledgements

We are grateful to Dr T. Sminia for his most helpful comments during the preparation of the manuscript. We are indebted to Drs R. Dikkeboom, E.S. Loker, J.J. Mellink, T.A.J. Reader, D. Rondelaud and V.R. Southgate for providing information. We thank R. Haynes for typing this manuscript.

References

Boray, J.C. (1966). Studies on the relative susceptibility of some lymnaeids to infection with *Fasciola hepatica* and *F. gigantica* and on the adaptation of *Fasciola* spp. *Ann. trop. Med. Parasit.* **60**: 114–24.

Boray, J.C. (1969). Experimental fascioliasis in Australia. *Adv. Parasit.* 7: 95–210.

Bourns, T.K.R. (1963). Larval trematodes parasitizing *Lymnaea stagnalis appressa* Say in Ontario with emphasis on multiple infections. *Can. J. Zool.* **41**: 937–41.

Buys, J., Wever, R., Van Stigt, R. & Ruitenberg, E.J. (1981). The killing of newborn larvae of *Trichinella spiralis* by eosinophil peroxidase *in vitro*. *Eur. J. Immunol.* **11**: 843–5.

Damian, R.T. (1979). Molecular mimicry in biological adaptation. In *Host-parasite interfaces*: 103–26. (Ed. Nickol, B.B.). Academic Press, New York.

Dikkeboom, R., Mulder, E.C. & Tijnagel, J.M.G.H. (In press). Evidence that blood cells of *Lymnaea stagnalis* generate active oxygen species. *Trop. geogr. Med.*

Dikkeboom, R., Van der Knaap, W.P.W., Maaskant, J.J. & De Jonge, A.J.R. (1985). Different subpopulations of haemocytes in juvenile, adult and *Trichobilharzia ocellata* infected *Lymnaea stagnalis*: a characterization using monoclonal antibodies. *Z. Parasitenk.* **71**: 815–9.

Dikkeboom, R., Van der Knaap, W.P.W., Meuleman, E.A. & Sminia, T. (1984). Differences between blood cells of juvenile and adult specimens of the pond snail *Lymnaea stagnalis*. *Cell Tissue Res.* **238**: 43–7.

Dikkeboom, R., Van der Knaap, W.P.W., Meuleman, E.A. & Sminia, T. (1985). A comparative study on the internal defence system of juvenile and adult *Lymnaea stagnalis*. *Immunology* **55**: 547–53.

Dondero, T.J., Ow-Yang, C.K. & Lie, K.J. (1977). Failure of irradiated *Echinostoma audyi* and *Hypoderaeum dingeri* to sensitize *Lymnaea rubiginosa* snails. *Southeast Asian J. trop. Med. publ. Hlth* **8**: 359–63.

Jeong, K.H., Lie, K.J. & Heyneman, D. (1980). Leucocytosis in *Biomphalaria glabrata* sensitized and resensitized to *Echinostoma lindoense*. *J. Invert. Path.* **35**: 9–13.

Jeong, K.H., Lie, K.J. & Heyneman, D. (1983). The ultrastructure of the amebocyte-producing organ in *Biomphalaria glabrata*. *Devl comp. Immunol.* 7: 217–28.

Joky, A., Matricon-Gondran, M. & Benex, J. (1985). Response of the amoebocyte-producing organ of sensitized *Biomphalaria glabrata* after exposure to *Echinostoma caproni* miracidia. *J. Invert. Path.* **45**: 28–33.

Jong, E.C., Mahmoud, A.A.F. & Klebanoff, S.J. (1981). Peroxidase-mediated toxicity to schistosomula of *Schistosoma mansoni*. *J. Immunol.* **126**: 468–71.

Klebanoff, S.J. (1968). Myeloperoxidase-halide-hydrogen peroxide antibacterial system. *J. Bacteriol.* **95**: 2131–8.

Klühspies, G. (1983). Angeborene und erworbene Resistenz von *Lymnaea stagnalis* und *Radix ovata* gegen Miracidien-Invasion. *Z. Parasitenk.* **69**: 591–611.

Lie, K.J. (1982). Survival of *Schistosoma mansoni* and other trematode larvae in the snail *Biomphalaria glabrata*. A discussion of the interference theory. *Trop. geogr. Med.* **34**: 111–22.

Lie, K.J., Lim, H.-K. & Ow-Yang, C.K. (1973). Synergism and antagonism between two trematode species in the snail *Lymnaea rubiginosa*. *Int. J. Parasit.* **3**: 729–33.

Loker, E.S. (1978). *Schistosomatium douthitti*: Exposure of *Lymnaea catascopium* to irradiated miracidia. *Expl Parasit.* **46**: 134–40.

Loker, E.S. (1979). Pathology and host responses induced by *Schistosomatium douthitti* in the freshwater snail *Lymnaea catascopium*. *J. Invert. Path.* **33**: 265–73.

Loker, E.S. & Bayne, C.J. (1982). *In vitro* encounters between *Schistosoma mansoni* primary sporocysts and hemolymph components of susceptible and resistant strains of *Biomphalaria glabrata*. *Am. J. trop. Med. Hyg.* **31**: 999–1005.

McReath, A.M. (1979). *Some aspects of the host-parasite relationship between* Fasciola hepatica *and various* Lymnaea *species*. Ph.D. Thesis: Portsmouth Polytechnic.

McReath, A.M., Reader, T.A.J. & Southgate, V.R. (1982). The development of the host-response in juvenile *Lymnaea palustris* to invasion by *Fasciola hepatica*. *Z. Parasitenk.* **67**: 175–84.

Mellink, J.J. & Van den Bovenkamp, W. (1985). *In vitro* culture of intramolluscan stages of the avian schistosome *Trichobilharzia ocellata*. *Z. Parasitenk.* **71**: 337–51.

Mellink, J.J., Van den Bovenkamp, W. & Van der Knaap, W.P.W. (1986). Parasite products possibly affecting host defences in the schistosome-snail model *Trichobilharzia ocellata–Lymnaea stagnalis*. *Devl comp. Immunol.* **10**: 140.

Meuleman, E.A. & Te Velde, A.A. (1984). Age-dependent susceptibility of *Lymnaea stagnalis* to *Trichobilharzia ocellata*. *Trop. geogr. Med.* **36**: 387–8.

Renwrantz, L., Schäncke, W., Harm, H., Erl, H., Liebsch, H. & Gercken, J. (1981). Discriminative ability and function of the immunobiological recognition system of the snail *Helix pomatia*. *J. comp. Physiol.* **141**: 477–88.

Roder, J.C., Bourns, T.K.R. & Singhal, S.K. (1977). *Trichobilharzia ocellata*: cercariae masked by antigens of the snail, *Lymnaea stagnalis*. *Expl Parasit.* **41**: 206–12.

Rondelaud, D. & Barthe, D. (1981). The development of the amoebocyte-producing organ in *Lymnaea truncatula* Müller infected by *Fasciola hepatica* L. *Z. Parasitenk.* **65**: 331–41.

Rondelaud, D. & Barthe, D. (1982). Relationship of the amoebocyte-producing organ with the generalized amoebocytic reaction in *Lymnaea truncatula* Müller infected by *Fasciola hepatica* L. *J. Parasit.* **68**: 967–9.

Sminia, T. (1972). Structure and function of blood and connective tissue cells of the fresh water pulmonate *Lymnaea stagnalis* studied by electron microscopy and enzyme histochemistry. *Z. Zellforsch. mikrosk. Anat.* **130**: 497–526.

Sminia, T. (1974). Hematopoiesis in the freshwater snail *Lymnaea stagnalis* studied by electron microscopy and autoradiography. *Cell Tissue Res.* **150**: 443–54.

Sminia, T. (1981). Phagocytic cells in molluscs. In *Aspects of developmental and comparative immunology* **1**: 125–32. (Ed. Solomon, J.B.). Pergamon Press, Oxford & New York.

Sminia, T. & Barendsen, L. (1980). A comparative morphological and enzyme histochemical study on blood cells of the freshwater snails *Lymnaea stagnalis*, *Biomphalaria glabrata*, and *Bulinus truncatus*. *J. Morph.* **165**: 31–9.

Sminia, T., Borghart-Reinders, E. & Van de Linde, A.W. (1974). Encapsulation of foreign materials experimentally introduced into the freshwater snail *Lymnaea stagnalis*. An electron microscopic and autoradiographic study. *Cell Tissue Res.* **153**: 307–26.

Sminia, T., Van der Knaap, W.P.W. & Edelenbosch, P. (1979). The role of serum factors in phagocytosis of foreign particles by blood cells of the freshwater snail *Lymnaea stagnalis*. *Devl comp. Immunol.* **3**: 37–44.

Sminia, T., Van der Knaap, W.P.W. & Kroese, F.G.M. (1979). Fixed phagocytes in the freshwater snail *Lymnaea stagnalis*. *Cell Tissue Res.* **196**: 545–8.

Sudds, R.H. (1960). Observations on schistosome miracidial behavior in the presence of normal and abnormal snail hosts and subsequent tissue studies of these hosts. *J. Elisha Mitchell sci. Soc.* **76**: 121–33.

Sullivan, J.T., Cheng, T.C. & Howland, K.H. (1984). Mitotic responses of the anterior pericardial wall of *Biomphalaria glabrata* (Mollusca) subjected to challenge. *J. Invert. Path.* **44**: 114–6.

Van der Knaap, W.P.W., Boerrigter-Barendsen, L.H., Van den Hoeven, D.S.P. & Sminia, T. (1981). Immunocytochemical demonstration of a humoral defence factor in blood cells (amoebocytes) of the pond snail, *Lymnaea stagnalis*. *Cell Tissue Res.* **219**: 291–6.

Van der Knaap, W.P.W., Boots, A.M.H., Meuleman, E.A. & Sminia, T. (1985). Search for shared antigens in the schistosome-snail combination *Trichobilharzia ocellata-Lymnaea stagnalis*. *Z. Parasitenk.* **71**: 219–26.

Van der Knaap, W.P.W., Boots, A.M.H., Van Asselt, L.A. & Sminia, T. (1983). Specificity and memory in increased defence reactions against bacteria in the pond snail *Lymnaea stagnalis*. *Devl comp. Immunol.* **7**: 435–43.

Van der Knaap, W.P.W., Doderer, A., Boerrigter-Barendsen, L.H. & Sminia, T. (1982). Some properties of an agglutinin in the haemolymph of the pond snail *Lymnaea stagnalis*. *Biol. Bull. mar. biol. Lab. Woods Hole* **162**: 404–12.

Van der Knaap, W.P.W., Meuleman, E.A. & Mellink, J.J. (1985). Schistosome-snail immunological compatibility: shared antigens in the model *Trichobilharzia ocellata-Lymnaea stagnalis. Devl comp. Immunol.* **9**: 172.

Van der Knaap, W.P.W., Meuleman, E.A. & Sminia, T. (1985). *Trichobilharzia ocellata* in *Lymnaea stagnalis*: immunological aspects of schistosome-snail compatibility. *Devl comp. Immunol.* **9**: 178.

Van der Knaap, W.P.W., Sminia, T., Kroese, F.G.M. & Dikkeboom, R. (1981). Elimination of bacteria from the circulation of the pond snail *Lymnaea stagnalis. Devl comp. Immunol.* **5**: 21–32.

Van der Knaap, W.P.W., Sminia, T., Schutte, R. & Boerrigter-Barendsen, L.H. (1983). Cytophilic receptors for foreignness, and some factors which influence phagocytosis by invertebrate leucocytes: *in vitro* phagocytosis by amoebocytes of the snail *Lymnaea stagnalis. Immunology* **48**: 377–83.

Van der Knaap, W.P.W., Tensen, C.P., Kroese, F.G.M. & Boerrigter-Barendsen, L.H. (1982). Adaptive defence reactions against bacteria in the pond snail *Lymnaea stagnalis. Devl comp. Immunol.* **6**: 775–80.

Symp. zool. Soc. Lond. (1986) No. 56 199–220

Immunity to trematode larvae in the snail *Biomphalaria*

Eric S. LOKER
Department of Biology, The University of New Mexico, Albuquerque, New Mexico 87131, U.S.A.

and Christopher J. BAYNE
Department of Zoology, Oregon State University, Corvallis, Oregon 97331, U.S.A.

Synopsis

Inbred strains of *Biomphalaria glabrata* and *Schistosoma mansoni*, and several species of echinostome flukes, have been used to explore the immunobiology of snail-trematode associations. The likelihood of success of a trematode sporocyst in *Biomphalaria* depends on its genetically-determined level of infectivity relative to the host's innate resistance. In compatible associations, the host fails to mount an effective immune response. Living trematode larvae secrete substances, as yet uncharacterized, that selectively interfere with the ability of host haemocytes to encapsulate and destroy trematodes. Other tactics of immune evasion such as molecular mimicry and acquisition of disguising host molecules may also be employed, but the extent to which trematodes rely on such phenomena has not been determined. In some host strains, haemocytes successfully destroy trematode larvae. The level of this innate resistance can often be enhanced by exposure of snails to irradiation-attenuated larvae. Haemocytes of resistant snails are by themselves capable of killing sporocysts; however, humoral factors from such hosts apparently promote sporocyst-killing responses from ordinarily benign haemocytes from compatible hosts. Such factors may act directly on compatible haemocytes, e.g. inducing a generalized state of activation, or they may act on sporocysts, rendering them susceptible to recognition and killing.

Introduction

Pulmonate snails of the genus *Biomphalaria* serve as obligatory intermediate hosts for *Schistosoma mansoni*, the most widespread and well-studied of the blood flukes infecting humans. Several other digenetic trematodes also use *Biomphalaria* for their larval development. The *Biomphalaria glabrata–S. mansoni* system has become the most intensively studied model for investigating the cellular and molecular basis of snail susceptibility and

ZOOLOGICAL SYMPOSIUM No. 56
ISBN 0-19-854004-3

resistance and trematode infectivity. This endeavour has been greatly facilitated by the development of several inbred strains of snail and schistosome. These strains are excellent experimental subjects for the immunologist interested in the mechanistic basis of host resistance and in the tactics used by larval flukes in subverting the gastropod internal defence system.

The benefits that accrue from such studies are many. They serve to heighten our awareness that trematode transmission in nature may depend critically on the degree of compatibility of local snail and trematode strains. Also, characterization of the genes and their products involved in resistance could be exploited for development of exquisitely specific methods of control. Studies with *Biomphalaria* will serve to illuminate the phenomenon of strict host specificity that typifies most trematode-gastropod relationships. Finally, the study of flukes and snails can provide insights into the molecular basis of 'self-non-self' recognition in invertebrates, and parasite-induced suppression and enhancement of invertebrate immune responses.

The snail species referred to throughout this review is *B. glabrata*, with specific inbred strains indicated where possible. The Puerto Rican 1 and 2 strains of *S. mansoni*, designated PR-1 and PR-2, are also discussed extensively. The derivation of host and parasite strains is described by Richards & Merritt (1972) and Richards (1975a,b,c). Also considered are several echinostome flukes, the most important being *Echinostoma paraensei* and *Echinostoma lindoense*. Earlier reviews by Basch (1976), Lie (1982) and Bayne (1983) provide important details and perspectives for this paper.

The internal defence system of *Biomphalaria*—general characteristics

Cellular components

Sminia (1981) reviewed the available literature on gastropod haematology and discussed the divergent opinions regarding the number of haemocyte types in snails. An update, as it relates to *Biomphalaria*, is pertinent in light of recent studies.

Two haemocyte types, granulocytes and hyalinocytes, have been described from *B. glabrata* blood (Cheng 1975; Cheng & Auld 1977; Schoenberg & Cheng 1980, 1981). Granulocytes, also referred to as type A cells, generally comprise more than 90% of all the haemocytes, and spread rapidly on glass (Schoenberg & Cheng 1980, 1981). As originally defined, *B. glabrata* hyalinocytes are spherical cells which attach to glass but do not spread appreciably (Cheng 1975). Such cells are comparable to electron-dense 'round cells' described by Sminia & Barendsen (1980).

In a later, modified description, hyalinocytes were characterized as 'pancake' shaped cells that spread on glass (Schoenberg & Cheng 1981).

Although not directly stated, it is implied that cells fitting this modified hyalinocyte description are the same as the 'type B' cells earlier identified by the same authors (Schoenberg & Cheng 1980). In addition to morphological dissimilarities, type A and B haemocytes respond differently to lectins. Concanavalin A (Con A) causes type A cells to round up, and increases their ability to bind erythrocytes; type B cells are unaffected (Schoenberg & Cheng 1981).

Yoshino & Granath (1985), using a monoclonal antibody as a surface probe, have recently characterized two haemocyte types designated BGH− and BGH+. Cells of the BGH− type closely resemble granulocytes or type A cells. BGH+ cells spread on glass, resemble type B cells, and therefore fit the modified hyalinocyte description of Schoenberg & Cheng (1981). However, Yoshino & Granath (1985) considered BGH+ cells to be a distinct set of granulocytes because BGH+ cells do not conform to the original description of the hyalinocyte as a non-spreading cell (Cheng 1975) and because they comprise a higher percentage of the haemocyte population (29%) than do hyalinocytes (approx. 10%).

'Classic' granulocytes (type A cells, BGH− cells) are the cells most directly involved in destruction of trematode larvae (Cheng & Garrabrant 1977; Loker, Bayne, Buckley & Kruse 1982; Jeong, Lie & Heyneman 1984). Granulocytes are actively phagocytic (Jeong & Heyneman 1976; Sminia & Barendsen 1980), and have pronounced ecto- and endoplasm, extensive rough endoplasmic reticulum, Golgi bodies, and numerous lysosome-like structures, which contain enzymes typically associated with lysosomes (Sminia & Barendsen 1980; Morona, Jourdane & Aeschlimann 1984). Cheng (1983) has reviewed the relevant literature on molluscan haemocyte enzymes.

One enzyme of particular interest is peroxidase, as its presence may indicate the existence of oxygen-dependent cytotoxic mechanisms comparable to those found in vertebrate effector cells (Klebanoff 1982). McKerrow, Jeong & Beckstead (1985) observed no peroxidase activity in *B. glabrata* haemocytes but other authors have noted that at least 50% of such cells are peroxidase-positive (Carter & Bogitsh 1975; Sminia & Barendsen 1980; Granath & Yoshino 1983a).

Although we know little about *Biomphalaria* haemocyte metabolism, one study indicates that the energy for phagocytosis is derived from glycolysis (Abdul-Salam & Michelson 1980a). Cheng (1976) observed that *Mercenaria* haemocytes actively phagocytosing bacteria exhibited no metabolic burst and lacked a peroxidase-hydrogen peroxide-halide antimicrobial system. Particularly in view of the progress made in understanding cytotoxic mechanisms in vertebrate effectors, additional study of molluscan haemocyte metabolism is sorely needed.

Progressively more diverse and sophisticated probes have been used to delineate the surface properties of *B. glabrata* haemocytes. Haemocytes specifically bind Con A, wheat germ agglutinin and *Ricinus* agglutinin (Schoenberg & Cheng 1980; Yoshino 1981a, 1983b), and antibodies to human fibronectin, murine Thy-1 antigen, and cell-free M line plasma (Yoshino 1983a,b). Monoclonal antibodies to haemocyte surfaces have revealed the BGH binding site (Yoshino & Granath 1985) and distinguish at least four other antigenically distinct subpopulations of glass-adherent haemocytes (Yoshino & Granath 1983). Heterogeneity among haemocytes has also been reported by Joky, Matricon-Gondran & Benex (1983) who, on the basis of size, morphology, lectin-binding specificities and surface charge, recognize four categories of 'granulocytes.' Quantitative studies of the enzyme content of spread haemocytes have also revealed several discrete haemocyte sub-populations (Granath & Yoshino 1983a).

The factors governing the ability of molluscan haemocytes to adhere to foreign objects, including trematode larvae, are poorly understood. One study indicates that if specific PR-1 *S. mansoni* sporocyst surface antigens are obscured by pretreatment with anti-sporocyst antibodies, *B. glabrata* haemocytes can no longer adhere to the sporocysts. This result suggests that antibody blocks a specific interaction between parasite antigen(s) and antigen binding sites on the surface of haemocytes (Bayne, Loker, Yui & Stephens 1984).

Biomphalaria blood cells originate in a haematopoietic organ which contains a blood sinus that allows proliferating cells direct access to the vena cava and the heart (Lie, Heyneman & Yau 1975). This organ contains clusters of primary and secondary blast cells that divide and eventually differentiate into granulocytes (Jeong, Lie & Heyneman 1983). Also, as dividing blast cells and haemocytes have been observed in the vena cava and the heart, this organ may not be the exclusive site of haematopoiesis (Jeong, Lie *et al.* 1983).

Humoral components

The means whereby gastropod haemocytes recognize a surface as 'non-self' may be dependent upon, or enhanced by, humoral factors. Plasma of some gastropod species contains opsonic substances (Sminia, van der Knaap & Edelenbosch 1979; Renwrantz, Schancke, Harm, Erl, Liebsch & Gercken 1981) but comparable substances have not yet been found in *B. glabrata* blood (Abdul Salam & Michelson 1980b). However, a number of agglutinins have been reported from *Biomphalaria*, the known properties of which are reviewed by Boswell & Bayne (1984). Although some molluscan agglutinins have been shown to be opsonins (van der Knaap, Doderer, Boerrigter-Barendsen & Sminia 1982; Renwrantz & Stahmer 1983), and to be intrinsic components of haemocyte plasma membranes (Vasta, Sullivan,

Cheng, Marchalonis & Warr 1982; Renwrantz & Stahmer 1983), no comparable agglutinins have yet been identified in *Biomphalaria*.

Humoral factors associated with trematode infections will be discussed later in this paper.

Trematode infection in *Biomphalaria*

Possible outcomes

The successful establishment of a larval trematode infection is dependent on the susceptibility of the host and on parasite infectivity (Richards 1975a,b; Basch 1975; Wright & Southgate 1976; Michelson & DuBois 1978). Both characteristics are under genetic control and exhibit considerable variability, resulting in potentially diverse interactions. Some of the phenomena noted when trematode miracidia penetrate *Biomphalaria* snails are:

1. *Unsuitability*. In snails of the 13141 strain exposed to PR-2 miracidia, sporocysts fail to develop, even though only a slight host response is mounted. Prior immunosuppression of 13141 snails does not enhance the ability of PR-2 sporocysts to develop, suggesting these snails are physiologically unsuitable for this parasite strain (Sullivan & Richards 1981; Sullivan, Richards, Lie & Heyneman 1981).

2. *Susceptibility*. PR-1 develops normally, without delay, in M line snails. In the absence of a conspicuous immune response, such hosts are considered to be susceptible to infection (Newton 1955; Richards & Merritt 1972). In some susceptible hosts, such as the 13152-142 strain, PR-1 sporocysts remain undeveloped for several months, but eventually resume normal development that culminates in cercaria production (Richards 1984).

3. *Resistance*. Within hours after penetration of 10-R2 or 13-16-R1 snails by PR-1 miracidia, the parasites are encapsulated and destroyed. Interference with the internal defence systems of these snails permits PR-1 sporocysts to develop, indicating these snails are physiologically suitable, but resistant (Lie, Heyneman & Richards 1977b). In some combinations, such as PR-2 *S. mansoni* and 93375, M line or 243432 *B. glabrata*, sporocysts are surrounded by thin haemocyte capsules but may persist for as long as three weeks; no cercariae are produced. Such hosts are known to be suitable, but express resistance characterized by minimal cellular reaction to the sporocyst. Unidentified humoral factors may also play a role in suppressing sporocyst development (Sullivan & Richards 1981; Sullivan, Richards *et al.* 1981).

4. *Interference*. If 10-R2 snails are first exposed to *E. paraensei*, their natural resistance to *S. mansoni* is impaired, allowing development of schistosome sporocysts (Lie, Heyneman & Richards 1977a,b). This suppressive ability has been termed interference (Lie & Heyneman 1976b; Lie 1982).

5. *Enhanced resistance.* By exposing 442132 strain snails to irradiation-attenuated PR-1 miracidia, the level of innate resistance can be boosted, so that the snails destroy normal, challenge miracidia (Lie, Jeong & Heyneman 1983).

6. *Self-cure.* Primary and secondary PR-1 sporocysts, permitted to develop in 10-R2 snails as a result of echinostome-mediated interference, may later be encapsulated and destroyed. This 'self-cure' may occur any time after the attenuated echinostomes have died, but may be delayed until after *S. mansoni* cercariae have been produced (Lie, Jeong & Heyneman 1980a).

Factors modifying outcomes

The actual path followed in any one infection can be markedly influenced by the age of the snail at the time of exposure (Richards 1984). Snails non-susceptible as juveniles remain so throughout their lives, whereas snails susceptible as juveniles may become non-susceptible at maturity. The susceptibility status of juvenile snails is determined by a complex of four or more genetic factors, the interactions of which are not clearly understood (Richards & Merritt 1972). Adult non-susceptibility is in some cases determined by a single gene, possibly with several alleles, with non-susceptibility dominant (Richards 1973, 1984). In other cases, adult non-susceptibility appears to involve additional genes, or incomplete dominance (Richards 1984). Also, snails susceptible as juveniles sometimes become non-susceptible as young adults and revert to susceptibility in old age (Richards 1973, 1977).

The fate of individual primary sporocysts can be influenced also by the number of other sporocysts simultaneously present in the snail. In some susceptible snails exposed to many schistosome miracidia, only four to five primary sporocysts develop normally; the remainder are destroyed by encapsulation responses. Exposure of control snails to monomiracidial infections indicates that most miracidia are potentially infective (Richards 1975c; Kassim & Richards 1979). Using different strains, Sullivan & Richards (1981) noted no obvious correlation between the number of miracidia penetrating and the proportion of sporocysts killed. These authors speculated that different snail genotypes have characteristic 'thresholds' for specific schistosome strains.

Infection of the susceptible host

Why do some snails support normal trematode development without mounting an efficacious immune response? Do such snails have a generalized defect in their internal defence system? Studies of *B. glabrata* show this is not so, insofar as some host strains are susceptible to one *S. mansoni* strain but resistant to others (Kassim & Richards 1979). Also, snails susceptible as juveniles frequently exhibit strong resistance as adults (Richards 1984). Finally, by exposing haemocytes from susceptible M line snails to Con A-treated PR-1 sporocysts (Boswell & Bayne 1985), or to plasma factors from resistant hosts (Bayne, Buckley & DeWan 1980b; Loker & Bayne 1982; Granath & Yoshino 1984), the cytotoxic potential of these haemocytes for *S. mansoni* sporocysts can be demonstrated. Susceptibility is a specific phenomenon, exhibited by otherwise immunocompetent hosts.

Responses of susceptible hosts to infection

Although unable to kill a compatible parasite, the susceptible snail apparently does respond to infection, at least in some cases. For example, upon infection of *B. glabrata* with compatible strains of *S. mansoni* or *E. paraensei*, the haematopoietic organ exhibits markedly enhanced mitotic activity (Sullivan, Cheng & Howland 1984). In M line snails harbouring viable *E. paraensei* larvae, the number of circulating haemocytes is significantly increased by two days post-exposure (dpe), and remains so until 15 dpe (Loker in preparation).

Stumpf & Gilbertson (1978, 1980) and Granath & Yoshino (1983b) note that in compatible *B. glabrata*, *S. mansoni* elicits a transient leucocytosis that subsides within 24 hpe. Haemocyte subpopulations delineated on the basis of enzyme content remain largely unchanged in M line snails infected with *S. mansoni* (Granath & Yoshino 1983b). Abdul-Salam & Michelson (1980c) reported that haemocyte counts in *S. mansoni*-infected snails rise steadily during the course of infection; by four to six weeks post-infection, counts are significantly elevated. Haemocytes from infected snails also exhibit qualitative differences; they are more likely to be clumped, and to cover larger areas when spread on glass (Abdul-Salam & Michelson 1980c).

Non-cellular components of the blood of susceptible snails are also altered upon infection. Haemolymph of echinostome-infected snails contains ‘miracidia-immobilizing substance,’ which is essentially absent from the blood of uninfected or *S. mansoni*-infected snails (Lie, Jeong & Heyneman 1980b). Within minutes of being placed in haemolymph containing this substance, echinostome (but not *S. mansoni*) miracidia exhibit slowed ciliary and swimming activity. Their cilia become clumped by a granular substance (Lie, Jeong *et al.* 1980b). A particulate material has been observed with phase contrast optics in the blood of echinostome-infected snails (Loker in preparation). This material is in some cases so abundant as to obscure spread haemocytes. As with the immobilizing substance described by Lie, Jeong *et al.* (1980b), this material is first evident at 1 dpe, reaches its maximum 10–15 dpe, and declines in abundance levels thereafter.

Blood of snails with echinostome infections also contains elevated titres of an agglutinin for calf erythrocytes (Jeong, Sussman, Rosen, Lie & Heyneman 1981). Significantly elevated titres do not appear until 21 dpe, suggesting this agglutinin is different from miracidia-immobilizing substance (Jeong, Sussman *et al.* 1981). Although *S. mansoni* infection is known to alter the composition of *B. glabrata* plasma (e.g. Bayne 1980), the functional relevance of the affected molecules is unknown.

Trematode evasion of host responses

How do compatible trematodes evade destruction in what is otherwise an immunologically competent host? One hypothesis, referred to as 'molecular mimicry', postulates that trematode larvae produce molecules that mimic those of the host and thereby evade detection as 'non-self' (e.g. Yoshino & Cheng 1978; Yoshino & Bayne 1983). According to a second hypothesis, trematode larvae acquire host antigens and become invested in a disguise of host molecules, as is believed to occur with adult schistosomes (McLaren 1984). These two hypotheses are discussed in this volume by Yoshino & Boswell (this volume p. 221). An essential element of both is parasite avoidance of recognition as non-self.

A third hypothesis, the interference hypothesis, postulates an aggressive role for the trematode, which actively suppresses the internal defence system of the host. Lie (1982) has provided an authoritative account of interference, so only selected details will be presented here. Although the molecular basis for this phenomenon has yet to be elucidated, apparently the larval stages of some, and perhaps all, trematodes secrete substances that become systemically distributed in the snail and selectively interfere with haemocyte function (Lie & Heyneman 1976b; Lie, Heyneman & Jeong 1976; Lie & Heyneman 1979b; Lie 1982).

Presence of even a single echinostome sporocyst can result in substantial interference (Lie, Heyneman & Jeong 1976), but the effect increases with the number of sporocysts present (Lie & Heyneman 1979b). One consequence is that the infected host's responsiveness to other trematodes is also drastically reduced (Lie, Heyneman & Richards 1977a,b). Both juvenile and adult innate resistance can be diminished (Lie, Heyneman & Richards 1977a), as can the resistance provoked by exposing snails to irradiated miracidia (Lie & Heyneman 1977).

In vitro studies of this phenomenon have now been undertaken. Approximately 25% of PR-1 sporocysts exposed *in vitro* to haemolymph from *E. paraensei*-infected 10-R2 snails are killed during a 24-h incubation period, as compared to a 90% killing rate in control tubes containing normal 10-R2 haemolymph (Loker, Bayne & Yui in press). In the presence of plasma from echinostome-infected snails, haemocytes from uninfected 10-R2 snails still kill PR-1 sporocysts, whereas haemocytes from infected snails when co-incubated with plasma from uninfected snails have depressed killing ability. The results imply haemocytes are the targets of echinostome-induced interference.

Inclusion of viable *E. paraensei* daughter rediae in tubes containing PR-1 sporocysts and normal 10-R2 haemolymph also diminishes sporocyst killing, suggesting that rediae secrete substances that can demonstrably affect haemocytes within the 24-h incubation period of the experiment (Loker, Bayne & Yui in press).

Haemocyte counts in *E. paraensei*-infected snails are elevated relative to controls, indicating that interference is not mediated by a parasite-induced leucopaenia. Also, haemocytes of infected snails retain the ability to encapsulate latex spheres and nematode larvae, to phagocytose cast miracidial plates, and to repair wounds (Lie, Jeong & Heyneman 1981). The selective nature of interference is in the parasite's best interest as a total abrogation of immune competence would increase the host's vulnerability to other pathogens (Lie 1982).

Abdul-Salam & Michelson (1980c) report that haemocytes from snails infected for four to six weeks with *S. mansoni* have significantly reduced ability to phagocytose sheep red blood cells. This may indicate that interference is not always strictly selective, or that snails with older infections suffer a general physiological deterioration that affects haemocytes.

According to Lie (1982), the interfering capability of the trematode relative to the level of natural resistance expressed by the host will determine the outcome of the specific host-parasite encounter. Even in compatible associations such an interplay would occur. Susceptibility therefore does not necessarily indicate a lack of resistance, or of ability to recognize the parasite's presence. Rather, it implies the snail has a low degree of innate resistance relative to the interfering capacity of the specific parasite to which it is susceptible (Lie 1982).

The principal question posed by the interference hypothesis relates to the length of time required by some trematodes to effectively interfere with haemocyte function. Insofar as haemocytes can attack sporocysts within a few hours post-exposure, it would seem that production of interfering factors would have to begin immediately after penetration. For some trematodes such as *E. paraensei*, this apparently happens (Lie, Heyneman & Richards 1977a), but for others, such as *S. mansoni*, it takes at least three days to observe fully developed, systemic interference (Lie, Heyneman & Jeong 1976). Do *S. mansoni* sporocysts rely on a fundamentally different evasive strategy in the meantime?

Lie (1982) notes that development of Lc-1 *S. mansoni* sporocysts is better in the immediate surroundings of the protective PR-1 sporocyst than at a distance from it, yet Kassim & Richards (1979) observe that double infection of M line snails with these two strains results in selective destruction of Lc-1 sporocysts. They provide photographs of encapsulated sporocysts lying adjacent to normal sporocysts. Clearly, comparative studies of the secretions of trematode sporocysts and of the temporal pattern of their production are needed.

Infection of the resistant host

General characteristics

Haemocyte responses

Upon infection of a resistant host, a trematode larva typically is surrounded by granulocytic haemocytes which cohere within hours to form a multilayered capsule. Similar encapsulation responses can also be observed in *in vitro* cultures containing haemocytes and sporocysts (Fig. 1). Phagocytosis of portions of the larval tegument is a prominent feature of encapsulation (Cheng & Garrabrant 1977; Lie, Jeong *et al.* 1980a; Loker, Bayne, Buckley *et al.* 1982; Jeong, Lie *et al.* 1984). It is not known if extracellular release of lysosomal enzymes or other lytic factors plays a prominent role in the demise of the parasite. The encapsulated sporocyst is typically destroyed in two to four days, but this rate varies considerably, presumably depending on the evasive capacity of the parasite and the host's level of resistance (Lie 1982).

In snails exposed to attenuated echinostome sporocysts, in concert with encapsulation, the haematopoietic organ becomes hyperplastic within 1 dpe, and a marked leucocytosis ensues. This response subsides one to five days after the encapsulated parasites are destroyed in the snail's heart (Jeong, Lie & Heyneman 1980; Joky, Matricon-Gondran & Benex 1985).

PR-2 sporocysts are slowly encapsulated and destroyed in M line snails, and elicit a haematopoietic response in the process (Sullivan, Cheng *et al.* 1984), yet 10-R2 snails infected with PR-1 *S. mansoni* exhibit a decrease in circulating haemocyte numbers at 1 hpe, followed by a gradual increase in numbers to control levels (Granath & Yoshino 1983b). Haemocyte subpopulations and plasma acid phosphatase content of 10-R2 snails fluctuate considerably after exposure to *S. mansoni*, whereas M line haemolymph shows no comparable response (Granath & Yoshino 1983b).

The changes observed in 10-R2 haemocytes may represent a trematode-induced 'activation' (Granath & Yoshino 1983b), a phenomenon possibly analogous to that exhibited by vertebrate macrophages. Jeong, Lie *et al.* (1983) note that haemocytes in activated haematopoietic organs are significantly larger than those seen in normal organs, a property also characteristic

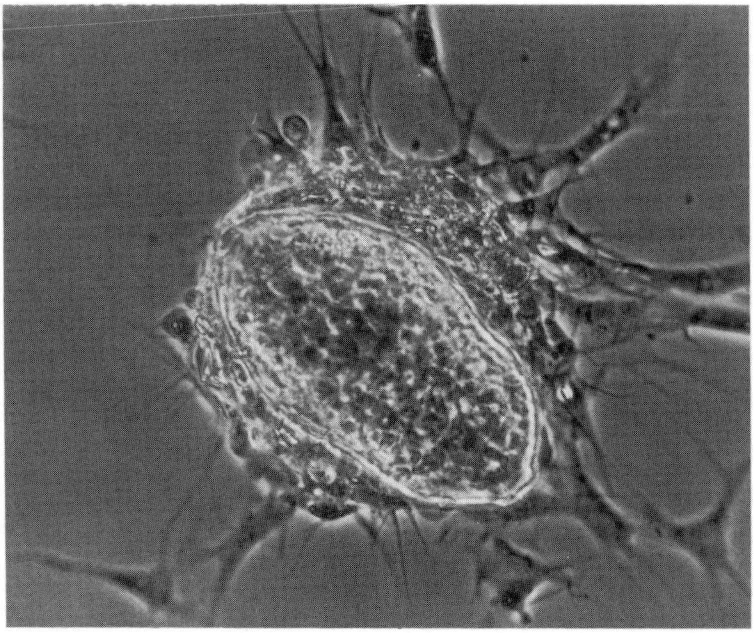

Fig. 1. *In vitro* encapsulation of PR–1 *Schistosoma mansoni* sporocysts by haemocytes of the 10–R2 strain of *Biomphalaria glabrata*. × 500, phase contrast. (From Bayne, Loker, Yui & Stephens 1984.)

of an activated state. Similarly, the enhanced spreading capacity of haemocytes from *S. mansoni*-infected snails may be indicative of an activated state (Abdul-Salam & Michelson 1980c).

Enhanced resistance

In host-parasite combinations in which low levels of either juvenile or adult resistance are expressed, it is possible, by using appropriate exposure methods, to boost the degree of resistance. This phenomenon has been termed 'acquired resistance', but perhaps more accurately could be considered 'enhanced resistance' (Lie, Jeong *et al.* 1983). For example, juvenile *B. glabrata* exposed to irradiated *E. paraensei* miracidia destroy such parasites in ventricular capsules within 10 days, and in the process become fully resistant to subsequent infections with normal miracidia (Lie, Heyneman & Lim 1975). Whereas virtually all juvenile snails exposed to irradiated miracidia develop enhanced resistance, only approximately 10% of juveniles sensitized with normal miracidia do so (Lie & Heyneman 1975); such snails apparently express a high level of innate resistance and destroy even non-irradiated parasites. A single parasite is sufficient to induce enhanced resistance (Lie, Jeong & Heyneman 1982). In either case, as a prerequisite for induction and retention of enhanced resistance, all sporocysts of the sensitizing exposure must be destroyed (Lie & Heyneman 1979a). The faster the encapsulation of irradiated miracidia takes place, the more the resistance is enhanced (Lie & Heyneman 1979a).

The resistance induced by irradiated parasites can persist for at least three weeks (Sullivan, Richards, Lie & Heyneman 1982). It may persist until snails subsequently express the adult resistance ordinarily apparent in this host-parasite combination (Lie & Heyneman 1975; Lie, Jeong *et al.* 1982, 1983). Attempts to induce resistance in snails susceptible as adults have not been successful (Lie, Jeong *et al.* 1983), suggesting that the ontogenesis of resistance is somehow hastened by exposure to irradiated parasites.

When snails with enhanced resistance are re-exposed to irradiated miracidia, a more rapid increase in the number of circulating haemocytes is observed than after the initial exposure (Jeong, Lie *et al.* 1980). There is also a stronger suppression of parasite migratory capacity and a greater tendency of haemocytes to flatten while attacking parasites (Lie & Heyneman 1976a).

In some cases, enhanced resistance is strongest when challenge is by the homologous parasite (Lie, Jeong *et al.* 1982), but in other cases, it is best expressed against a heterologous, but closely related echinostome species. Lie, Jeong *et al.* (1982) conclude that although the response of snails with enhanced resistance to a challenge exposure is relatively specific, the stimuli for inducing that resistance may be relatively non-specific. Generally, enhanced resistance induced by echinostomes fails to protect snails from *S. mansoni* (Lie & Heyneman 1975), although Sullivan, Richards *et al.* (1982) noted that resistance induced by the psilostomatid *Ribeiroia marini* provided a slight, but significant, degree of protection against *S. mansoni*.

Unlike results obtained with echinostomes, induction of enhanced resistance by exposure to irradiated *S. mansoni* miracidia has been difficult to demonstrate and has succeeded only in juvenile 442132, a strain that exhibits strong adult resistance to PR-1 (Lie, Jeong *et al.* 1983). Even in this case, the rate at which irradiated PR-1 sporocysts are destroyed is slower, and the degree to which resistance is enhanced is lower than is the case with echinostomes.

In another attempt to induce resistance to *S. mansoni*, MRLc snails were first exposed to normal miracidia of incompatible Lc-1 *S. mansoni* which were encapsulated within two to three days. Despite this, such snails remained fully susceptible to PR-1 (Lie, Heyneman & Richards 1979; Kassim & Richards 1979). Again, the inability of MRLc to exhibit enhanced resistance to PR-1 may be fundamentally related to the complete lack of innate resistance, even in adult snails, to this parasite strain (Lie, Jeong *et al.* 1983).

Although we have much to learn about resistance and how it is boosted, it seems clear that it exhibits specificity. This is implied by the above-mentioned studies with MRLc snails, and by study of certain 10-R2 stocks which, for reasons that are not clear, lose their resistance to PR-1 (Lie, Heyneman & Richards 1979). Nonetheless, these snails retain unaltered resistance to Lc-1 *S. mansoni*, and juvenile susceptibility-adult resistance to echinostomes, suggesting that resistance to different parasites may be controlled by different genes (Lie, Heyneman & Richards 1979).

Mechanisms of resistance

Comparison of susceptible and resistant host strains

The molecular basis of resistance in *Biomphalaria*, or in any other molluscan species, is unknown. Currently, studies of this subject focus on the 10-R2 and M line strains of *B. glabrata*. At all ages, the former strain is resistant and the latter susceptible to PR-1 *S. mansoni* (Richards & Merritt 1972).

Although the plasmas of M line and 10-R2 snails exhibit differences in protein composition (e.g. Bayne 1980), both allow survival of PR-1 sporocysts for at least four days (Bayne, Buckley & DeWan 1980a; LoVerde, Shoulberg & Gherson 1984), implying that resistance cannot be attributed solely to constitutive differences in levels of soluble plasma factors. Haemocytes from the two strains do clearly differ in killing ability *in vitro* because, when plasma is replaced by culture medium, 10-R2 haemocytes still kill PR-1 sporocysts whereas M line haemocytes do not (Bayne, Buckley *et al.* 1980a).

The lectin-binding specificities of M line and 10-R2 haemocytes differ only slightly (Schoenberg & Cheng 1980). Haemocytes from both strains exhibit patch and cap formation upon exposure to Con A (Yoshino 1981b), and bind a polyvalent anti-M line plasma antiserum (Yoshino 1983a,b). One of five monoclonal antibodies raised against M line haemocytes revealed significant differences in binding patterns between M line and 10-R2 haemocytes; it binds 74.2% of the former and 43.7% of the latter (Yoshino & Granath 1983). Granath & Yoshino (1983a) note that 83.7% of 10-R2 haemocytes are positive for acid phosphatase and 86.3% are positive for non-specific esterases; the comparable figures for M line haemocytes are 75.4% and 79.9%, respectively. Also, 10-R2 snails have an average of 221 ± 25.2 haemocytes/mm^3 as compared to 112.1 ± 12.3 for M line snails (Granath & Yoshino 1983a). These authors suggest that the greater number and higher enzymatic content of 10-R2 haemocytes may play a role in innate resistance to *S. mansoni*.

Employing a different monoclonal antibody, Yoshino & Granath (1985) note that 40% of M line haemocytes and only 10% of 10-R2 haemocytes are positive for the BGH surface marker. Because BGH+ cells have lower acid phosphatase content and lower phagocytic capability, they conclude that M line snails are potentially less able to mount haemocyte responses to foreign objects. Subsequently, it was noted that treatment of M line snails with urethane, a molluscan anaesthetic, induces a temporary three-fold increase in circulating haemocytes, increases the prevalence of acid phosphatase-positive cells, and decreases the percentage of BGH+ haemocytes (Granath & Yoshino 1985). Such snails remain fully susceptible to *S. mansoni*, prompting the conclusion that induction of resistance involves more than acquisition of lysosomal enzymes. These authors also observe that factors other than the relative numbers of BGH− haemocytes determine the relative susceptibility of *B. glabrata* to *S. mansoni*.

The role of plasma factors

Soluble factors in *B. glabrata* blood may influence the parasite's chances of survival. Although sporocysts are not killed by *in vitro* exposure to plasma from naive, resistant hosts, production of toxic factors may be induced by infection. This could explain the *in vivo* histological observations of Sullivan & Richards (1981) and of Jourdane (1982) who noted *S. mansoni* sporocyst degeneration without obvious participation of haemocytes.

Also, exposure of PR-1 sporocysts to M line haemocytes and 10-R2 plasma results in a degree of sporocyst damage comparable to that produced by 10-R2 haemocytes and plasma (Bayne, Buckley *et al.* 1980b). Sporocysts held for a comparable 24-h period with M line haemocytes and plasma are not damaged. Loker & Bayne (1982), after further study, noted that for sporocysts co-incubated for 24 h with plasma from resistant 10-R2 or 13-16-R1 strains and M line haemocytes, the predominant trend was sporocyst destruction in five replicates and lack of substantial damage in six. Of a total of 222 sporocysts subjected to this treatment, 64 (28.8%) became abnormal in appearance. Sporocysts pre-incubated in 10-R2 plasma for up to 6 h and then transferred to assay tubes containing M line haemocytes and plasma were not destroyed after a 24-h incubation period (Loker & Bayne 1982). The pattern of results suggest that, to be effective, the 10-R2 factor(s) must be continuously present with haemocytes and sporocysts during the 24-h incubation.

Using similar *in vitro* methods but a different susceptible host strain (the 641 strain), LoVerde *et al.* (1984) noted that 641 haemocytes do not kill sporocysts in the presence of 10-R2 plasma. However, a strict comparison of results is not possible because, unlike M line haemocytes, 641 cells will not bind to sporocysts in the presence of homologous plasma. Interestingly, exposure of 641 haemocytes to 10-R2 plasma promotes adherence to sporocysts (LoVerde *et al.* 1984).

Independent confirmation of the existence of soluble factors in 10-R2 blood that can mediate sporocyst destruction by M line haemocytes is provided by Granath & Yoshino (1984). Injection of 10-R2 plasma into M line snails induces complete protection from a primary infection with PR-1 *S. mansoni* in 60.4% of the snails. Inoculation of homologous M line plasma has no such effect. However, when sporocysts pre-incubated in 10-R2 or M line plasma for 1 h are injected into M line snails, more than 80% of the snails of either group become infected (Granath & Yoshino 1984).

One hypothesis consistent with both *in vitro* and *in vivo* results is that 10-R2 plasma possesses factors that activate M line haemocytes to destroy sporocysts. This response can be elicited by providing 10-R2 plasma to M line haemocytes as they encounter sporocysts, but not by presenting haemocytes with sporocysts pre-incubated in resistant plasma. This phenomenon may be directly comparable to the development of enhanced resistance resulting from exposure of snails to irradiated miracidia.

Subsequent to these studies, Loker, Yui & Bayne (1984) reported that glutaraldehyde-fixed PR-1 sporocysts become agglutinated in the plasma of resistant 10-R2 and 13-16-1 strains, but not in plasma from susceptible M line, MRLc, or L-311 strains. Living sporocysts are readily clumped in the plasma of all five strains. The observed agglutination of living sporocysts may be indicative of a nutritive response, or of modulation of the sporocyst tegument upon exposure to host body fluids. That the sporocyst surface is capable of modulation, upon exposure to host plasma or to Con A, is implied by experiments of Bayne, Loker & Yui (in press) and Boswell & Bayne (1986), respectively. Agglutination of metabolically inert, fixed sporocysts suggests that the plasma of resistant hosts may contain multivalent binding factors with specificity for sporocyst surface receptors (Loker, Yui *et al.* 1984).

Western blot analysis of proteins extracted from fixed sporocysts that have been incubated in plasma reveals several snail proteins common to both strains, and at least one protein which is more concentrated in the extract of sporocysts held in resistant plasma (Bayne, Loker *et al.* in press). If an opsonin-like molecule is present in resistant plasma, why does pre-incubation in resistant plasma fail to facilitate sporocyst killing when sporocysts are later transferred to M line haemolymph? Although the answer is far from clear, one possibility is that the 10-R2 factor is rapidly sloughed or degraded once bound to the living tegument. Our immunoblot experiments suggest that most antigens acquired from snail plasma are no longer detectable in tegumental extracts prepared 12 h after transfer of sporocysts to culture medium (Bayne, Loker *et al.* in press).

Concluding remarks

It is apparent that many fundamental questions remain regarding the immune response of *Biomphalaria* and other gastropods to digenetic trematodes. Given that different parasites may evoke different resistance mechanisms in the same host (Lie, Heyneman & Richards 1979), we may also speculate that different parasite species or strains rely on diverse evasive tactics. Differences between echinostomes and schistosomes with respect to the degree of host specificity, interference capability and ease with which resistance can be artificially enhanced, may be indications of divergent evasive strategies.

Results to date imply that larval trematode secretions can both stimulate haemocyte production and induce specific immunosuppression. An important priority for future research is to identify the responsible molecules and to elucidate their specific effects on haemocytes. *S. mansoni* antigen has been detected on the surface of, and within, haemocytes from infected snails (Abdul-Salam & Michelson 1983), raising the possibility that trematode antigens block or alter haemocyte receptor molecules.

The molecular basis of resistance also awaits clarification. Although we are conditioned to believe that resistant animals possess specific opsonins or haemocyte activators that susceptible hosts lack, they may simply be relatively unaffected by suppressive factors released by the parasite.

Finally, most current studies of gastropod-trematode immunobiology have used inbred laboratory strains representing a limited number of species. Additional study of a diversity of snail and fluke species freshly collected from natural habitats is needed to confirm the validity and generality of our laboratory models. In any case, the long co-evolutionary interplay between trematode and gastropod, the legacy of which is a remarkable degree of host specificity, will continue to provide us with new challenges and insights for many years to come.

References

Abdul-Salam, J.M. & Michelson, E.H. (1980a). *Biomphalaria glabrata* amoebocytes: assay of factors influencing *in vitro* phagocytosis. *J. Invert. Path.* **36**: 52–9.

Abdul-Salam, J.M. & Michelson, E.H. (1980b). Phagocytosis by amoebocytes of *Biomphalaria glabrata*: absence of opsonic factor. *Malac. Rev.* **13**: 81–3.

Abdul-Salam, J.M. & Michelson, E.H. (1980c). *Biomphalaria glabrata* amoebocytes: effects of *Schistosoma mansoni* infection on *in vitro* phagocytosis. *J. Invert. Path.* **35**: 241–8.

Abdul-Salam, J.M. & Michelson, E.H. (1983). *Schistosoma mansoni*: Immunofluorescent detection of its antigen reacting with *Biomphalaria glabrata* amoebocytes. *Expl Parasit.* **55**: 132–7.

Basch, P.F. (1975). An interpretation of snail-trematode infection rates: specificity based on concordance of compatible phenotypes. *Int. J. Parasit.* **5**: 449–52.

Basch, P.F. (1976). Intermediate host specificity in *Schistosoma mansoni*. *Expl Parasit.* **39**: 150–69.

Bayne, C.J. (1980). Humoral factors in molluscan parasite immunity. In *Aspects of developmental and comparative immunology* **1**: 113–24. (Ed. Solomon, J.B.). Pergamon Press, Oxford.

Bayne, C.J. (1983). Molluscan immunobiology. In *The Mollusca* **5**: 407–86. (Eds Saleuddin, A.S.M. & Wilbur, K.M.). Academic Press, San Diego.

Bayne, C.J., Buckley, P.M. & DeWan, P.C. (1980a). Macrophagelike hemocytes of resistant *Biomphalaria glabrata* are cytotoxic for sporocysts of *Schistosoma mansoni in vitro*. *J. Parasit.* **66**: 413–9.

Bayne, C.J., Buckley, P.M. & DeWan, P.C. (1980b). *Schistosoma mansoni*: cytotoxicity of hemocytes from susceptible snail hosts for sporocysts in plasma from resistant *Biomphalaria glabrata*. *Expl Parasit.* **50**: 409–16.

Bayne, C.J., Loker, E.S. & Yui, M.A. (In press). Interactions between the plasma proteins of *Biomphalaria glabrata* (Gastropoda) and the sporocyst tegument of *Schistosoma mansoni* (Trematoda). *Parasitology*.

Bayne, C.J., Loker, E.S., Yui, M.A. & Stephens, J.A. (1984). Immune-recognition of *Schistosoma mansoni* primary sporocysts may require specific receptors on *Biomphalaria glabrata* hemocytes. *Parasite Immunol.* **6**: 519–28.

Boswell, C.A. & Bayne, C.J. (1984). Isolation, characterization and functional assessment of a hemagglutinin from the plasma of *Biomphalaria glabrata*, intermediate host of *Schistosoma mansoni*. *Devl comp. Immunol.* **8**: 559–68.

Boswell, C.A. & Bayne, C.J. (1985). *Schistosoma mansoni*: lectin-dependent cytotoxicity of hemocytes from susceptible host snails, *Biomphalaria glabrata*. *Expl Parasit.* **60**: 133–8.

Boswell, C.A. & Bayne, C.J. (1986). Lectin-dependent cell-mediated cytotoxicity in an invertebrate model: Con A does not act as a bridge. *Immunology* **57**: 261–4.

Carter, O.S. & Bogitsh, B.J. (1975). Histologic and cytochemical observations of the effects of *Schistosoma mansoni* on *Biomphalaria glabrata*. *Ann. N. Y. Acad. Sci.* **266**: 380–93.

Cheng, T.C. (1975). Functional morphology and biochemistry of molluscan phagocytes. *Ann. N. Y. Acad. Sci.* **266**: 343–79.

Cheng, T.C. (1976). Aspects of substrate utilization and energy requirement during molluscan phagocytosis. *J. Invert. Path.* **27**: 263–8.

Cheng, T.C. (1983). The role of lysosomes in molluscan inflammation. *Am. Zool.* **23**: 129–44.

Cheng, T.C. & Auld, K.R. (1977). Hemocytes of the pulmonate gastropod *Biomphalaria glabrata*. *J. Invert. Path.* **30**: 119–22.

Cheng, T.C. & Garrabrant, T.A. (1977). Acid phosphatase in granulocytic capsules formed in strains of *Biomphalaria glabrata* totally and partially resistant to *Schistosoma mansoni*. *Int. J. Parasit.* **7**: 467–72.

Granath, W.O., Jr. & Yoshino, T.P. (1983a). Characterization of molluscan phagocyte subpopulations based on lysosomal enzyme markers. *J. exp. Zool.* **226**: 205–10.

Granath, W.O., Jr. & Yoshino, T.P. (1983b). Lysosomal enzyme activities in susceptible and refractory strains of *Biomphalaria glabrata* during the course of infection with *Schistosoma mansoni*. *J. Parasit.* **69**: 1018–26.

Granath, W.O., Jr. & Yoshino, T.P. (1984). *Schistosoma mansoni*: passive transfer of resistance by serum in the vector snail, *Biomphalaria glabrata. Expl Parasit.* **58**: 188–93.

Granath, W.O., Jr. & Yoshino, T.P. (1985). *Biomphalaria glabrata* (Gastropoda): effect of urethane on the morphology and function of hemocytes, and on susceptibility to *Schistosoma mansoni* (Trematoda). *J. Invert. Path.* **45**: 324–330.

Jeong, K.H. & Heyneman, D. (1976). Leukocytes of *Biomphalaria glabrata*: morphology and behaviour of granulocytic cells *in vitro. J. Invert. Path.* **28**: 357–62.

Jeong, K.H., Lie, K.J. & Heyneman, D. (1980). Leucocytosis in *Biomphalaria glabrata* sensitized and resensitized to *Echinostoma lindoense. J. Invert. Path.* **35**: 9–13.

Jeong, K.H., Lie, K.J. & Heyneman, D. (1983). The ultrastructure of the amebocyte-producing organ in *Biomphalaria glabrata. Devl comp. Immunol.* **7**: 217–28.

Jeong, K.H., Lie, K.J. & Heyneman, D. (1984). An ultrastructural study on ventricular encapsulation reactions in *Biomphalaria glabrata* exposed to irradiated echinostome parasites. *Int. J. Parasit.* **14**: 127–33.

Jeong, K.H., Sussman, S., Rosen, S.D., Lie, K.J. & Heyneman, D. (1981). Distribution and variation of hemagglutinating activity in the hemolymph of *Biomphalaria glabrata. J. Invert. Path.* **38**: 256–63.

Joky, A., Matricon-Gondran, M. & Benex, J. (1983). Fine structural differences in the amoebocytes of *Biomphalaria glabrata. Devl comp. Immunol.* **7**: 669–72.

Joky, A., Matricon-Gondran, M. & Benex, J. (1985). Response of the amoebocyte-producing organ of sensitized *Biomphalaria glabrata* after exposure to *Echinostoma caproni* miracidia. *J. Invert. Path.* **45**: 28–33.

Jourdane, J. (1982). Studies of mechanisms of elimination in a non-compatible snail-schistosome combination when infected naturally or by microsurgical transplantations of parasitical stages. *Acta trop.* **39**: 325–35. (In French: English summary.)

Kassim, O.O. & Richards, C.S. (1979). Host reactions in *Biomphalaria glabrata* to *Schistosoma mansoni* miracidia, involving variations in parasite strains, numbers and sequence of exposures. *Int. J. Parasit.* **9**: 565–70.

Klebanoff, S.J. (1982). Oxygen-dependent cytotoxic mechanisms in phagocytes. In *Advances in host defense mechanisms* **1**: 111–48. (Eds Gallin, J.I. & Fauci, A.S.). Raven Press, New York.

Lie, K.J. (1982). Survival of *Schistosoma mansoni* and other trematode larvae in the snail *Biomphalaria glabrata.* A discussion of the interference hypothesis. *Trop. geogr. Med.* **34**: 111–22.

Lie, K.J. & Heyneman, D. (1975). Studies on resistance in snails: a specific tissue reaction to *Echinostoma lindoense* in *Biomphalaria glabrata* snails. *Int. J. Parasit.* **5**: 621–5.

Lie, K.J. & Heyneman, D. (1976a). Studies on resistance in snails. 3. Tissue reactions to *Echinostoma lindoense* sporocysts in sensitized and resensitized *Biomphalaria glabrata. J. Parasit.* **62**: 51–8.

Lie, K.J. & Heyneman, D. (1976b). Studies on resistance in snails. 6. Escape of *Echinostoma lindoense* sporocysts from encapsulation in the snail heart and subsequent loss of the host's ability to resist infection by the same parasite. *J. Parasit.* **62**: 298–302.

Lie, K.J. & Heyneman, D. (1977). *Schistosoma mansoni, Echinostoma lindoense,* and *Paryphostomum segregatum*: interference by trematode larvae with acquired resistance in snails, *Biomphalaria glabrata*. *Expl Parasit*. **42**: 343–7.

Lie, K.J. & Heyneman, D. (1979a). Acquired resistance to echinostomes in four *Biomphalaria glabrata* strains. *Int. J. Parasit.* **9**: 533–7.

Lie, K.J. & Heyneman, D. (1979b). Capacity of irradiated echinostome sporocysts to protect *Schistosoma mansoni* in resistant *Biomphalaria glabrata*. *Int. J. Parasit.* **9**: 539–43.

Lie, K.J., Heyneman, D. & Jeong, K.H. (1976). Studies on resistance in snails. 7. Evidence of interference with the defense reaction in *Biomphalaria glabrata* by trematode larvae. *J. Parasit.* **62**: 608–15.

Lie, K.J., Heyneman, D. & Lim, H.K. (1975). Studies on resistance in snails: specific resistance induced by irradiated miracidia of *Echinostoma lindoense* in *Biomphalaria glabrata* snails. *Int. J. Parasit.* **5**: 627–31.

Lie, K.J., Heyneman, D. & Richards, C.S. (1977a). Studies on resistance in snails: interference by nonirradiated echinostome larvae with natural resistance to *Schistosoma mansoni* in *Biomphalaria glabrata*. *J. Invert. Path.* **29**: 118–25.

Lie, K.J., Heyneman, D. & Richards, C.S. (1977b). *Schistosoma mansoni*: temporary reduction of natural resistance in *Biomphalaria glabrata* induced by irradiated miracidia of *Echinostoma paraensei*. *Expl Parasit.* **43**: 54–62.

Lie, K.J., Heyneman, D. & Richards, C.S. (1979). Specificity of natural resistance to trematode infections in *Biomphalaria glabrata*. *Int. J. Parasit.* **9**: 529–31.

Lie, K.J., Heyneman, D. & Yau, P. (1975). The origin of amoebocytes in *Biomphalaria glabrata*. *J. Parasit.* **61**: 574–6.

Lie, K.J., Jeong, K.H. & Heyneman, D. (1980a). Tissue reactions induced by *Schistosoma mansoni* in *Biomphalaria glabrata*. *Ann. trop. Med. Parasit.* **74**: 157–66.

Lie, K.J., Jeong, K.H. & Heyneman, D. (1980b). Inducement of miracidia-immobilizing substance in the hemolymph of *Biomphalaria glabrata*. *Int. J. Parasit.* **10**: 183–8.

Lie, K.J., Jeong, K.H. & Heyneman, D. (1981). Selective interference with granulocyte function induced by *Echinostoma paraensei* (Trematoda) larvae in *Biomphalaria glabrata* (Mollusca). *J. Parasit.* **67**: 790–6.

Lie, K.J., Jeong, K.H. & Heyneman, D. (1982). Further characterization of acquired resistance in *Biomphalaria glabrata*. *J. Parasit.* **68**: 529–31.

Lie, K.J., Jeong, K.H. & Heyneman, D. (1983). Acquired resistance in snails. Induction of resistance to *Schistosoma mansoni* in *Biomphalaria glabrata*. *Int. J. Parasit.* **13**: 301–4.

Loker, E.S. (In preparation). *Alterations in the hemolymph of* Biomphalaria glabrata *infected with* Echinostoma paraensei.

Loker, E.S. & Bayne, C.J. (1982). *In vitro* encounters between *Schistosoma mansoni* primary sporocysts and hemolymph components of susceptible and resistant strains of *Biomphalaria glabrata*. *Am. J. trop. Med. Hyg.* **31**: 999–1005.

Loker, E.S., Bayne, C.J., Buckley, P.M. & Kruse, K.T. (1982). Ultrastructure of encapsulation of *Schistosoma mansoni* mother sporocysts by hemocytes of juveniles of the 10-R2 strain of *Biomphalaria glabrata*. *J. Parasit.* **68**: 84–94.

Loker, E.S., Bayne, C.J. & Yui, M.A. (In press). *Echinostoma paraensei:*

Biomphalaria glabrata hemocytes are targets of echinostome-mediated interference with host snail resistance to *Schistosoma mansoni*. *Expl Parasit.*

Loker, E.S., Yui, M.A. & Bayne, C.J. (1984). *Schistosoma mansoni*: Agglutination of sporocysts, and formation of gels on miracidia transforming in plasma of *Biomphalaria glabrata*. *Expl Parasit.* **58**: 56–62.

LoVerde, P.T., Shoulberg, N. & Gherson, J. (1984). Role of cellular and humoral components in the encapsulation response of *Biomphalaria glabrata* to *Schistosoma mansoni* sporocysts *in vitro*. *Progr. clin. biol. Res.* **157**: 17–29.

McKerrow, J.H., Jeong, K.H. & Beckstead, J.H. (1985). Enzyme histochemical comparison of *Biomphalaria glabrata* amebocytes with human granuloma macrophages. *J. Leukocyte Biol.* **37**: 341–7.

McLaren, D.J. (1984). Disguise as an evasive stratagem of parasitic organisms. *Parasitology* **88**: 597–611.

Michelson, E.A. & DuBois, L. (1978). Susceptibility of Bahian populations of *Biomphalaria glabrata* to an allopatric strain of *Schistosoma mansoni*. *Am. J. trop. Med. Hyg.* **27**: 782–6.

Morona, D., Jourdane, J. & Aeschlimann, A. (1984). Study of the *in vitro* evolution of the hemolymph granulocytes of *Biomphalaria glabrata*. Study by scanning electron microscopy. *Annls Parasit. hum. comp.* **59**: 467–75. (In French.)

Newton, W.L. (1955). The establishment of a strain of *Australorbis glabratus* which combines albinism and high susceptibility to infection with *Schistosoma mansoni*. *J. Parasit.* **41**: 526–8.

Renwrantz, L., Schancke, W., Harm, H., Erl, H., Liebsch, H. & Gercken, J. (1981). Discriminative ability and function of the immunobiological recognition system of the snail *Helix pomatia*. *J. comp. Physiol.* **141**: 477–88.

Renwrantz, L. & Stahmer, A. (1983). Opsonizing properties of an isolated hemolymph agglutinin and demonstration of lectin-like recognition molecules at the surface of hemocytes from *Mytilus edulis*. *J. comp. Physiol.* **149**: 535–46.

Richards, C.S. (1973). Susceptibility of adult *Biomphalaria glabrata* to *Schistosoma mansoni* infection. *Am. J. trop. Med. Hyg.* **22**: 748–56.

Richards, C.S. (1975a). Genetic studies on variation in infectivity of *Schistosoma mansoni*. *J. Parasit.* **61**: 233–6.

Richards, C.S. (1975b). Genetic factors in susceptibility of *Biomphalaria glabrata* for different strains of *Schistosoma mansoni*. *Parasitology* **70**: 231–41.

Richards, C.S. (1975c). Genetic studies of pathologic conditions and susceptibility to infection in *Biomphalaria glabrata*. *Ann. N. Y. Acad. Sci.* **266**: 394–410.

Richards, C.S. (1977). *Schistosoma mansoni*: susceptibility reversal with age in the snail host *Biomphalaria glabrata*. *Expl Parasit.* **42**: 165–8.

Richards, C.S. (1984). Influence of snail age on genetic variation in susceptibility of *Biomphalaria glabrata* for infection with *Schistosoma mansoni*. *Malacologia* **25**: 493–502.

Richards, C.S. & Merritt, J.W., Jr. (1972). Genetic factors in the susceptibility of juvenile *Biomphalaria glabrata* to *Schistosoma mansoni* infection. *Am. J. trop. Med. Hyg.* **21**: 425–34.

Schoenberg, D.A. & Cheng, T.C. (1980). Lectin-binding specificities of hemocytes from two strains of *Biomphalaria glabrata* as determined by microhemadsorption assays. *Devl comp. Immunol.* **4**: 617–28.

Schoenberg, D.A. & Cheng, T.C. (1981). The behaviour of *Biomphalaria glabrata* (Gastropoda: Pulmonata) hemocytes following exposure to lectins. *Trans. Am. microsc. Soc.* **100**: 345–54.

Sminia, T. (1981). Gastropods. In *Invertebrate blood cells* **1**: 191–232. (Eds Ratcliffe, N.A. & Rowley, A.F.). Academic Press, London.

Sminia, T. & Barendsen, L. (1980). A comparative morphological and enzyme histochemical study on blood cells of the freshwater snails *Lymnaea stagnalis*, *Biomphalaria glabrata*, and *Bulinus truncatus*. *J. Morph.* **165**: 31–9.

Sminia, T., van der Knaap, W.P.W. & Edelenbosch, P. (1979). The role of serum factors in phagocytosis of foreign particles by blood cells of the freshwater snail *Lymnaea stagnalis*. *Devl comp. Immunol.* **3**: 37–44.

Stumpf, J.L. & Gilbertson, D.E. (1978). Hemocytes of *Biomphalaria glabrata*: factors affecting variability. *J. Invert. Path.* **32**: 177–81.

Stumpf, J.L. & Gilbertson, D.E. (1980). Differential leukocytic responses of *Biomphalaria glabrata* to infection with *Schistosoma mansoni*. *J. Invert. Path.* **35**: 217–8.

Sullivan, J.T., Cheng, T.C. & Howland, K.H. (1984). Mitotic responses of the anterior pericardial wall of *Biomphalaria glabrata* (Mollusca) subjected to challenge. *J. Invert. Path.* **44**: 114–6.

Sullivan, J.T. & Richards, C.S. (1981). *Schistosoma mansoni*, NIH-SM-PR-2 strain, in susceptible and nonsusceptible stocks of *Biomphalaria glabrata*: comparative histology. *J. Parasit.* **67**: 702–8.

Sullivan, J.T., Richards, C.S., Lie, K.J. & Heyneman, D. (1981). *Schistosoma mansoni*, NIH-SM-PR-2 strain, in non-susceptible *Biomphalaria glabrata*: protection by *Echinostoma paraensei*. *Int. J. Parasit.* **11**: 481–4.

Sullivan, J.T., Richards, C.S., Lie, K.J. & Heyneman, D. (1982). *Ribeiroia marini*: irradiated miracidia and induction of acquired resistance in *Biomphalaria glabrata*. *Expl Parasit.* **53**: 17–25.

van der Knaap, W.P.W., Doderer, A., Boerrigter-Barendsen, L.H. & Sminia, T. (1982). Some properties of an agglutinin in the haemolymph of the pond snail *Lymnaea stagnalis*. *Biol. Bull. mar. biol. Lab. Woods Hole* **162**: 404–12.

Vasta, G.R., Sullivan, J.T., Cheng, T.C., Marchalonis, J.J. & Warr, G.W. (1982). A cell membrane-associated lectin of the oyster hemocyte. *J. Invert. Path.* **40**: 367–77.

Wright, C.A. & Southgate, V.R. (1976). Hybridization of schistosomes and some of its implications. *Symp. Br. Soc. Parasit.* **14**: 55–86.

Yoshino, T.P. (1981a). Comparison of Concanavalin A-reactive determinants on hemocytes of two *Biomphalaria glabrata* snail stocks: receptor binding and redistribution. *Devl comp. Immunol.* **5**: 229–39.

Yoshino, T.P. (1981b). Concanavalin A-induced receptor redistribution on *Biomphalaria glabrata* hemocytes: characterization of capping and patching responses. *J. Invert. Path.* **38**: 102–12.

Yoshino, T.P. (1983a). Surface antigens of *Biomphalaria glabrata* (Gastropoda) hemocytes: occurrence of membrane-associated hemolymph-like factors antigenically related to snail hemoglobin. *J. Invert. Path.* **41**: 310–20.

Yoshino, T.P. (1983b). Lectins and antibodies as molecular probes of molluscan hemocyte surface membranes. *Devl comp. Immunol.* 7: 641–4.

Yoshino, T.P. & Bayne, C.J. (1983). Mimicry of snail host antigens by miracidia and primary sporocysts of *Schistosoma mansoni*. *Parasite Immunol.* **5**: 317–28.

Yoshino, T.P. & Cheng, T.C. (1978). Snail host-like antigens associated with the surface membranes of *Schistosoma mansoni* miracidia. *J. Parasit.* **64**: 752–4.

Yoshino, T.P. & Granath, W.O., Jr. (1983). Identification of antigenically distinct hemocyte subpopulations in *Biomphalaria glabrata* (Gastropoda) using monoclonal antibodies to surface membrane markers. *Cell Tissue Res.* **232**: 553–64.

Yoshino, T.P. & Granath, W.O., Jr. (1985). Surface antigens of *Biomphalaria glabrata* (Gastropoda) hemocytes: functional heterogeneity in cell subpopulations recognized by a monoclonal antibody. *J. Invert. Path.* **45**: 174–86.

Symp. zool. Soc. Lond. (1986) No. 56 221–238

Antigen sharing between larval trematodes and their snail hosts: how real a phenomenon in immune evasion?

Timothy P. YOSHINO
and Carl A. BOSWELL

Department of Zoology,
University of Oklahoma,
Norman,
Oklahoma 73019
USA

Synopsis

The parasitic life style is obviously a successful one, yet survival is dependent upon circumventing the host's potentially devastating immune response. One of the means by which parasites might escape destruction is by presenting a surface that appears like host tissue, thus avoiding stimulation of the host's immune system. There is substantial evidence that schistosome parasites display significant amounts of host-like antigens on their surface. Some of these antigens appear to be of parasite origin and represent examples of molecular mimicry as defined by Damian (1964), while others are acquired from the host by direct adsorption of host molecules onto the parasite's tegumental surface. That the vertebrate stages of blood flukes share antigens with their mammalian host has been known for some time, and evidence is now accumulating that the intramolluscan stages also display antigens which are serologically crossreactive with snail tissues. Evidence for snail host-like antigens on larval trematodes actually being relevant to the parasites' survival, however, has not been demonstrated. It may be that the methodologies so far employed have simply not identified the relevant parasite or host molecules that allow survival in susceptible snails, or that antigen sharing as an immune evasion mechanism is not operative in the schistosome-mollusc system. Further research into other trematode-mollusc systems is needed before the question of immune evasion by antigen-sharing mechanisms can be fully resolved.

Introduction

The concept of immune evasion by parasites is not a new one, and its existence is probably an underlying assumption in associations in which parasites are relatively long-lived within their hosts. In vertebrate host systems, with their diversified and highly adaptive immunological responses, natural selection has favoured mechanisms adopted by their parasites to reduce that responsive capacity through a variety of means including immunosuppres-

ZOOLOGICAL SYMPOSIUM No. 56
ISBN 0-19-854004-3

sion, antigenic variation or modulation, intracellular parasitism, molecular mimicry and antigen masking (Brown 1976; Houba 1976; Capron & Camus 1979; Mitchell 1982; Damian 1984).

Although invertebrate hosts lack the adaptive system of lymphoid cells responsible for specific immunoglobulin production, cell-mediated reactions and immune regulatory functions, they have been shown to be highly capable of distinguishing between 'self' and 'non-self' materials (Chorney & Cheng 1980), and possess systems of circulating and fixed blood cells which are effective in eliminating foreign intruders through processes such as phagocytosis, encapsulation or cytotoxicity (Lackie 1980; Ratcliffe & Rowley 1981). Gastropod molluscs, the principal intermediate host of larval digenetic trematodes and the focus of the present review, also possess systems of circulating blood cells (haemocytes) and fixed phagocytes which represent formidable 'barriers' to pathogens gaining entry into these animals (reviewed by Bayne 1983; Cheng 1984). Yet, despite this seemingly effective system of internal defence, many species of larval Digenea can still be found in an unmolested state within equally numerous gastropod species, suggesting that these parasites have evolved means of either circumventing or disrupting the host's potentially destructive defence system.

Mechanisms of parasite immune evasion

The extensive knowledge of the cellular and biochemical basis for immune responsiveness in vertebrates has provided a firm foundation for investigations into the mechanisms used by protozoan and helminth parasites to evade or suppress both regulatory and effector immune pathways in these animals. However, unlike vertebrate host models, our rudimentary understanding of the fundamental mechanisms of immune function in invertebrates has permitted little more than the establishment of working hypotheses for possible evasion mechanisms. For the purpose of discussion of some of these mechanisms, we have categorized them into two main groups: *Active* evasion, which denotes the active biochemical participation of the parasite (e.g., production and secretion of molecules) in countering a normally aggressive cellular host response, and *passive* evasion, where parasites, by virtue of their genetically-endowed or acquired surface chemistry, are rendered unrecognizable by elements of the host defence system ('non-recognition', Lackie 1980). Mechanisms responsible for active evasion may be further divided into additional subcategories: namely *direct*, in which the parasite destroys haemocytes attempting encapsulation reactions through a cytolytic process (Collin 1970; Ubelaker, Cooper & Allison 1970; Yoshino 1976), and *indirect*, where host haemocytes are rendered incapable of recognizing and subsequently reacting against the parasite through, presumably, production and secretion of soluble, parasite-derived interfering or suppressive factors (Salt 1970; Lie, Heyneman & Jeong 1976). The mechanisms of active evasion are discussed in several other contributions to this symposium (see chapters by Lackie and Loker & Bayne) and, therefore, will not be covered further. Instead, we would like to concentrate on the strategy of antigen sharing as a possible mechanism for evasion by parasites of immune recognition in their invertebrate host, with special emphasis on gastropod molluscs and their larval trematodes. It is the intent of this paper to evaluate the evidence for the sharing of common molecules between these parasites and their snail hosts, and to determine whether a case can be made for antigen sharing as a passive form of immune evasion in this group of helminths.

Antigen sharing in passive immune evasion

The hypothesis that antigen sharing may represent a possible means of avoiding host recognition grew out of earlier ideas of Sprent (1959) and Dineen (1963a) who suggested that long-lived parasites infecting immunocompetent hosts would undergo a process of immunological selection leading eventually to a reduction in antigenic disparity between parasite and host. Damian (1964) further hypothesized that, as a consequence of convergent evolution, parasites in their mammalian hosts might be avoiding immune recognition by expressing surface antigens which mimicked those of their hosts. Mimicked antigens are those which are encoded within the genome of the parasite and expressed as constitutive components of the organism's external surface. A highly adaptive immune system capable of efficiently eliminating parasites exhibiting foreign (i.e. 'non-self') antigens has been suggested as the selective driving force behind the molecular mimicry hypothesis (Damian 1964). Two modifications of Damian's molecular mimicry hypothesis were later suggested to explain various immunobiological aspects of the schistosome in its mammalian host. Capron, Biguet, Vernes & Afchain (1968) theorized that schistosome larvae, upon entry into a particular mammalian host, could be induced to synthesize host species-specific antigens, thus accounting for the wide range of mammalian species which are susceptible to infection by these worms. Antigen masking, in which juvenile worms after entering their host take up and incorporate disguising molecules of host origin into their surface membranes, was suggested as a possible explanation for the phenomenon of concomitant immunity observed in schistosome infections (Smithers & Terry 1969; Smithers, Terry & Hockley 1969). Investigators interested in the question of how larval helminths escape recognition and encapsulation in their invertebrate hosts quickly incorporated these ideas into hypotheses suggesting that mechanisms similar to molecular mimicry or antigen masking may also be important in regulating the interactions of invertebrate hosts with their parasites (Heyneman, Faulk & Fudenberg 1971; Wright 1971; Basch 1976; Stein & Basch 1979; Lackie 1980; Bayne 1983).

The innate resistance or susceptibility of snails to larval trematode infections has a complex genetic basis which involves the interaction of characteristics inherent in both the host (presence or absence of susceptibility or resistance factors) and the invading parasite (infectivity factors). As Basch (1975) has pointed out, host infection will result only when there exists a concordance of relevant 'compatibility' genes in both the snail and the parasite. Perhaps one of the most obvious manifestations of genetic discordance in the expression of compatibility factors is the rapid and effective cellular reaction elicited by entry of 'foreign' miracidia into a host. Numerous examples of such reactions have been reported with parasite strains infecting snail strains from different geographic regions (e.g., Newton 1952; Barbosa & Barreto 1960; Boray 1966; Jourdane 1982) or parasites entering host strains which have been genetically selected for resistance (Newton 1954; Richards 1973, 1975).

Implicit in the reasoning upon which molecular mimicry and antigen acquisition hypotheses are based is that circulating haemocytes, the principal host effector cells, recognize both self and non-self through the interaction of their surface membrane components with their external chemical environment (Chorney & Cheng 1980). Since all snails, parasite-susceptible and -resistant strains alike, possess similar systems of circulating haemocytes, one of the possible 'breakdowns' in immune defence in susceptible hosts would appear to be the inability of these cells to recognize invading larval forms as foreign, thus abrogating the inductive phase of the cellular response. Molecular mimicry or antigen masking could serve as mechanisms responsible for the lack of parasite recognition by host haemocytes observed in susceptible snails. What is the evidence for either of these passive evasion mechanisms operating in trematode/snail relationships?

Molecular mimicry

As Damian (1979) has pointed out, two essential criteria for claiming molecular mimicry as an immune evasive mechanism are (1) that surface determinants expressed by the parasite originate from the parasite (i.e. are not acquired from the host) and (2) that, as a consequence of possessing mimicked determinants, the parasite is protected from an otherwise lethal host reaction. Meeting both criteria is not an easy task. Numerous early reports claimed a sharing of antigens between snails and larval trematodes (Capron, Biguet, Rose & Vernes 1965; Heyneman *et al.* 1971; Benex & Tribouley 1974; Kemp, Greene & Damian 1974; Jackson 1976; Jackson & DeMoor 1976; Roder, Bourns & Singhal 1977). However, in all cases, because the parasite stages used in these experiments had prior contact with the snail host, it could not be determined whether antibodies reactive with shared antigens were directed against parasite-derived mimicked antigens or antigens acquired from the host.

Some of the first indications that larval trematodes might be expressing mimicked snail-like determinants came from molecular probe investigations using carbohydrate-binding lectins (Yoshino, Cheng & Renwrantz 1977). This study showed that miracidia and primary sporocysts of *Schistosoma mansoni* possessed sugar-binding sites for several lectins (concanavalin A, *Dolichos* agglutinin, or anti-H eel serum agglutinin) which were previously shown to be reactive with soluble plasma components found in the snail host, *Biomphalaria glabrata* (Stanislawski, Renwrantz & Becker 1976). Sporocysts, in this case, were derived by the *in vitro* cultivation of miracidia in media (DiConza & Basch 1974) devoid of snail materials. The sharing of carbohydrate determinants was not surprising since glycoconjugates such as glycoproteins, glycolipids or polysaccharides are common constituents of eucaryotic plasma membranes (Nicholson 1974). However, results of this early study were viewed as important for two reasons: first, it demonstrated that transformation of the miracidium to primary sporocyst was accompanied by dramatic changes in the parasite's surface chemistry and it was thought that any potential for altering surface components during this crucial stage of establishing host infections could be important in immune evasion. Second, the successful application of membrane probe reagents suggested that the use of other probes with greater sensitivity or specificity would be a promising line of investigation. Emphasis was placed on the early intramolluscan stages of the parasite since they represent the primary targets of haemocyte-mediated reactions in resistant host strains and, therefore, would be most likely to exhibit shared antigens involved in immune avoidance.

In follow-up experiments, polyclonal antibodies raised in rabbits to soluble haemolymph (plasma) components of the snail, *B. glabrata*, were employed as probe reagents to determine whether miracidia and culture-derived sporocysts of *S. mansoni* shared crossreactive antigens with their intermediate host. Strong serological crossreactivity was observed on miracidial epidermal plate membranes, ciliary membranes and, to a lesser degree, on the surface of intercellular ridges (Yoshino & Cheng 1978; Yoshino & Bayne 1983). Similarly, primary sporocysts, transformed *in vitro* in the absence of snail components, possessed antigenic determinants which were crossreactive with anti-snail plasma antibodies, although sporocyst reactivity was clearly reduced in comparison to miracidia (Yoshino & Bayne 1983). Davis (1983) confirmed the existence of mimicked antigens on the parasite surface in reciprocal experiments in which four mouse monoclonal antibodies (IIIB12, IIH11, VH6, IID12), produced against schistosome miracidia, were crossreactive with methanol-fixed haemocytes or soluble plasma components of *B. glabrata*. Similar experiments by Bayne &

Stephens (1983), demonstrating the crossreactivity of a sporocyst-specific antiserum (polyclonal) with snail haemocytes, provide another reciprocal combination and independent confirmation that, at least in the *Schistosoma/Biomphalaria* system, parasites appear to be engaged in some form of host molecular mimicry. Although most of the evidence supporting antigen mimicry has come from the *S. mansoni/B. glabrata* model, this phenomenon may be more widespread in trematode/snail systems. For example, serological crossreactivity was recently found between miracidial stages of the avian schistosome *Trichobilharzia ocellata* and plasma antigens of its snail host *Lymnaea stagnalis* (van der Knaap, Meuleman, Boots & de Cleen 1984). However, in contrast to the previously reported findings involving *S. mansoni*, the anti-snail plasma antibodies used by these investigators were not crossreactive with primary sporocysts *in situ* (van der Knaap, Boots, Meuleman & Sminia 1985). These differing results indicate that, although similar in some respects, subtle variations in the extent of antigen sharing between the snail host and different larval stages may exist.

As pointed out previously, the second criterion for claiming molecular mimicry as a mechanism for immune evasion is that possession of such shared molecules confers on the parasite protection from aggressive host defences (Damian 1979). In mammalian host systems evidence supporting an immune evasive role has been mainly correlative. For instance, Dineen (1963b) demonstrated that extracts from the nematode *Haemonchus contortus* elicited an antibody response in a non-permissive host, the rabbit, that was qualitatively distinct from what resulted after immunization of the permissive, sheep, host. Adult and larval worms elicited only stage-specific antibody responses in sheep, while similarly immunized rabbits produced not only the same stage-specific antibodies, but additional ones which crossreacted between the parasite stages. These crossreactive epitopes were considered 'fitness characteristics' by Dineen because their recognition in rabbits was correlated with aborted development of larval forms, whereas their non-recognition in the natural sheep host was correlated with normal parasite development. It might be speculated that the so-called 'fitness' antigens not recognized in sheep represent products of molecular mimicry. However, direct evidence to support this notion is not available. Damian, Greene & Hubbard (1973) established that adult *S. mansoni* possessed a tegumental molecule which mimicked mouse α_2-macroglobulin (α_2M) but not baboon α_2M. They then immunized baboons against mouse α_2M in an attempt to elicit an immunoprotective response against adult mouse worms subsequently introduced into α_2M-sensitized animals. Results of this experiment showed that challenge worm burdens recovered from α_2M-sensitized animals were similar to those recovered from non-sensitized controls indicating that, even though antibodies against mouse α_2M were produced in baboons and were found to be crossreactive with α_2M-like epitopes on challenge parasites, these mimicked antigens were of little protective immune relevance in the mouse host.

Likewise, in trematode/mollusc systems there is no evidence for a parasite-protective role of mimicked molecules since many of the difficulties inherent in designing meaningful experiments in mammalian host systems (Damian 1979) also apply to molluscs. Some studies have attempted to correlate the degree of sharing of mimicked molecules with the host susceptibility state. On the assumption that early developmental stages of trematode larvae might be expected to share a greater proportion of mimicked antigens with a susceptible snail host than with a resistant one, Yoshino & Bayne (1983) measured the antibody titres to mimicked determinants in antisera produced against plasma components from a susceptible and a resistant strain of *B. glabrata*. Using miracidia and *in vitro*-transformed primary sporocysts of *S. mansoni* as test specimens in an immunofluorescence assay, they found that the titre of anti-susceptible antiserum was higher against both the miracidial and sporocyst stages than the anti-resistant snail antiserum. This result suggested a greater degree of mimicked antigen sharing between schistosomes and their susceptible host than the resistant host strain. However, as cautioned by the authors, since the anti-snail plasma antisera were produced in different rabbits, this difference in antibody stength may also be a reflection of individual variation in antibody-producing ability of the immunized rabbits.

In reciprocal experiments using the same schistosome/snail system, anti-miracidial monoclonal antibodies (Davis 1983) and anti-sporocyst antiserum (Bayne & Stephens 1983) crossreacted equally well with haemocytes from both susceptible and resistant host strains. These findings suggest that either mimicked molecules are of little or no consequence in the establishment of compatible relationships or the methods (primarily serological ones) for assessing the degree of antigen sharing are inadequate for identifying relevant molecules. A more sensitive way of looking at the molecular similarity between snails and their trematode parasites is to begin comparisons at the genomic level using tools of molecular biology. With these powerful techniques, combined with effective *in vitro* culture methods now available, information useful in assessing the possible role of molecular mimicry in regulating parasite/host compatibility should be forthcoming.

Acquisition of host antigens

Antigen masking has been suggested as an alternative hypothesis to molecular mimicry as a mechanism of parasite immune evasion, although it should be noted that acquisition of host antigens by parasites may be facilitated by parasite mimicry of host receptor molecules such as Fc (Kemp, Merritt, Bogucki, Rosier & Seed 1977; Torpier, Capron & Ouaissi 1979) or complement (Santoro, Ouaissi, Pestel & Capron 1980) receptors. Whether through receptor-mediated uptake or non-specific adsorption, schistosomula and adults of *S. mansoni* have been shown to acquire a variety of mammalian host components including blood group antigens (Smithers *et al.* 1969; Dean 1974; Goldring, Kusel & Smithers 1977), immunoglobulins (Kemp, Merritt, Bogucki *et al.* 1977; Kemp, Merritt & Rosier 1978), major histocompatibility complex (MHC) antigens (Sher, Hall & Vadas 1978; Gitter & Damian 1982), complement (Santoro *et al.* 1980) and others. Moreover, acquisition of mammalian host antigens by schistosomes can aid in the protection of these parasites from host immune attack (Smithers *et al.* 1969; McLaren & Terry 1982), thus indicating a functional role for acquired antigens in immune evasion.

In invertebrate host/parasite systems there is also considerable evidence supporting the ability of larval trematodes to acquire snail host antigens and express them at their tegumental surfaces. Most of the literature has concentrated on the later stages of trematode development (primarily the cercaria) in which incorporation of snail antigens has been clearly demonstrated (see Damian 1979, for review). However, since successful infection of the gastropod host is dependent upon survival and normal development of the primary sporocyst, one might expect this stage to be able to bind host molecules if immune avoidance was based on a chemical masking mechanism. Starting with this premise, Stein & Basch (1979) demonstrated that culture-transformed sporocysts of *S. mansoni* bind a naturally-occurring plasma haemagglutinin (HA) from the snail *B. glabrata*, and since the plasma source for the HA was a susceptible host strain, they hypothesized that this material might be involved in avoidance of recognition by host haemocytes. Similarly, plasma agglutinins from different strains of *B. glabrata* have subsequently been shown to react with living and aldehyde-fixed sporocysts (Loker, Yui & Bayne 1984). Plasma agglutinins for living parasites occurred in both susceptible and resistant snail strains, while those reactive with fixed sporocysts were found only in resistant snails. It is clear from this and other studies (Michelson & Dubois 1977; Boswell & Bayne 1984) that plasma HAs comprise a heterogeneous molecular population in this snail species and may be diversified enough in their sugar-binding properties to serve in parasite immunorecognition processes (Bayne, Boswell, Loker & Yui 1985). However, if antigen masking is playing a role in immune evasion, the fact that similar HAs with reactivity to living sporocysts appear in both resistant and susceptible snails would suggest that plasma HAs are not directly involved in the avoidance process.

Using an anti-snail plasma antiserum, Bayne *et al.* (1985) and Bayne, Loker & Yui (in preparation) showed in subsequent experiments that, in addition to plasma HAs, *in vitro*-transformed sporocysts acquire a variety of other soluble plasma components. Plasma antigens were taken up rapidly (within 20 min of *in vitro* sporocyst incubation in fresh plasma). However,

Table 1. *In vitro* acquisition of snail plasma antigens by culture-transformed primary sporocysts of *Schistosoma mansoni* detected by monoclonal anti-plasma antibodies (mAB).

mAB No.	Age of sporocysts in culture (days)	Sporocyst surface reactivity (IFA)[a] Plasma incubation time (min) 0	15	30	60	120
VI–16	1	–[b]	+/++	+/++	++	++
	4	–	+/++	+/++	++	++
	8	–	ND	+/++	+/++	++
VB10.3Bg	1	–	±/+	±/+	+	+/++
	4	–	±	±/+	+	+/++
	8	–	ND	ND	+	+/++
II–B.1.5	1	–	±	+	+	+
	4	–	–	–	±	+
	8	–	ND	–	±	±
IV–H.1.2	1	–	–	ND	–	ND
	4	–	–	–	–	–

[a] Plasma-treated and untreated (0 min) sporocysts were exposed to mAB for 20 min, washed with phosphate buffered saline (Yoshino 1981), and incubated in FITC-conjugated goat anti-mouse Ig antiserum for an additional 20 min. The plasma used to treat sporocysts was obtained from a susceptible strain of snail, *Biomphalaria glabrata* (M-line).
[b] Immunofluorescence (IFA) scoring criteria: – = no surface fluorescence, ± = weak, irregular distribution of surface fluorescence, + = continuous surface staining, ++ intense surface fluorescence. ND = not determined.

following tegumental adsorption, soluble haemolymph antigens were gradually lost from the sporocyst surface within 3 h of incubation in plasma-free culture medium. Western blot analysis of plasma-incubated sporocysts suggested that the uptake of plasma molecules was also selective. Using another trematode-snail system, *Trichobilharzia*/*Lymnaea*, van der Knaap, Meuleman *et al.* (1984) and van der Knaap, Boots *et al.* (1985) showed immunohistochemically that young primary sporocysts and cercariae, but not older mother sporocysts or daughter sporocysts (25 days post-infection), were able to acquire host antigens. It was suggested that the host-like antigens displayed by miracidia might help to protect this stage upon entry into the snail host, whereas the primary sporocyst might be protected immediately following transformation by antigens acquired from the host (van der Knaap, Meuleman *et al.* 1984).

In recent experiments, using anti-snail plasma monoclonal antibodies (mABs), we have confirmed an earlier preliminary observation (Yoshino 1983) and those of Bayne *et al.* (1985), that plasma antigens were taken up rapidly and in a selective fashion (Yoshino, Boswell & Watson in preparation). In these experiments, groups of *S. mansoni* primary sporocysts (NMRI strain) which had been transformed and subsequently incubated in a modified DiConza & Basch (1974) medium for one, four, or eight days were exposed for 15, 30, 60, or 120 min to *B. glabrata* plasma (M-line), washed three times with serum-free medium, fixed in 4% paraformal-

Table 2. Serological reactivity of various monoclonal antibodies (mAB) to snail plasma antigens (Enzyme-linked Immunosorbent Assay, ELISA; Western blot) and plasma-exposed *Schistosoma mansoni* primary sporocysts (Immunofluorescence Assay, IFA).

mAb No.	Plasma reactivity (ELISA OD_{405})[a]	Plasma components in Western blots (No. of bands)	Sporocyst surface reactivity (IFA)[b]		
			Untreated	Trypsinized (0.1%)	Endo H-treated (0.02 U)
VI–16	0.472	32	++[c]	±/+	++
II–D2.6	0.338	ND	++	+	++
II–B.1.5	0.382	3	+/++	ND	+/++
III–E.1	0.339	37	++	ND	±/+
IV–H.1.2	0.293	20	–	ND	ND

[a] For the enzyme-linked immunosorbent assay all monoclonals were ammonium sulphate fractionated and standardized to 0.5 mg/ml phosphate buffered saline (PBS). Polystyrene multi-well plates were sensitized with snail plasma (1:20 diluted with distilled water) which served as the test antigen.
[b] To perform the immunofluorescence assays, test sporocysts were pre-incubated in snail plasma (M-line) for 2 h, washed with PBS, and fixed with 4% buffered paraformaldehyde prior to enzyme treatment. Sporocysts were treated for 20 min at 22°C with trypsin, Endo H (endo-ß-N-acetylglucosaminidase H), or were left untreated (positive control). All sporocysts were exposed to mAB and FITC-labelled goat anti-mouse Ig antibody prior to assessment of surface staining using epifluorescence microscopy.
[c] Immunofluorescence scoring criteria were as described in Table 1.

dehyde and finally washed with a reaction buffer containing 0.5% bovine serum albumin (BSA) in snail phosphate buffered saline, pH 7.2 (PBS; Yoshino 1981). One-, four-, and eight-day sporocysts incubated in plasma-free culture medium served as negative controls. As illustrated by mABs VI-16 and VB10.3Bg in Table 1, the uptake of plasma antigens recognized by these antibodies was rapid (usually within 15 min of plasma incubation). Antigen uptake was also qualitatively selective as indicated by plasma components recognized by mAB IV-H.1.2 (Table 2). In this case, at least 20 plasma components were detected by Western blot analysis, yet plasma-exposed sporocysts did not display any mAB-reactive materials on their tegumental surface (Table 1). Furthermore, as sporocysts matured their ability to acquire certain plasma components appeared to change. For example, one-day sporocysts bound a plasma antigen(s) recognized by mAB II-B.1.5 within 15 min, while eight-day sporocysts took up very little of this same material after 60 min of plasma incubation (Table 1). This finding suggests that as sporocysts age, in addition to chemical changes occurring as a result of acquiring host antigens, the tegumental surface may also be undergoing innate ontogenetic modifications. A similar age-dependent ability of primary sporocysts to acquire host antigens has been reported in natural infections of *Trichobilharzia* in *L. stagnalis* (van der Knaap, Meuleman *et al.* 1984). Although not directly pertinent to the discussion of antigen masking, this observation raises an interesting question of whether sporocysts, as they mature, normally become less sensitive to haemocyte-mediated reactions in resistant snails. This appears to be the case with schistosomula, a stage which has been shown, when cultured in serum-free media, to become innately more resistant to antibody-mediated (Dean 1977) or antibody-dependent cell-mediated (Dessein *et al.* 1981) killing activity. With the availability of sensitive *in vitro* (Boswell & Bayne 1985) and *in vivo* (Granath & Yoshino 1984) assays, answers to this question are certain to be forthcoming.

Finally, in an attempt to learn more about the chemical nature of the adsorbed host molecules we subjected plasma-exposed sporocysts to mild protease (0.1% trypsin) or carbohydrase (Endo H) treatment to determine the effects of these enzymes on the epitopes recognized by several anti-plasma mABs. Endo H (endo-β-N-acetylglucosaminidase H; Sigma) is an endoglycosidase from *Streptomyces* which has a specificity for internal asparagine-linked N-acetylchitobiose sugars (Tai *et al.* 1975). Most of the antigenic determinants of acquired plasma components were associated with protease-sensitive structures since trypsin treatment reduced significantly the capacity of several mABs to react to plasma-exposed, trypsin-treated sporocysts (Table 2). Treatment of plasma-incubated sporocysts with this endoglycosidase had no apparent influence on monoclonal anti-plasma antibody activity except in one case involving mAB III-E.1. Endo H-treatment of these sporocysts reduced the binding capacity of III-E.1 (Table 2) suggesting that the antibody-binding epitope is probably carbohydrate in nature. The generalized reactivity of this mAB with numerous plasma components in Western blots supports this finding since mAB-reactive sugar moieties might be expected to be quite common constituents of numerous glycoconjugates found in snail plasma (Stanislawski *et al.* 1976). The finding of carbohydrate antibody-binding sites points out a source of potential misinterpretation of serological data since one cannot assume that antibodies produced against a complex protein mixture such as snail plasma are recognizing only specific amino acid sequences (Mattes & Steiner 1978; Warr, DeLuca & Griffin 1979).

Awareness of this potential source of error is particularly important in studies of molecular mimicry where serological crossreactivity is thought to be the result of expression of common protein determinants through genetic convergence. On the other hand, perhaps our view of chemical mimicry should be broadened (i.e. not necessarily restricted to just protein structural similarities), especially in view of the potential importance of carbohydrate-binding lectins (agglutinins) in mediating recognition of non-self invertebrates (Sharon 1984) and invertebrates (Renwrantz 1983; chapter by Renwrantz, this volume). The sharing of saccharide moieties between host and parasite as illustrated above and by Yoshino *et al.* (1977) raises the issue of whether selective pressure of the immune response in molluscs might have favoured parasite 'mimicry' of host carbohydrate components or the acquisition of glycoconjugated macromolecules of host origin.

It is clear that early developing sporocysts of the schistosome group are capable of selectively acquiring host molecules, and the potential certainly exists for using this mechanism to evade immune recognition in the susceptible host. However, as in the case of molecular mimicry, little evidence is available to support a functional (i.e. parasite-protective) role for acquired host components. Loker & Bayne (1982) found that pre-incubation of culture-transformed sporocysts of *S. mansoni* in plasma from a susceptible snail strain did not protect these parasites from cytotoxic killing by resistant snail haemocytes in an *in vitro* assay. In *in vivo* experiments a similar lack of parasite protection by acquired plasma factors has also been demonstrated (Granath & Yoshino 1984). In this study, sporocysts, pre-incubated in susceptible *B. glabrata* plasma and then injected into resistant snails, failed to develop patent infections, indicating an absence of functional masking components. Results of these *in vitro* and *in vivo* experiments do not support the idea of a parasite-protective role of acquired host components in innately resistant snails. However, it should be realized that in the experiments cited above, sporocysts were exposed to 'masking' plasma for relative short time intervals and only after these sporocysts had already been transformed and incubated in complex culture media for 24h prior to plasma treatments. Perhaps protective masking substances need to be acquired during the initial phase of miracidial-sporocyst transformation, since this represents a period of dynamic membrane change for the parasite (DiConza & Basch 1974; Yoshino *et al.* 1977; Meuleman, Lyaruu, Khan, Holzmann & Sminia 1978). Additional studies on the biochemical interaction of parasite and host molecules during this critical period of trematode development are needed before conclusions regarding the role of acquired host molecules in parasite immune evasion can be made.

Conclusions

What we can conclude from the currently available evidence is that the presence of host-derived or host-like molecules on the parasite's surface is no guarantee of protection for that parasite. The molecular interactions between host and parasite that determine a 'non-reaction' in susceptible hosts are possibly so subtle that the methods employed to date are not sensitive enough to detect them. Polyclonal and even monoclonal antibodies against snail haemolymph components recognize a plethora of molecules, both from the original antigenic stimulus (e.g. plasma) and from the source of crossreactivity (e.g. sporocyst tegument). Crossreactivity between snail molecules and larval trematodes isolated in the absence of snail materials has been considered to be a form of molecular mimicry. However, the possibility that at least some of the probe antibodies used to demonstrate mimicry may be reacting to carbohydrate epitopes indicates that not all crossreacting antigens may be considered mimicked molecules as originally defined by Damian (1964, 1979). Antibodies with binding specificity to common glycosyl moieties of parasite membrane or snail plasma molecules could account for the extensive crossreactivity observed in these host/parasite systems. In addition, it is clear that developing primary sporocysts of some trematode species, such as the schistosomes, do acquire in a selective fashion molecules of host origin, thus leading to speculation that such molecules may be serving a parasite-protective function in susceptible hosts. Yet the fact remains that in systems of trematodes and snails examined to date there is still no direct evidence to indicate that antigen sharing in any form represents an immune evasive mechanism. It is tempting to conclude that such a mechanism does not exist, but we would remind the reader that (1) not all experimental approaches have been exploited to examine this problem and (2) the majority of information we have comes from a limited number of host/parasite systems. We anticipate that with further research the question of antigen sharing in immune evasion will be resolved.

Acknowledgements

The authors gratefully acknowledge Bonnie Watson and Joe Newsome for their fine technical assistance, and Dr Fred Lewis (Biomedical Research Institute, Rockville, MD) for providing infected mouse livers. Also we thank Dr Christopher Bayne (Oregon State University) for providing us with a preprint of his most recent work for citation in this review. Preliminary original data given in this paper and previously published research by TPY was supported by the National Institutes of Health, NIAID, through Grant No. AI15503 and Research Career Development Award K04 AI00491–01.

References

Barbosa, F.S. & Barreto, A.C. (1960). Differences in susceptibility of Brazilian strains of *Australorbis glabratus* to *Schistosoma mansoni*. *Expl Parasit.* 9: 137–40.

Basch, P.F. (1975). An interpretation of snail-trematode infection rates: Specificity based on concordance of compatible phenotypes. *Int. J. Parasit.* 5: 449–52.

Basch, P.F. (1976). Intermediate host specificity in *Schistosoma mansoni*. *Expl Parasit.* 39: 150–69.

Bayne, C.J. (1983). Molluscan immunobiology. In *The Mollusca* 5: 407–86. (Eds Saleuddin, A.S.M. & Wilbur, K.M.). Academic Press, New York.

Bayne, C.J., Boswell, C.A., Loker, E.S. & Yui, M.A. (1985). Plasma components which mediate cellular defences in the gastropod mollusc *Biomphalaria glabrata*. *Devl comp. Immunol.* 9: 523–30.

Bayne, C.J. & Stephens, J.A. (1983). *Schistosoma mansoni* and *Biomphalaria glabrata* share epitopes: Antibodies to sporocysts bind host snail hemocytes. *J. Invert. Path.* **42**: 221–3.

Benex, J. & Tribouley, J. (1974). Mise en évidence à la surface de la cercaire de *Schistosoma mansoni* d'antigène possédant le caractère planorbe. *C. r. hebd. Séanc. Acad. Sci., Paris* **279**: 683–5.

Boray, J.C. (1966). Studies on the relative susceptibility of some lymnaeids to infection with *Fasciola hepatica* and *F. gigantica* and on the adaption of *Fasciola* spp. *Ann. trop. Med. Parasit.* **60**: 114–24.

Boswell, C.A. & Bayne, C.J. (1984). Isolation, characterization and functional assessment of a hemagglutinin from the plasma of *Biomphalaria glabrata*, intermediate host of *Schistosoma mansoni. Devl comp. Immunol.* **8**: 559–68.

Boswell, C.A. & Bayne, C.J. (1985). *Schistosoma mansoni*: Lectin-dependent cytotoxicity of hemocytes from susceptible host snails, *Biomphalaria glabrata. Expl Parasit.* **60**: 133–8.

Brown, K.N. (1976). Specificity in host-parasite interactions. *Receptors and Recognition* (Series A) **1**: 119–75.

Capron, A., Biguet, J., Rose, F. & Vernes, A. (1965). Les antigènes de *Schistosoma mansoni*. II. Étude immunoélectrophoretique comparée de divers stades larvaires et des adultes des deux sexes. Aspects immunologiques des relations hôte-parasite de la cercaire et de l'adulte de *S. mansoni. Annls Inst. Pasteur, Paris* **109**: 798–810.

Capron, A., Biguet, J., Vernes, A. & Afchain, D. (1968). Structure antigénique des helminthes. Aspects immunologiques des relations hôte-parasite. *Path. Biol. Paris* **16**: 121–38.

Capron, A. & Camus, D. (1979). Immunoregulation by parasite extracts. *Springer Semin. Immunopathol.* **2**: 69–77.

Cheng, T.C. (1984). A classification of molluscan hemocytes based on functional evidences. *Comp. Pathobiol.* **6**: 111–46.

Chorney, M.H. & Cheng, T.C. (1980). Discrimination of self and non-self in invertebrates. In *Contemporary topics in immunobiology* **9**: 37–54. (Eds Marchalonis, J.J. & Cohen, N.). Plenum Press, New York.

Collin, W.K. (1970). Electron microscopy of postembryonic stages of the tapeworm, *Hymenolepis citelli. J. Parasit.* **56**: 1159–71.

Damian, R.T. (1964). Molecular mimicry: Antigen sharing by parasite and host and its consequences. *Am. Nat.* **98**: 129–49.

Damian, R.T. (1979). Molecular mimicry in biological adaptation. In *Host-parasite interfaces*: 103-26. (Ed. Nickol, B.B.). Academic Press, New York.

Damian, R.T. (1984). Immunity in schistosomiasis: A holistic view. *Contemp. Top. Immunobiol.* **12**: 359–420.

Damian, R.T., Greene, N.D. & Hubbard, W.J. (1973). Occurrence of mouse α_2-macroglobulin antigenic determinants on *Schistosoma mansoni* adults, with evidence on their nature. *J. Parasit.* **59**: 64–73.

Davis, C.D. (1983). *Monoclonal antibody analysis of antigens from larval* Schistosoma mansoni. M.S. Thesis: University of Oklahoma, Norman, Oklahoma.

Dean, D.A. (1974). *Schistosoma mansoni*: Adsorption of human blood group A and B antigens by schistosomula. *J. Parasit.* **60**: 260–3.

Dean, D.A. (1977). Decreased binding of cytotoxic antibody by developing *Schistosoma mansoni*. Evidence for a surface change independent of host antigen adsorption and membrane turnover. *J. Parasit.* **63**: 418–24.

Dessein, A., Samuelson, J.C., Butterworth, A.E., Hogan, M., Sherry, B.A., Vadas, M.A. & David, J.R. (1981). Immune evasion by *Schistosoma mansoni*: Loss of susceptibility to antibody or complement-dependent eosinophil attack by schistosomula cultured in medium free of macromolecules. *Parasitology* **82**: 357–74.

DiConza, J.J. & Basch, P.F. (1974). Axenic cultivation of *Schistosoma mansoni* daughter sporocysts. *J. Parasit.* **60**: 757–63.

Dineen, J.K. (1963a). Immunological aspects of parasitism. *Nature, Lond.* **197**: 268–9.

Dineen, J.K. (1963b). Antigenic relationship between host and parasite. *Nature, Lond.* **197**: 471–2.

Gitter, B.D. & Damian, R.T. (1982). Murine alloantigen acquisition by schistosomula of *Schistosoma mansoni*: Further evidence for the presence of K, D, and I region gene products on the tegumental surface. *Parasite Immunol.* **4**: 383–93.

Goldring, O.L., Kusel, J.R. & Smithers, S.R. (1977). *Schistosoma mansoni:* Origin *in vitro* of host-like surface antigens. *Expl Parasit.* **43**: 82–93.

Granath, W.O., Jr. & Yoshino, T.P. (1984). *Schistosoma mansoni*: Passive transfer or resistance by serum in the vector snail, *Biomphalaria glabrata. Expl Parasit.* **58**: 188–93.

Heyneman, D., Faulk, W.P. & Fudenberg, H.H. (1971). *Echinostoma lindoense* : Larval antigens from the snail intermediate host, *Biomphalaria glabrata. Expl Parasit.* **29**: 480–92.

Houba, V. (1976). Pathophysiology of the immune response to parasites. In *Pathophysiology of parasitic infection*: 221–32. (Ed. Soulsby, E.J.L.). Academic Press, New York.

Jackson, T.F.H.G. (1976). Intermediate host antigens associated with the cercariae of *Schistosoma haematobium. J. Helminth.* **50**: 45–7.

Jackson, T.F.H.G. & DeMoor, P.P. (1976). A demonstration of the presence of anti-snail antibodies in individuals infected with *Schistosoma haematobium. J. Helminth.* **50**: 59–63.

Jourdane, J. (1982). Étude des mecanismes de rejet dans les couples mollusque-schistosome incompatibles à partir d'infestations par voie naturelle et par transplantations microchirurgicales de stades parasitaires. *Acta trop.* **39**: 325–35.

Kemp, W.M., Greene, N.D. & Damian, R.T. (1974). Sharing of Cercarienhüllen Reaktion antigens between *Schistosoma mansoni* cercariae and adults and uninfected *Biomphalaria pfeifferi. Am. J. trop. Med. Hyg.* **23**: 197–202.

Kemp, W.M., Merritt, S.C., Bogucki, M.S., Rosier, J.G. & Seed, J.R. (1977). Evidence for adsorption of heterospecific host immunoglobulin on the tegument of *Schistosoma mansoni. J. Immunol.* **119**: 1849–54.

Kemp, W.M., Merritt, S.C. & Rosier, J.G. (1978). *Schistosoma mansoni*: Identification of immunoglobulins associated with the tegument of adult parasites from mice. *Expl Parasit.* **45**: 81–7.

Lackie, A.M. (1980). Invertebrate immunity. *Parasitology* **80**: 393–412.

Lie, K.J., Heyneman, D. & Jeong, K.H. (1976). Studies on resistance in snails. 7. Evidence of interference with the defense reaction in *Biomphalaria glabrata* by trematode larvae. *J. Parasit.* **62**: 608–15.

Loker, E.S. & Bayne, C.J. (1982). *In vitro* encounters between *Schistosoma mansoni* primary sporocysts and hemolymph components of susceptible and resistant strains of *Biomphalaria glabrata. Am. J. trop. Med. Hyg.* **31**: 999–1005.

Loker, E.S., Yui, M.A. & Bayne, C.J. (1984). *Schistosoma mansoni* : Agglutination of sporocysts, and formation of gels on miracidia transforming in plasma of *Biomphalaria glabrata. Expl Parasit.* **58**: 56–62.

McLaren, D.J. & Terry, R.J. (1982). The protective role of acquired host antigens during schistosome maturation. *Parasite Immunol.* **4**: 129–48.

Mattes, M.J. & Steiner, L.A. (1978). Antiserum to frog immunoglobulins cross-react with a periodate-sensitive cell surface determinant. *Nature, Lond.* **273**: 761–3.

Meuleman, E.A., Lyaruu, D.M., Khan, M.A., Holzmann, P.J. & Sminia, T. (1978). Ultrastructural changes in the body wall of *Schistosoma mansoni* during the transformation of the miracidium into the mother sporocyst in the snail host *Biomphalaria pfeifferi. Z. Parasitenk.* **56**: 227–42.

Michelson, E.H. & Dubois, L. (1977). Agglutinins and lysins in the molluscan family Planorbidae: A survey of hemolymph, egg-masses, and albumen-gland extracts. *Biol. Bull. mar. biol. Lab. Woods Hole* **153**: 219–27.

Mitchell, G.F. (1982). Effector mechanisms of host-protective immunity to parasites and evasion by parasites. In *Parasites—their world and ours*: 24–33. (Eds Mettrick, D.F. & Desser, S.S.). Elsevier Biomedical Press, Amsterdam.

Newton, W.L. (1952). The comparative tissue reaction of two strains of *Australorbis glabratus* to infection with *Schistosoma mansoni. J. Parasit.* **38**: 362–6.

Newton, W.L. (1954). Tissue response to *Schistosoma mansoni* in second generation snails from a cross between two strains of *Australorbis glabratus. J. Parasit.* **40**: 352–5.

Nicholson, G.L. (1974). The interactions of lectins with animal cell surfaces. *Int. Rev. Cytol.* **39**: 89–190.

Ratcliffe, N.A. & Rowley, A.F. (Eds). (1981). *Invertebrate blood cells*, **1** & **2**. Academic Press, New York.

Renwrantz, L. (1983). Involvement of agglutinins (lectins) in invertebrate defense reactions: The immuno-biological importance of carbohydrate-specific binding molecules. *Devl comp. Immunol.* **7**: 603–8.

Richards, C.S. (1973). Susceptibility of adult *Biomphalaria glabrata* to *Schistosoma mansoni* infection. *Am. J. trop. Med. Hyg.* **22**: 748–56.

Richards, C.S. (1975). Genetic factors in susceptibility of *Biomphalaria glabrata* for different strains of *Schistosoma mansoni. Parasitology* **70**: 231–41.

Roder, J.C., Bourns, T.K.R. & Singhal, S.K. (1977). *Trichobilharzia ocellata*: Cercariae masked by antigens of the snail, *Lymnaea stagnalis. Expl Parasit.* **41**: 206–12.

Salt, G. (1970). *The cellular defence reactions of insects.* Cambridge University Press, Cambridge.

Santoro, F., Ouaissi, M.A., Pestel, J. & Capron, A. (1980). Interaction between *Schistosoma mansoni* and the complement system: Binding of Clq to schistosomula. *J. Immunol.* **124**: 2886–91.

Sharon, N. (1984). Surface carbohydrates and surface lectins are recognition determinants in phagocytosis. *Immunol. Today* **5**: 143–7.

Sher, A., Hall, B.F. & Vadas, M.A. (1978). Acquisition of murine major histocompatibility complex gene products by schistosomula of *Schistosoma mansoni. J. exp. Med.* **148**: 46–57.

Smithers, S.R. & Terry, R.J. (1969). Immunity in schistosomiasis. *Ann. N.Y. Acad. Sci.* **160**: 826–40.

Smithers, S.R., Terry, R.J. & Hockley, D.J. (1969). Host antigens in schistosomiasis. *Proc. R. Soc. Lond.* (B) **171**: 483–94.

Sprent, J.F.A. (1959). Parasitism, immunity and evolution. In *The evolution of living organisms*: 149–65. Melbourne University Press, Victoria.

Stanislawski, E., Renwrantz, L. & Becker, W. (1976). Soluble blood group reactive substances in the hemolymph of *Biomphalaria glabrata* (Mollusca). *J. Invert. Path.* **28**: 301–8.

Stein, P.C. & Basch, P.F. (1979). Purification and binding properties of hemagglutinin from *Biomphalaria glabrata. J. Invert. Path.* **33**: 10–18.

Tai, T., Yamashita, K., Ogata-Arakawa, M., Koide, N., Muramatsu, T., Iwashita, S., Inoue, Y. & Kobata, A. (1975). Structural studies of two ovalbumin glycopeptides in relation to the endo-β-N-acetylglucosaminidase specificity. *J. Biol. Chem.* **250**: 8569–75.

Torpier, G., Capron, A. & Ouaissi, M.A. (1979). Receptors for IgG (Fc) and human β_2-microglobulin on *S. mansoni* schistosomula. *Nature, Lond.* **278**: 447–9.

Ubelaker, J.E., Cooper, N.B. & Allison, V.F. (1970). Possible defensive mechanism of *Hymenolepis diminuta* cysticercoids to haemocytes of the beetle *Tribolium confusum. J. Invert. Path.* **16**: 310–2.

Van der Knaap, W.P.W., Boots, A.M.H., Meuleman, E.A. & Sminia, T. (1985). Search for shared antigens in the schistosome-snail combination *Trichobilharzia ocellata-Lymnaea stagnalis. Z. Parasitenk.* **71**: 219–26.

Van der Knaap, W.P.W., Meuleman, E.A., Boots, A.M.H. & de Cleen, M. (1984). Parasite-host immunological compatibility: shared antigens in the schistosome-snail model *Trichobilharzia ocellata-Lymnaea stagnalis. Trop. geogr. Med.* **36**: 386–7. (Abstract.)

Warr, G.W., DeLuca, D. & Griffin, B.R. (1979). Membrane immunoglobulin is present on thymic and splenic lymphocytes of the trout *Salmo gairdneri. J. Immunol.* **123**: 910–7.

Wright, C.A. (1971). *Flukes and snails.* George Allen and Unwin, London.

Yoshino, T.P. (1976). Encapsulation response of the marine prosobranch *Cerithidea californica* to natural infections of *Renicola buchanani* sporocysts (Trematoda: Renicolidae). *Int. J. Parasit.* **6**: 423–31.

Yoshino, T.P. (1981). Comparison of concanavalin A-reactive determinants on hemocytes of two *Biomphalaria glabrata* snail stocks; Receptor binding and redistribution. *Devl comp. Immunol.* **5**: 229–39.

Yoshino, T.P. (1983). Selective acquisition of snail host antigens by larval *Schistosoma mansoni*: Detection by monoclonal antibodies to snail hemocytes and hemolymph. *Devl comp. Immunol. Suppl.* **3**: 225. (Abstract.)

Yoshino, T.P. & Bayne, C.J. (1983). Mimicry of snail host antigens by miracidia and primary sporocysts of *Schistosoma mansoni*. *Parasite Immunol.* **5**: 317–28.

Yoshino, T.P. & Cheng, T.C. (1978). Snail host-like antigens associated with the surface membranes of *Schistosoma mansoni* miracidia. *J. Parasit.* **64**: 752–4.

Yoshino, T.P., Cheng, T.C. & Renwrantz, L.R. (1977). Lectin and human blood group determinants of *Schistosoma mansoni*: Alteration following *in vitro* transformation of miracidium to mother sporocyst. *J. Parasit.* **63**: 818–24.

Symp. zool. Soc. Lond. (1986) No. 56 239–274

Genetic variability in resistance to parasitic invasion: population implications for invertebrate host species

Roy M. ANDERSON

Department of Pure and Applied Biology, Imperial College, London University, London SW7 2BB

Synopsis

Recent studies suggest that parasites may influence the abundance of their invertebrate host populations and, concomitantly, exert strong selective pressures. This paper combines components of immunology, ecology and population genetics to examine the population consequences of genetically based variability in invertebrate host resistance to infectious diseases. Particular emphasis is placed on the mechanisms that promote stable coexistence between host and pathogen and maintain polymorphic host populations. Simple mathematical models of the invertebrate immune system, the interaction between host and pathogen populations and changes in gene frequencies are developed and their properties explored. For microparasitic organisms that exhibit epidemic patterns of population behaviour, frequency-and density-dependent selective pressures are of particular importance in maintaining host genetic diversity. In the case of macroparasites, which are endemic in character, heterozygous advantage, migration and sex linkage are likely to be of greater significance. Where possible theoretical predictions are compared with observed trends.

Introduction

The members of an invertebrate population characteristically differ in their vulnerability to infectious disease agents. This heterogeneity may arise as a result of a wide variety of factors which include host age and sex, past experience of infection, host nutritional status and the genetic constitution of an individual. Within invertebrates, genetic factors may control a variety of processes relevant to host acquisition and survival of infection. Of particular interest in the context of invertebrate immunity is the genetic control of the host's ability to recover from infection. Laboratory selection experiments clearly demonstrate that resistance or vulnerability to infection are genetically determined traits (Macdonald 1962; Richards 1975). Whether resistance is a consequence of host defence mechanisms, or simply due to an inadequate physiological environment, is less certain. Some of the most

ZOOLOGICAL SYMPOSIUM No. 56
ISBN 0-19-854004-3

detailed studies concern the arthropod vectors of infectious diseases of agricultural, veterinary or medical significance. Ward (1963), in experiments involving the selection of strains of the mosquito *Aedes aegypti* resistant to infection by a malarial parasite, *Plasmodium gallinaceum* (Fig. 1), found that in 26 generations of laboratory selection, strain vulnerability to infection was decreased by roughly 98%. Other examples are provided by studies of bacterial-virus (bacteriophage) interactions (E.S. Anderson 1957). In chemostat studies of viral and bacterial population growth, phage-resistant bacteria often arise by mutation and enjoy a significant selective advantage over the vulnerable strain (Fig. 2) (see E.S. Anderson 1957; Levin & Lenski 1983). Surprisingly, in many such studies involving a wide variety of invertebrate species, resistance appears to be a simple genetic character controlled by a single or small group of genes. Invariably the resistance trait is partially or completely dominant.

These experimental observations raise a number of interesting questions. First, and most important, is the issue of whether genetically controlled resistance is associated with the invertebrate host's defence system. Secondly, if resistant hosts are at a selective advantage in an infected population why do they not totally replace the vulnerable strain? In natural habitats, the widely observed occurrence of vulnerable and resistant strains in the same population implies the maintenance of balanced polymorphisms. Lastly, given that we have some answers to the previous two questions, how important is variability in resistance to infection in maintaining a stable interaction between host and parasite populations?

This paper examines certain aspects of these questions, placing particular emphasis on the population genetics and dynamics of host-parasite associations. Throughout, use is made of simple mathematical models but, where possible, reference is made to appropriate biological examples. The paper concludes with a discussion of theoretical predictions in the light of observed patterns.

Immunity in invertebrates

Vertebrate immune systems, with their ability to distinguish between self and non-self, are characterized by specificity, dissemination, amplification and memory (Roitt 1984). These features mean that a second or later exposure to a particular infection will, in general, evoke an accelerated response, even at a site remote from that of the primary infection. The ability to mount an enhanced response on second exposure to a specific antigen is termed acquired immunity and is thought to arise via 'immunological memory' generated by the sensitization and clonal selection of small, long-lived lymphocytes which recognize specific antigens. Invertebrates clearly have the ability to recognize 'non-self' or 'foreign' material, as discussed in

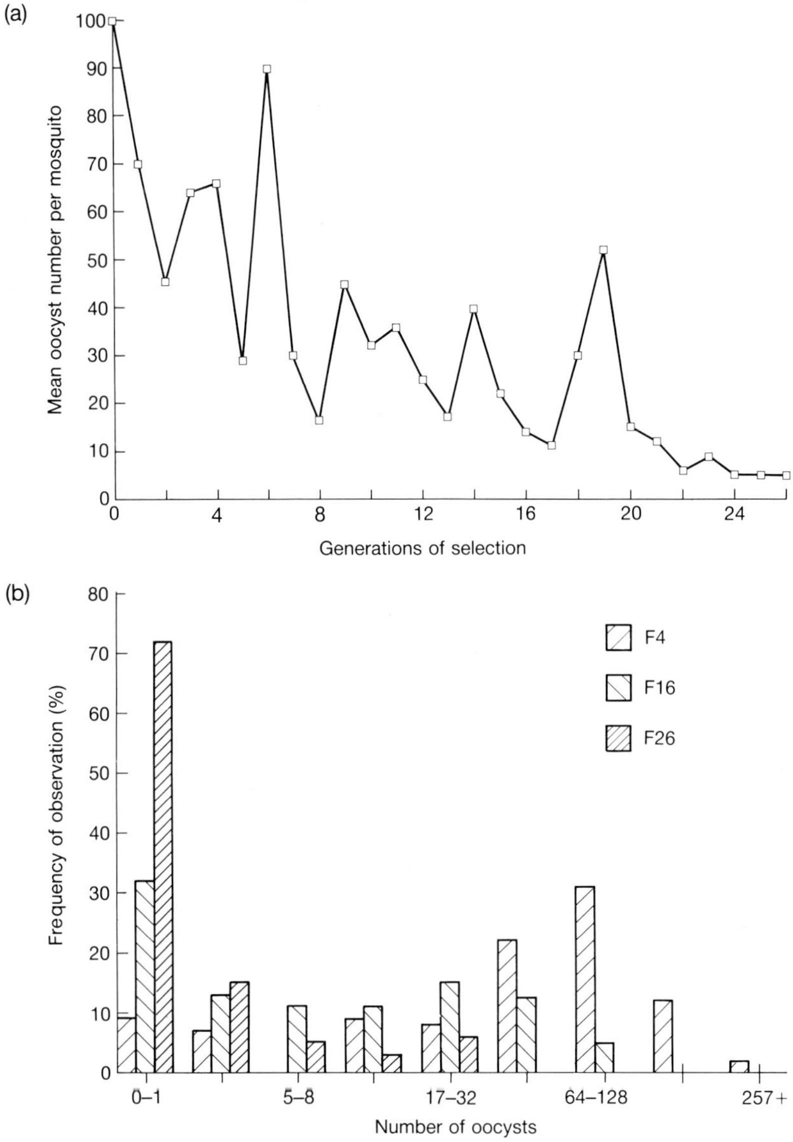

Fig. 1. Selection of a strain of *Aedes aegypti* resistant to infection by *Plasmodium gallinaceum* (data from Ward 1963).
(a). The mean oocyst count after successive generations of selection.
(b). The frequency distributions of malarial oocysts in hosts infected after varying intervals of selection. As the duration of selection progressed from generation 4 (F4) to generation 26 (F26) the distribution changed from an essentially unimodal form with a high mean oocyst count to a bimodal distribution with a low mean oocyst count.

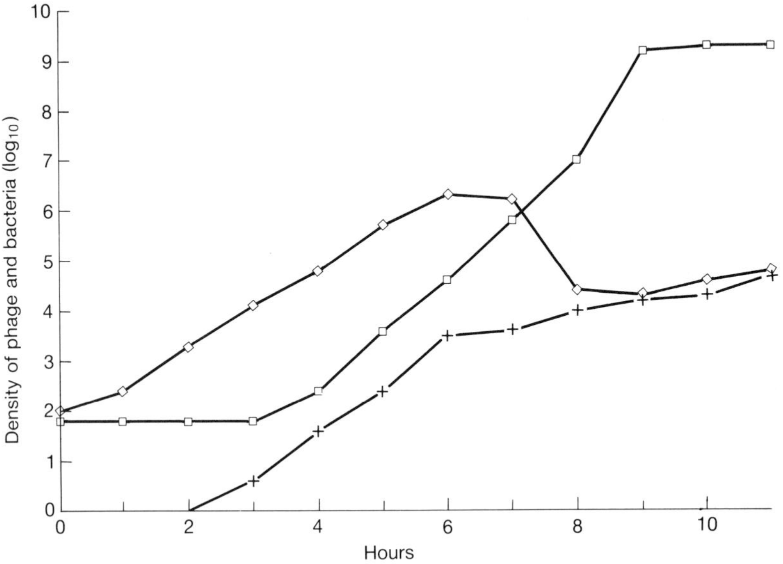

Fig. 2. The evolution of a strain of *Salmonella typhi* (V_i–type *A*) resistant to a bacteriophage (data from E.S. Anderson 1957). Note that the phage population does not begin to increase in size until the density of bacteria is in excess of 10^4 — 10^5 cells. Thereafter the resistant bacterium evolves and is selected for so that by time 10 h it dominates the population and halts phage population growth.
◇, susceptible bacteria; □, free phage; +, resistant bacteria.

other chapters in this volume. They are able to mount cellular or humoral responses to parasitic invasion which are often very effective in promoting host recovery from infection. The cellular response in insects, for example, involves encapsulation by haemocytes (Salt 1970) while in molluscs amoebocytes and fibroblasts play an important role. Following infection there is often a rapid increase in the number of effector cells (haemocytes, phagocytic amoebocytes and fibroblastic cells) and, as such, the defence system involves a component of amplification. Dissemination also occurs via the circulatory blood cells in molluscs and insects which represent the primary effector component involved in the internal defence mechanisms. These cells are known to participate in a variety of defence-related processes, including endocytosis (phagocytosis and pinocytosis), cytotoxic reactions and capsule formation (Bang 1975; Lackie 1980; Ratcliffe & Rowley 1981; Bayne 1983). Recent research suggests that in certain invertebrates a degree of specificity is involved (Yoshino & Granath 1985). For example, in molluscs the population of circulatory cells appears to be composed of molecularly distinct cell subpopulations (each expressing distinctive surface epitopes) capable of some degree of functional compartmentalization (Renwrantz & Cheng 1977a,b).

The invertebrate defence system is therefore qualitatively similar to that of vertebrates in its ability to recognize 'non-self' and the characteristics of amplification, dissemination and, perhaps, specificity. The missing character, however, is thought to be memory (Anderson & May 1981). For example, the experiments of Bayne & Kime (1970) involving repeated exposure of the land snail (*Helix pomatia*) to a bacterial infection of the haemolymph suggested that the rate of elimination of the pathogen was similar in first, second and third exposures to infection. For short-lived invertebrate species a memory component to the defence system is of limited value to the fitness of the species provided short life expectancy is not entirely a consequence of mortality induced by infectious agents (Anderson 1981). For long-lived species, however, such as some decapod crustaceans which only attain reproductive maturity after a lengthy period of growth and development, a memory component would seem to be essential to the reproductive success of the species (McKay & Jenkin 1969, 1970a,b). It therefore appears likely that certain higher invertebrates have evolved a memory component to the defence systems, perhaps via long-lived subpopulations of circulatory cells specific to particular classes of antigens. The presence or absence of memory is of great significance as a determinant of the population dynamics of host-pathogen systems (see Anderson & May 1979, 1981; Anderson 1981).

In this paper the word *immunity* is used to denote a defence mechanism with the capabilities of recognition of foreign antigen, of amplification and of dissemination. The degree of specificity to the antigen is not regarded as a prerequisite for immunity, since non-specific responses can clearly be very effective in the elimination of an infection. The term *acquired immunity* is used to imply the presence of a memory component in the defence system. The words *resistant* and *vulnerable* are employed to denote a genetic basis to the effectiveness of the host defences.

Pathogen populations within individual hosts

The significance of invertebrate immunity to the transmission success of pathogens within populations of hosts is in part determined by the course of infection within an individual host. This section briefly examines the dynamics of the interaction between pathogen replication within a host and the effector cells of the invertebrate defence system.

A very simple interpretation of invertebrate defences and pathogen replication can be captured by two equations denoting changes in the densities of 'effector cells' (haemocytes in the blood of the host), $E(t)$, and parasites, $P(t)$, within the host. It is assumed that the density of cells is determined by a constant input term, Λ (representing recruitment to the population in a manner analogous to the production of primordial lymphocytes in mammals by the reticuloendothelial system), a mortality term, μ, and an amplification or proliferation term arising from contacts between cells and antigen (the parasite). Proliferation is assumed to be directly proportional to the densities of cells and parasites (a mass action term) with the constant of proportionality defined as γ. The parasite is assumed to replicate at a net (births minus natural mortality) per capita rate, r, and to be removed (by phagocytosis or encapsulation etc.) at a net rate directly proportional to the densities of cells and parasites with contact coefficient β. These assumptions lead to the following pair of differential equations.

$$dE/dt = \Lambda - \mu E + \gamma PE \tag{1}$$

$$dP/dt = rP - \beta PE \tag{2}$$

The pathogen killing term in Equation 2 (e.g. βPE) can be taken to represent cellular or humoral factors, or both combined. Humoral responses may, in certain instances, be triggered by cell recognition of foreign antigen and, as such, their severity is likely to be associated with cell density.

The simple model of the invertebrate immune response to a replicating antigen, defined in Equations 1 and 2, has some interesting properties. In the absence of infection in a naive host ($P = 0$), the system settles to an equilibrium state, E^*, where effector cell density is set at Λ/μ (immigration rate/death rate). When a naive host acquires infection the parasite population increases in density provided the following condition is satisfied;

$$r > \Lambda\beta/\mu \tag{3}$$

If not, the initial inoculum of the pathogen (P_0) decays to zero as time progresses. Equation 3 simply states that the replication rate of the pathogen, r, must exceed the product of the equilibrium density of effector cells ($E^* = \Lambda/\mu$) times the rate of contact between cells and parasites, β. Given that Equation 3 is satisfied the system settles in the long term, to an equilibrium state where the densities of effector cells, E^*, and parasites, P^*, are given by;

$$E^* = r/\beta \tag{4}$$

$$P^* = [r - \Lambda\beta/\mu][\mu/(r\gamma)]. \tag{5}$$

Interestingly, the parasite persists within the host (provided Equation 3 is satisfied) irrespective of the magnitude of the effector cell proliferation parameter, γ. The model system is stable to small perturbations but has a marked tendency to exhibit oscillatory behaviour. The system exhibits weak damping to the stable point defined by Equations 4 and 5. The interval, T, between peak parasitaemias is approximately given by

$$T \simeq 2\pi(\mu r)^{-1/2}, \tag{6}$$

provided β is small in comparison to the values of μ or r. Illustrations of the various patterns of behaviour exhibited by the model are displayed in Fig. 3. The oscillatory behaviour of the system is of particular interest; temporal patterns show 'epidemic' peaks of parasite abundance in the host followed by peaks in effector cell density. Between peak parasitaemias the density of the parasite is held at a very low level (virtually eliminated) by high densities of effector cells (Fig. 3a). If the life expectancy of these cells is long relative to the life span of the host, following the initial explosion of the parasite population the host is virtually immune to reinfection as a consequence of the elevated levels of effector cells triggered by the first upsurge in parasite density (Fig. 3b). In these cases, the predicted interval between parasitaemias is longer than the life expectancy of the host. The defence system therefore has a memory component as a consequence of the long-lived circulatory cells (μ small relative to the intrinsic reproductive rate of the pathogen, r).

The time lag between the point of infection and the time when effector cell density expands rapidly is a consequence of a 'threshold effect' implicit in the structure of the model. From Equation 1 it can be seen that effector cell density only begins to expand rapidly once the parasite exceeds a critical density, P_T where

$$P_T \simeq \mu/\gamma \tag{7}$$

Clearly if P_T is larger than the density of parasites required to kill the host, then the defence system is ineffectual in dealing with the pathogen (Fig. 3d). The critical parameters in this context are the life expectancy of the effector cells ($1/\mu$) and the coefficient of cell proliferation γ. An effective defence system requires γ large and μ small, relative to the reproductive potential of the parasite, r. Innate resistance or vulnerability to a specific infection are therefore determined by a variety of parameters as defined in Equations 3 and 4. A totally resistant host would be one in which the condition defined in Equation 3 is not met. If this condition is satisfied the host is vulnerable and whether it survives infection or not depends on the relative magnitudes of the density of parasites required to kill the host and the threshold parasite density necessary to trigger effector cell proliferation (Equation 7). The effectiveness of the immune response depends on the values of the parameters γ (cell proliferation) and β (cytotoxicity), while acquired immunity depends on the life expectancy of the sensitized effector cells ($1/\mu$).

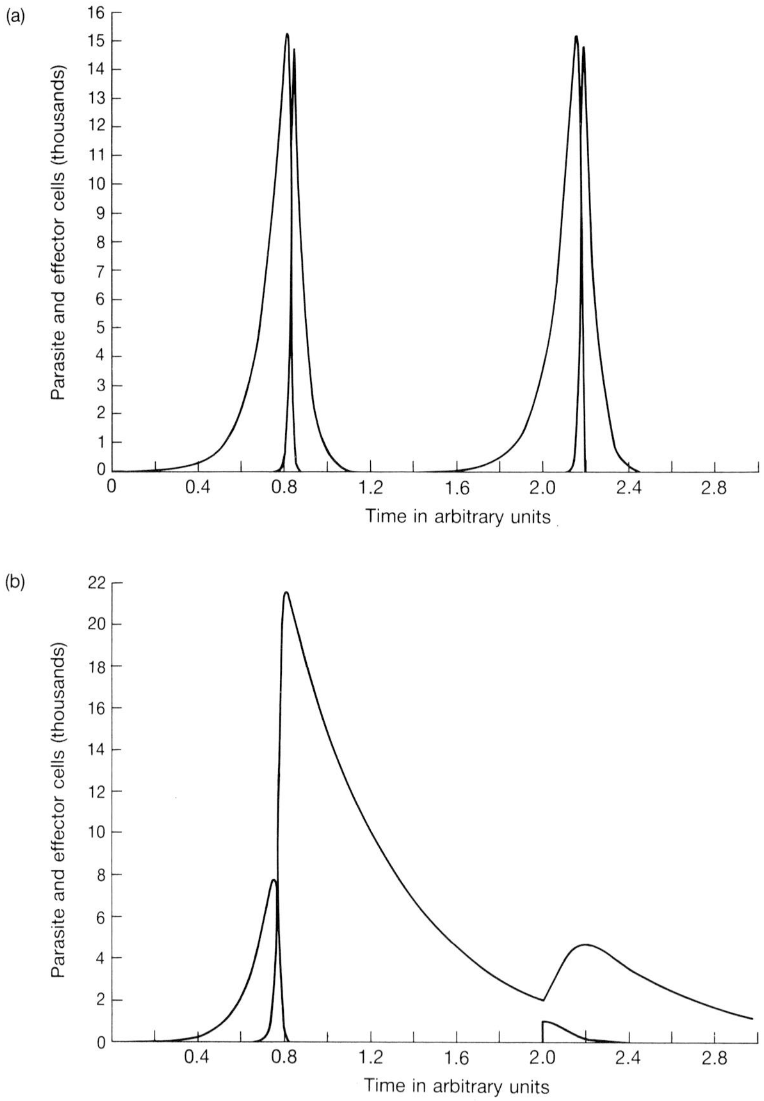

Fig. 3. Patterns of dynamical change in parasite and effector cell densities through time as predicted by Equations 1 and 2 in the main text.
(a). Oscillatory changes in parasite and effector cell densities (the leading peak is parasite density) induced by short-lived effector cells (no "memory").
Parameter values: $\Lambda = 1$, $\mu = 20$, $\gamma = 0.01$, $r = 10$, $\beta = 0.01$, $P(o) = 10$, $E(o) = \Lambda/\mu$.
(b). Similar to (a) but representing changes in parasite and effector cell densities under the assumption that effector cells are long-lived (a component of 'memory'). A 'challenge' infection introduced at time $t = 2$ is rapidly eliminated. Parameter values as for (a) except $\Lambda = 2$, $\mu = 2$, $\beta = 0.005$. Parasite peaks are in advance of peak effector cell density.

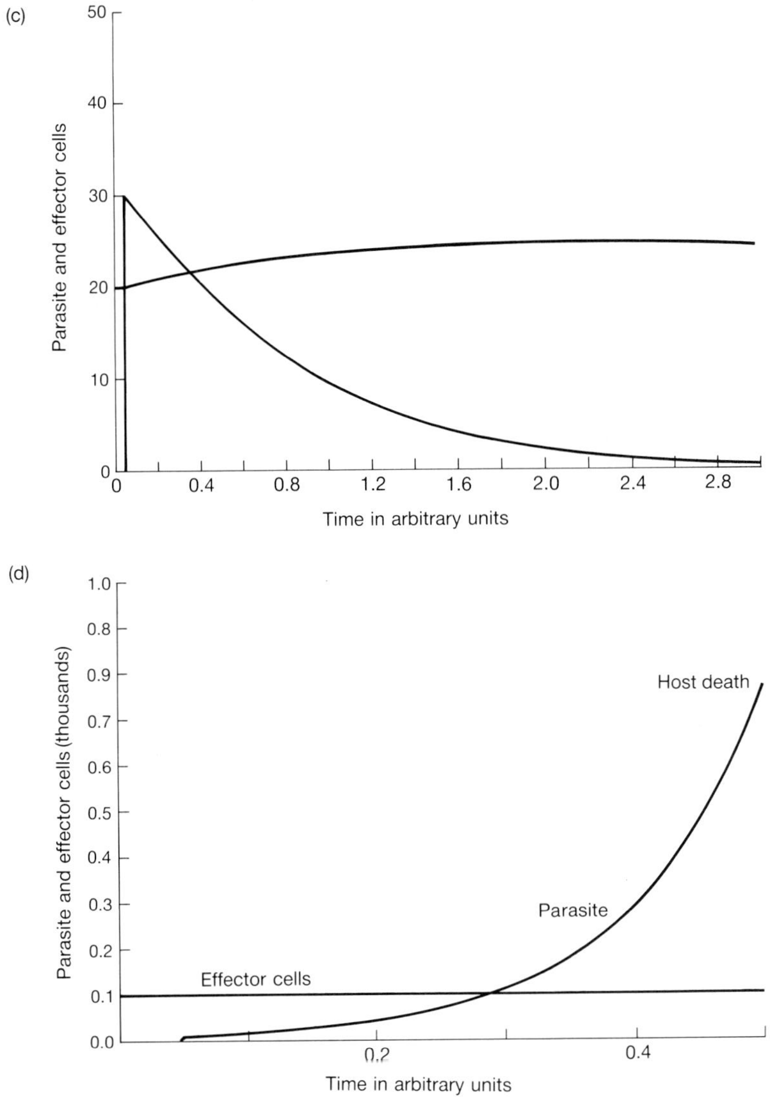

(c). Parasite reproductive rate too low to overcome the 'innate resistance' of the host so that after infection the parasite population decays to extinction. Effector cell density remains approximately constant. Parameter values: $\Lambda = 2$, $\mu = 0.1$, $\gamma = 0.01$, $r = 1$, $\beta = 0.1$, $P(o) = 30$, $E(o) = \Lambda/\mu$.

(d). Host response insufficient to suppress parasite population with the outcome that parasite density grows until the host is killed. Parameter values: $\Lambda = 2$, $\mu = 0.02$, $\gamma = 0.00001$, $r = 10$, $\beta = 0.001$, $P(o) = 10$, $E(o) = \Lambda/\mu$.

The incubation (time from infection to the point of infectiousness to other hosts) and infectious (duration of infectiousness) periods will be determined by the growth and decay patterns of parasite abundance in the host. For example, in Fig. 3a, the time period between the point of infection and the 'epidemic' in parasite abundance is equivalent to the incubation period, and the length of the epidemic is equivalent to the host's infectious period, given that infectiousness is in part determined by parasite density.

In summary, this very simple model of the invertebrate immune response to a replicating pathogen captures, in qualitative terms, certain important characteristics of host-parasite interactions: 1. innate resistance or vulnerability to infection (Equation 3); 2. recovery from infection (Fig. 3b); 3. incubation and infectious periods; 4. host death (Equation 7); and 5. a memory component to the defence system. The predicted pattern of parasite elimination following a first infection is very reminiscent of observed events as illustrated in Fig. 4. If the host population is heterogeneous with respect to the parameter values (defined in Equations 1 and 2) operating within an individual, then a specific parasite will create varied responses. These may include innate host-resistance, host death and host recovery, which may or may not result in acquired immunity. In future work on invertebrate immunity more attention should be focused on the demographic properties of effector cell populations, their genetic control and their interaction with the demography of the parasite population within the host.

The interaction between pathogen and host populations

The dynamical interactions between populations of hosts and parasites have been the subject of much research in recent years (see Anderson & May 1978, 1979; May & Anderson 1978, 1979). Some of this attention has been focused on invertebrate hosts (see Anderson & May 1981) and the principal conclusions in the context of invertebrate immunity are well illustrated by reference to a simple population model of the interaction between an invertebrate host and a microparasite (virus, bacterium or protozoan which undergoes direct replication within an individual host). We consider a compartmental model in which the host population is divided into susceptible and infectious individuals of density $X(t)$ and $Y(t)$ at time t respectively. The equations are:

$$dX/dt = aX - bX - \beta XY + \gamma Y \tag{8}$$

$$dY/dt = \beta XY - (b + \alpha + \gamma)Y \tag{9}$$

Here a is the per capita host birth rate, b is the per capita natural death rate, β is a transmission coefficient, α is the disease-induced death rate and γ

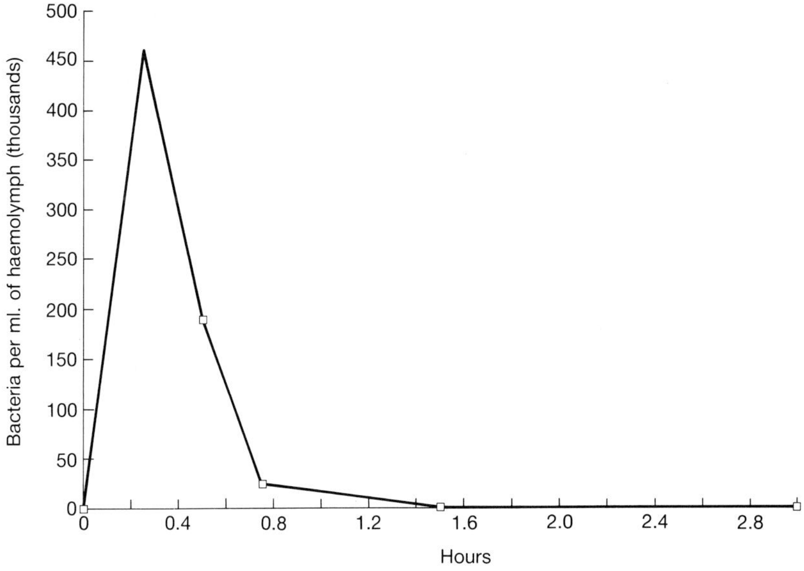

Fig. 4. Elimination of bacteria from the land snail *Helix pomatia* by circulatory amoebocytes (from Bayne & Kime 1970). Note how the defence mechanisms rapidly overcome pathogen population growth.

is the per capita recovery rate of infected hosts. This model assumes that infected hosts do not reproduce, that the disease increases host mortality and that infected hosts are able to recover from infection but not acquire immunity to reinfection. With respect to the model discussed in the previous section, which described the course of parasite infection in an individual host, recovery is equivalent to the reduction of parasitaemia to a very low level (see Fig. 3a). For simplicity, the incubation period of infection (hosts infected but not infectious) is ignored in Equations 8 and 9.

An important general concept to emerge from this simple model concerns the existence of a critical threshold density of susceptible hosts, X_T, below which the infection is unable to persist within the host community. The value of X_T is given from Equation 9 as

$$X_T = (b + \alpha + \gamma)/\beta \qquad (10)$$

The critical density is low if transmission efficiency is high (β large) and high if the rates of disease-induced mortality and recovery from infection are high. The component $1/(b + \alpha + \gamma)$ denotes the average infectious period of an infected host and if this is short ($(b + \alpha + \gamma)$ large) then the critical density for disease persistence is high. In a totally susceptible population of density X, the average number of secondary cases of infection generated by one primary case (the basic reproductive rate or transmission potential of the infection), R_o, is defined as

$$R_o = \beta X/(\alpha + b + \gamma) = X/X_T \quad (11)$$

Clearly R_o must exceed unity in value for disease persistence, which implies that $X > X_T$.

In the absence of the parasite the host population grows exponentially with parameter $r = (a-b)$ given that $a > b$. When the parasite is present the host population is regulated to a stable equilibrium state, provied $R_o > 1$, where

$$X^* = (\alpha + b + \gamma)/\beta = X_T \quad (12)$$

$$Y^* = (a - b)X^*/(b + \alpha) \quad (13)$$

and

$$N^* = X^* + Y^* = X^*(a + \alpha)/(b + \alpha) \quad (14)$$

If $\gamma = 0$, the system is neutrally stable and identical in structure and behaviour to the classical Lotka-Volterra equations of predator-prey interactions. The equilibrium density of susceptibles, X^*, is equal to the critical threshold density X_T for disease persistence. An interesting test of this prediction is provided by the observations of Benz (1962) on a nuclear polyhedrosis viral infection of the forest tent caterpillar *Malacosoma disstria*. The densities of susceptible hosts in areas of endemic infection were similar to those of virus-free populations at lower altitudes in the Canadian forest habitat.

The density of hosts, N^*, is related to the rate of recovery of infected individuals (i.e. the ability of the hosts to mount an effective defence to parasitic invasion). The total density N^* decays as the rate of recovery, γ, declines (Fig. 5a) and the proportion of infected hosts that recover is determined by the ratio $\gamma/(b + \alpha + \gamma)$. The major conclusion to emerge from this model is that high rates of host recovery from infection (an effective immunological defence system) increase the critical density of susceptible hosts necessary for the persistence of the infection within the host population. If the parasite is able to persist the equilibrium density of hosts is directly related to the speed of recovery from infection (Anderson & May 1981).

Parasite persistence within the host population is made less likely if those hosts that recover acquire a degree of immunity to reinfection. The simple model defined by Equations 8 and 9 can be modified to illustrate this point. If $Z(t)$ is the density of immune hosts at time t, then the new model is of the form

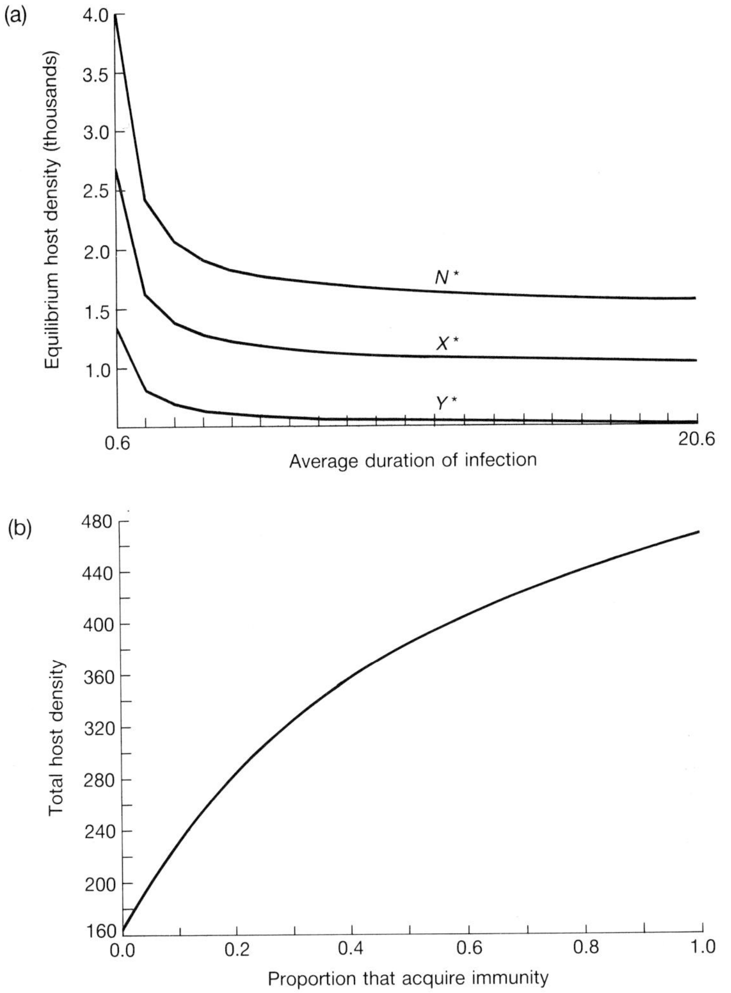

Fig. 5. (a). The impact of the rate of recovery from infection, γ, (average duration of infection $= 1/\gamma$) on equilibrium host density (see text for details). Predictions of model defined by Equations 8 and 9.
N^* = total density, X^* = density of susceptibles,
Y^* = density of infecteds. Parameter values: $a = 1$,
$b = 0.05$, $\beta = 0.001$, $\alpha = 0.5$, $\sigma = 0$.
(b). The impact of acquired immunity on the equilibrium host density (see text for details). Predictions of Equations 15–17 in the main text.

$$dX/dt = aX - bX - \beta XY + \gamma(1-p)Y + \sigma z \qquad (15)$$

$$dY/dt = XY - (b + \alpha + \gamma)Y \qquad (16)$$

$$dZ/dt = p\gamma Y - (b + \sigma)z \tag{17}$$

Here γ denotes the rate of recovery from infection of hosts, p denotes the proportion that acquire immunity to reinfection and σ records the rate of loss of immunity. The critical density of susceptibles necessary for disease persistence (and hence the equilibrium density X^*) remains as defined in Equation 10. The equilibrium densities of infected, Y^*, immune, Z^*, and all hosts, N^*, are;

$$Y^* = (a - b)X^*/[(b + \alpha + \gamma p) - \sigma p\gamma/(b + \sigma)] \tag{18}$$

$$Z^* = p\gamma Y^*/(b + \sigma) \tag{19}$$

$$N^* = X^* + Y^* + Z^* \tag{20}$$

The system is again stable but exhibits weakly damped oscillatory behaviour following perturbation from the equilibrium state (Equations 18 to 20). Of particular interest, in the context of immunity to infection, is the influence of the parameter p on the equilibrium density N^*. This is illustrated graphically in Fig. 5b where it can be seen that host density rises in a non-linear manner as the proportion that acquire immunity increases.

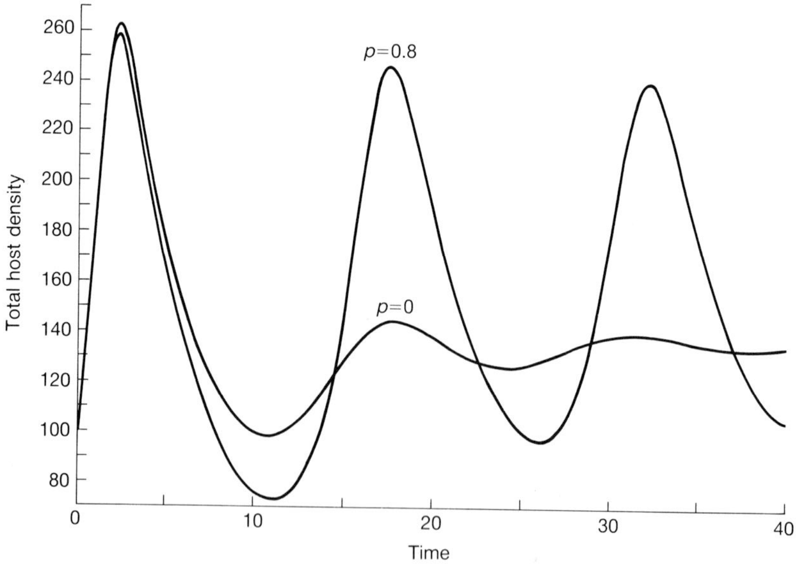

Fig. 6. Changes in host density ($N(t)$) through time as predicted by the model defined in Equations 15–17 in the main text for different values of the parameter p that determines the proportion of hosts that acquire immunity or recover from infection. Parameter values: $a = 1$, $b = 0.01$, $\alpha = 0.1$, $\gamma = 0.05$, $\beta = 0.01$.

The general point to emerge from these models is that the abilities to recover from infection and acquire immunity to reinfection on recovery both decrease the degree to which the parasite can suppress host density below the level that would be achieved in the absence of infection. A good defence system, which facilitates rapid recovery from infection, decreases the parasite's ability to persist in low-density host populations (high γ values create high X_T values, see Equation 10). Acquired immunity has no influence on the critical host density for pathogen persistence (X_T), but enhances the oscillatory behaviour (recurrent epidemic cycles) of the interaction between host and parasite populations (Fig. 6).

Population genetics of host-pathogen interactions

Host and parasite populations are genetically heterogeneous. Differences in phenotype between individuals in a population may be due to differences in genotype, or they may reflect non-heritable effects imposed on the individuals by environmental factors. Whether an observed trait is heritable or non-heritable is of particular importance in the study of host-parasite interactions. The resistance of a host to infection, for example, may reflect either genetic background or past exposure to a pathogen. A further problem arises when assessing the importance of observed variation to the fitness of the species. Many inherited characteristics are thought to be neutral in the context of natural selection (Crow & Kimura 1970). As such, even though host and parasite populations exhibit a great deal of phenotypic variation, only a proportion of this may be inherited, and of this proportion a smaller fraction will be of relevance to the fitness of the species.

Laboratory experiments provide a wealth of information on the significance of genetic factors as determinants of host responses to infection. The most intensively studied systems are those involving interactions between parasites and the laboratory mouse (Skamene, Kongshavn & Landy 1980). Vertebrate host resistance to infection appears in general to be a dominant character (partial or complete), irrespective of whether the pathogen is a virus, bacterium, protozoan or helminth (Table 1). However, there are a few exceptions to this trend such as the recessive control of acquired resistance to *Leishmania donovani* in mice (Blackwell, Freeman & Bradley 1980). The definition of resistance is complicated by the many facets of the mammalian immune system and by the often complex development cycles of pathogens within the host. Different genes, or groups of genes, control different host defence mechanisms such as non-specific immunity, specific cellular or humoral responses or acquired immunity. The genetic basis of the mammalian immune response is receiving a great deal of attention today (Wakelin

Table 1. Examples of dominance (partial or complete) in the genetic trait of resistance to parasitic infection.

	Host	Parasite	Author(s)
Invertebrates	*Culex pipiens*	*Plasmodium cathemerium*	Huff (1931)
	Aedes aegypti	*Plasmodium gallinaceum*	Kilama & Craig (1969)
	,, ,,	*Dirofilaria immitis*	McGreevy *et al.* (1974)
	,, ,,	*Brugia malayi*	Macdonald (1962)
	,, ,,	*Brugia pahangi*	Macdonald & Ramachandran (1965)
	Cicadulina mbila	Maize streak disease virus	Storey (1932)
	Biomphalaria glabrata	*Schistosoma mansoni*	Richards (1975)
Vertebrates	Laboratory mouse	*Leishmania donovani* (innate resistance)	Blackwell *et al.* (1980)
	,, ,,	*Leishmania tropica*	Skamene *et al.* (1980)
	,, ,,	*Salmonella typhimurium*	Skamene *et al.* (1980)
	,, ,,	*Trypanosoma rhodesiense*	Greenbatt, Rosenstreich & Diggs (1980)
	,, ,,	Measles virus	Rager-Zisman, Neighbour, Grace & Bloom (1980)
	,, ,,	Herpes simplex virus (type 2)	Kirchner, Engler, Zawatzky & Schroder (1980)
	,, ,,	*Trichinella spiralis*	Wakelin (1984)
	,, ,,	*Trichuris muris*	Wakelin (1984)
	,, ,,	*Taenia taeniformis*	Wakelin (1984)
	,, ,,	*Rickettsia tsutsugamushi*	Groves, Rosenstreich & Osterman (1980)
	,, ,,	*Listeria monocytogenes*	Cheers, McKenzie, Mandel & Yu Yu Chan (1980)

1984; Bodmer 1980), but our knowledge of invertebrate systems is very restricted at present. Invertebrate populations are genetically heterogeneous in their ability to withstand parasitic infection and the resistant trait is usually partially or completely dominant (Table 1) (Curtis & Graves 1983).

The genetic basis of susceptibility of mosquitoes to infection with filarial nematodes has been established for a variety of specific associations including *Culex fatigans* and *Wuchereria bancrofti*, *Aedes aegypti* and *Dirofilaria immitis*, and *A. aegypti* and *Brugia malayi* (Macdonald 1962; Thomas & Ramachandran 1970; McGreevy, McClelland & Lavoipierre 1974). In all these examples the resistant trait is dominant. Present understanding of snail susceptibility to schistosome infection is to a large extent due to the work of Richards (1970, 1973, 1975, 1977, 1984). Breeding and selection experiments have shown that the susceptibility of *Biomphalaria glabrata* to *Schistosoma mansoni* infection is under oligogenic control and that the resistance trait is probably controlled by a few genes with resistance dominant (Richards 1975; Rollinson & Southgate 1985). More generally, there is abundant evidence that individuals within natural populations differ in their resistance to viral, bacterial and fungal infections (L. Bailey 1973;

Maramorosch & Shope 1975). Martignoni (1957) found, for example, that the pathogenicity of a granulosis virus of the European larch budmoth, *Eucosoma griseana*, declined during an epidemic outbreak of infection in Switzerland, with larvae collected after the epidemic being much more resistant to infection than those collected at the onset of the outbreak. Similar changes were observed by Martignoni & Schmid (1961) in populations of the Californian oakworm, *Phryganidia californica*, infected with a nuclear-polyhedrosis virus.

The population implications of genetic variability in resistance to infection (the genetic control of the invertebrate host defence system) can be assessed by a variety of approaches, of which three are discussed below.

Heterogeneous host populations

The simplest approach to formulating a model of heterogeneity in resistance to infection is to consider the population dynamics of different strains of the host without reference to changes in gene frequency (see Anderson & May 1982; May & Anderson 1983). The simple model outlined in Equations 8 and 9 can be extended to mimic parasite transmission in a population of hosts containing a resistant and a vulnerable strain (denoted respectively by the subscripts 1 and 2) as follows:

$$dX_i/dt = r_i X_i - \sum_{j=1}^{2} \beta_{ij} X_i Y_j + \gamma_i Y_i \tag{21}$$

$$dY_i/dt = \sum_{j=1}^{2} \beta_{ij} X_i Y_j - d_i Y_i \tag{22}$$

where $d_i = (\alpha + b + \gamma_i)$ and $r_i = (a_i - b_i)$. It is assumed that the resistant (type 1) and vulnerable (type 2) hosts differ in their abilities to transmit infection (β_{ij}), recover from infection (γ_i) and reproduce (r_i). The transmission coefficient β_{ij} denotes transmission between susceptibles of type i and infectious hosts of type j. The resistant strain is assumed to be able to recover from infection much more rapidly than the vulnerable strain ($\gamma_1 > \gamma_2$).

The properties of Equations 21 and 22 have recently been examined by Holt & Pickering (in press) in the context of the impact of infectious diseases on competition between host species. However, the model is also appropriate for a simple description of heterogeneity in host vulnerability to infection (where the two types of host represent different strains as opposed to species). If we define the equilibrium densities of susceptibles and infecteds of resistant (type 1) and vulnerable (type 2) hosts as X_i^* and Y_i^*

when the total host population consists of only one strain, and as $\bar{X}_i$ and $\bar{Y}_i$ when both strains coexist, then at equilibrium

$$\frac{\bar{Y}_i}{\bar{X}_i} = \frac{r_i}{d_i - \gamma_i} = \frac{Y_i^*}{X_i^*} \tag{23}$$

Thus, for this model, it can be concluded that the equilibrium ratio of infected to susceptible individuals in each strain is independent of the presence of the alternative host strain. More generally, the equilibrium densities $\bar{Y}_i$ and $\bar{X}_i$ are given by

$$\bar{Y}_1 = \frac{Y_1^* - (\beta_{11}/\beta_{12})Y_2^*}{1 - (\beta_{12}/\beta_{11})(\beta_{21}/\beta_{22})} \tag{24}$$

$$\bar{Y}_2 = \frac{Y_2^* - (\beta_{21}/\beta_{22})Y_2^*}{1 - (\beta_{12}/\beta_{11})(\beta_{21}/\beta_{22})} \tag{25}$$

$$\bar{X}_i = \bar{Y}_i(d_i - \gamma_i)/r_i \tag{26}$$

where Y_i^* is given by Equations 12 and 13.

As shown by Holt & Pickering (in press), both strains only persist together (i.e. a heterogeneous host population or balanced polymorphism) provided

$$\beta_{11}\beta_{22} > \beta_{12}\beta_{21} \tag{27}$$

In other words, within-strain disease transmission must be greater than between-strain transmission. Note that the condition for stable coexistence does not depend on the parameter γ_i which denotes the rate of recovery from infection.

Aside from the question of coexistence, the model can provide insights into the question of the susceptibility of a single-strain population to invasion. Suppose a resistant strain evolves such that initially the resistant host is very rare. Under what circumstances will the strain increase and persist? First, either strain will only be able to establish provided Condition 27 is satisfied. If this is not the case, then the likelihood that the resistant strain will establish depends on the relative magnitudes of the inter- and intra-strain transmission coefficient (β_{ij}), the recovery rates (γ_i) and the intrinsic growth rate (r_i). The resistant strain will replace the vulnerable strain provided

$$\frac{r_1 d_1}{r_2 d_2} > \frac{\beta_{11}}{\beta_{21}} \text{ and } \frac{r_1 d_1}{r_2 d_2} > \frac{\beta_{12}}{\beta_{22}}. \tag{28}$$

Thus if the vulnerable strain enjoys a substantial reproductive advantage over the resistant strain ($r_2 >> r_1$) then the resistant strain will never be able to establish within the host population.

In summary, this model suggests that a stable polymorphism will always arise, provided within-strain transmission is greater than between-strain transmission. If stable coexistence is not possible then whether or not a resistant strain replaces the vulnerable strain depends critically on the reproductive rates of the two strains. In general, if resistance confers a substantial reproductive disadvantage then the vulnerable strain will always resist invasion by the resistant mutant. Fig. 7 illustrates these points by reference to numerical solutions of Equations 20 and 21 with different sets of parameter values. A fuller discussion of the properties of this model is given in Holt & Pickering (in press).

Changes in gene frequency

The model discussed in the previous section is concerned with changes in population density rather than changes in gene frequency (unless the host is haploid). Ideally, both should be considered within the same framework (May & Anderson 1983). However, a consideration of gene frequency models (which take no account of changes in population density) provides some insight into the mechanisms that maintain genetic diversity within host populations under selective pressures induced by infectious diseases. A central issue concerns the question of why host populations are not totally resistant to infection given that the resistant trait appears in general to be a dominant characteristic (such that in diploid organisms the heterozygous individuals exhibit resistance as opposed to vulnerability). Genetic polymorphism is defined as the occurrence together in the same locality of two or more strains of a species in such proportions that the rarest of them cannot be maintained merely by recurrent mutation (Ford 1975). It appears as though most host populations are polymorphic with respect to resistance to a defined infectious disease agent. Balanced polymorphism can be maintained in a variety of ways.

Heterozygous advantage

From the limited experimental evidence available at present, it appears as though the resistant trait may confer a disadvantage with respect to intrinsic reproductive potential, when compared with the trait of vulnerability. The most detailed quantitative study of this issue is that of Minchella & Loverde (1983) who demonstrated that strains of *Biomphalaria glabrata* vulnerable to infection by *Schistosoma mansoni* had, in the absence of infection, higher intrinsic reproductive potential (Fisher's R_o; see Fisher 1930) than resistant strains. The difference in the net reproductive rate, R_o, was approximately twofold. These authors, however, did not provide information on the

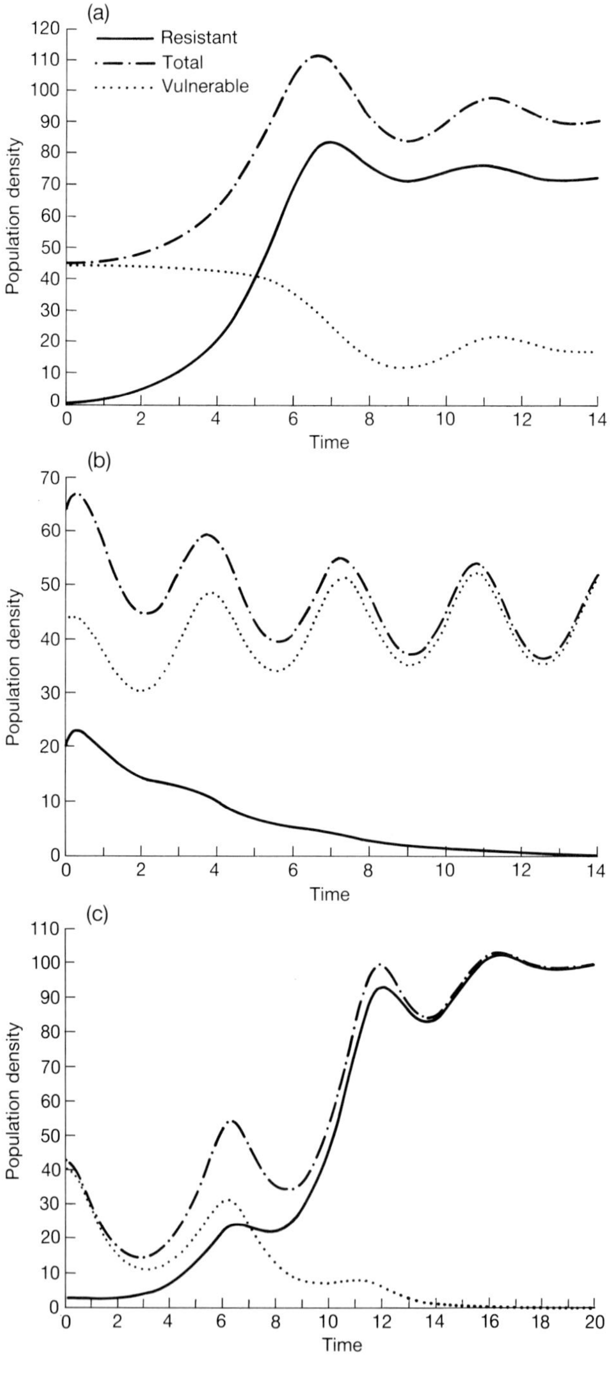
(a)
Resistant
Total
Vulnerable
Population density
Time
(b)
Population density
Time
(c)
Population density
Time

relative reproductive successes in the presence of infection of the resistant and vulnerable homozygotes and the resistant heterozygotes (assuming that resistance is conferred by a single or small group of linked genes and is dominant over vulnerability). It seems likely that the heterozygote may be at a selective advantage over both homozygotes given the concomitant influence of genes which confer resistance and genes that confer high reproductive potential. Further examples of heterozygous advantage are given by Macdonald (1962) and Townson (1971) for mosquitoes infected by filarial parasites. If we denote the resistant and vulnerable alleles as A and a respectively, and define the relative fitnesses of the genotypes AA, Aa and aa as $1-t$, 1 and $1-s$ respectively (where t and s lie between 0 and 1 and $t < s$ when the parasite is present) then in a large, randomly mating population, the equilibrium gene frequency, q, of the resistant genes A is given by

$$q = \frac{s}{s+t} \tag{29}$$

Ignoring the trivial equilibrium $q = 0$ and $q = 1$ (for situations in which the population only contains vulnerable or resistant genes respectively) then Equation 29 denotes a stable polymorphism with both the genes for resistance and vulnerability maintained within the population. Their relative frequencies will depend on the magnitude of the difference in the respective fitnesses of resistant and vulnerable hosts (e.g. if $t << s$, then q approaches unity in value).

This example of heterozygous advantage serves to illustrate one way in which genetic variability can be maintained. There is clearly a need, however, for detailed experimental studies to assess the frequency with which heterozygous individuals enjoy a relative fitness advantage over homozygous resistant and vulnerable strains (Townson 1971).

Migration

In a heterogeneous environment in which parasites are distributed unevenly between host patches so that some remain uninfected, migration of vulnerable host strains from uninfected to infected patches can maintain the stable coexistence of both strains in an infected population. Consider an example

---

Fig. 7. Heterogeneous host populations — invariant pathogen. Changes through time in the densities of vulnerable and resistant host strains and total population density (N) as predicted by Equations 21 and 22 in the main text.
(a). Stable coexistence of vulnerable and resistant strains (parameter values as for (b) except $\beta_{11} = 0.01$, $\beta_{12} = 0.01$, $\beta_{21} = 0.05$, $\beta_{22} = 0.1$).
(b). Vulnerable strain wins (parameter values:
$r_1 = 1$, $r_2 = 3$, $\alpha + b = 1$, $\beta_{11} = 0.01$, $\beta_{12} = 0.1$, $\beta_{21} = 0.05$, $\beta_{22} = 0.1$, $\gamma_1 = 1$, $\gamma_2 = 0.1$).
(c). Resistant strain wins (parameter values as for (b) except $r_2 = 1$, $\beta_{11} = 0.04$, $\beta_{12} = 0.1$, $\beta_{21} = 0.05$, $\beta_{22} = 0.1$).

in which two autosomal alleles segregate in a large randomly mating population in which the dominant resistant allele is denoted by A and the vulnerable recessive allele by a. The frequencies of the alleles A and a are denoted by q and p respectively and the three genotypes AA, Aa and aa are assumed to have the respective fitnesses of 1, $1–t$, and $1–s$ (vulnerability confers a selective disadvantage in the homozygote aa when the infection is present in the population). The parameters s and t adopt values between 0 and 1. Migration into the population is assumed to increase the frequency of the vulnerable genotype (aa) by a proportion m in each generation. After selection the frequencies of the genotypes AA, Aa and aa are respectively $q^2(1-m)$, $2pq(1-m)(1-t)$ and $p^2(1-s)(1-m) + m$. At equilibrium the frequency of the vulnerable allele, $\bar{p}$, is defined by the equation;

$$\bar{p}^3(2t - s) + \bar{p}^2(s - 3t) + \bar{p}(t + m) - m = 0 \qquad (30)$$

The positive real roots (in the interval 0–1) of this equation represent possible equilibria. An example of a stable polymorphic steady state is presented in Fig. 8. Migration is therefore able to maintain a stable polymorphism where the population consists of resistant and vulnerable genotypes. It is probable that migration is an important factor in maintaining genetic variability in vulnerability to infection within invertebrate host populations (it will be of particular relevance for highly mobile hosts, such as various species of winged insects (e.g. mosquitoes)).

Unequal selection in males and females plus sex linkage

In certain invertebrate host-parasite associations, infection is of unequal severity in male and female hosts and therefore applies an unequal selective pressure on the different sexes. In addition, the resistant trait may be sex-linked. A good example of both these effects is provided by the mosquito hosts of the filarial nematode parasites. The habit of blood feeding in these vector species is restricted to the females. The filarial parasites therefore have no influence on the survival or mating capabilities of the male organism. In addition, mosquito resistance to filarial infection has been shown by a variety of laboratory studies to be a sex-linked genetic trait (McGreevy *et al.* 1974; Macdonald 1962).

Both factors, acting alone or in combination, can maintain stable polymorphisms where both resistant and vulnerable hosts coexist (Haldane 1924, 1964; Haldane & Jayakar 1964; Wright 1969).

Frequency-dependent selection

Frequency-dependent selecion implies that the fitness of a genotype (sometimes of a gene) changes as its frequency alters within the population. This form of selection most commonly arises as a consequence of non-linear

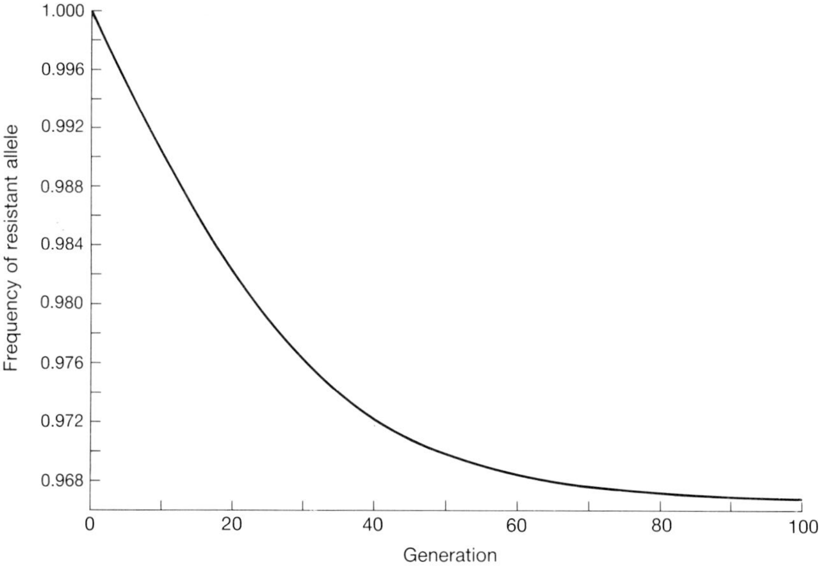

Fig. 8. Migration. Changes in the frequency of the resistant allele, A, through time following introduction of a vulnerable allele (by migration ($m = 0.001$, $s = 0.9$)) into a resistant population.

effects in two-species population interactions (Clarke 1976). Host-parasite associations are a good example (see Equations 11 and 12); parasite persistence appears to be highly dependent on host density. Below a critical density of susceptible hosts the parasite is unable to maintain itself and therefore exerts no selective pressure on the resistant and vulnerable genotypes. If vulnerable hosts have a greater reproductive potential in the absence of infection than resistant hosts, then the vulnerable genotype will increase in frequency when host density is too low to maintain parasite transmission. However, once host density rises above the critical level for parasite persistence, the resistant genotype will enjoy a selective advantage.

This advantage will continue until a high frequency of resistant hosts drives the parasite population to a low level to re-create a selective advantage for the vulnerable host. As such, we may envisage oscillatory fluctuations in gene frequencies corresponding with peaks or troughs in parasite abundance.

In the following example, the resistant allele is defined as A and is assumed to be completely dominant to the vulnerable allele defined as a. Let us assume that the fitnesses of the three genotypes AA, Aa and aa are $k[1-s_1(1-p^2)]$, $k[1-s_1(1-p^2)]$ and $k[1-s_2p^2]$ respectively where $p = (1-q)$ is the frequency of the vulnerable allele and k is a constant. The frequency-dependent selection functions are chosen here for mathematical simplicity rather than for their relevance to actual cases. The function for the genotype *AA* represents the assumption that the fitness of the resistant allele declines as its frequency increases in the population as a direct consequence of reduced parasite abundance. The vulnerable allele is assumed to confer a selective advantage over the resistant allele when the parasite is rare (Wright 1969). The recurrence relationship for the frequency of the vulnerable allele in generation $t+1$, p_{t+1}, is given by

$$p_{t+1} = \frac{p_t^2[1-s_2p_t^2] + p_t(1-p_t)[1-s_1(1-p_t^2)]}{(1-p_t^2)[1-s_1(1-p_t^2)] + p_t^2(1-s_2p_t^2)} \tag{31}$$

There is a stable polymorphic equilibrium point $\bar{p}$ (with both resistant and vulnerable alleles maintained in the population) at

$$\bar{p} = [s_1/(s_1 + s_2)]^{1/2} \tag{32}$$

provided $s_1/(s_1 + s_2)$ falls between 0 and 1, and both s_1 and s_2 are positive (Wright 1969; Clarke & O'Donald 1964). The behaviour of this model is not oscillatory but the choice of more complicated (and perhaps more realistic) frequency-dependent fitness functions can induce cyclic or chaotic changes in allele frequencies (see May & Anderson 1983). This topic is examined in the following section.

Population genetics and population dynamics

The previous section outlines various ways in which genetic variability in host resistance to infection can be maintained under the selective pressures exerted by a single species of pathogen. The models examined, however, make the assumption that factors other than pathogens ultimately determine the magnitude of the host population at the start of each generation. In other words, the host population is held to be constant and the fitness of the different genotypes is assumed to be independent of changes in host and pathogen densities. Such models are frequency-, but not density-dependent. The concomitant treatment of both population dynamic and population genetic factors raises some formidable problems. However, recent work has begun to focus on this issue and includes analyses of host-parasite and predator-prey interactions (see Gillespie 1975; Hamilton 1980; Roughgarden 1983; Anderson & May 1982; Kemper 1982; Beck 1984).

Discrete generations

Consider a diploid host population with discrete non-overlapping generations where resistance is again controlled by a single locus with two alleles, *A* (resistant) and *a* (vulnerable). Resistance is assumed to be completely dominant so that the heterozygote Aa has the same fitness as the homozygote AA. If p_t is the frequency of A and $q_t = 1-p_t$ the frequency of a, in generation t, the frequency p_{t+1} in the next generation (Crow & Kimura 1970) is given by

$$p_{t+1} = p_t W_{AA}/\bar{W}_t. \tag{33}$$

Here W_{AA} and W_{aa} are the fitnesses of the resistant and vulnerable genotypes respectively, and $\bar{W}_t$ is the average fitness in generation t,

$$\bar{W}_t = (1 - q_t^2)W_{AA} + q_t^2 W_{aa}. \tag{34}$$

It is now necessary to add to this description of gene frequency changes (Equations 33 and 34) a series of equations describing changes in the densities of the three genotypes AA, Aa and aa. For non-overlapping generations a convenient model is provided by an adaptation of the Reed-Frost discrete time epidemic equations (see Abbey 1952 and N.T.J. Bailey 1975). If X_{it} and Y_{it} (where $i = 1 \ldots 3$) are the densities of susceptible and infected genotypes of type i ($1 = AA$, $2 = Aa$ and $3 = aa$) at time t, then the difference equations for changes with respect to time are

$$X_{i,t+1} = X_{i,t} + a\theta_i \sum_{i=1}^{3} X_{i,t} - b_i X_{i,t} - X_{i,t}[1 - \prod_{j=1}^{3} (1 - \beta_{i,j})^{Y_{j,t}} \tag{35}$$

$$Y_{i,t+1} = Y_{i,t} + X_{i,t}[1 - \prod_{j=1}^{3} (1 - \beta_{i,j})^{Y_{j,t}}] - (b_i + \alpha_i)Y_{i,t} \tag{36}$$

It is here assumed that only susceptibles are able to reproduce (at a rate a, independent of genotype) and that no infecteds recover from infection. The term $\beta_{i,j}$ denotes the probability that a susceptible of genotype i has 'effective contact' (leading to infection) with an infective of genotype j (the value of $\beta_{i,j}$ must lie between 0 and 1). Random mating is assumed, so that the newly born susceptibles are apportioned among the three genotypes in the standard proportions $\theta_i = p^2, 2pq, q^2$ for $i = 1, 2, 3$, respectively. If only susceptibles reproduce then $\hat{p}$ denotes the frequency of gene A among susceptibles, namely

$$\hat{p} = (X_1 + X_2/2)/(X_1 + X_2 + X_3).$$

More generally the frequency of gene A in the total population is given by

$$p = (X_1 + X_2/2 + Y_1 + Y_2/2)/(X_1 + Y_1 + X_2 + Y_2 + X_3 + Y_3) \tag{37}$$

Changes in allele frequency and population density can be calculated via Equations 32 and 36.

Many simplifying assumptions have been made in the construction of this frequency- and density-dependent model but the structure remains complex. The properties of first-order non-linear difference equations have received much attention in recent years (e.g. May 1976). As non-linearities become more marked, stable equilibrium points give way, by a bifurcation process, to a cascade of period doublings and then to a regime of apparent chaos where the trajectories are effectively indistinguishable from the sample function of a random process.

The model defined by Equations 33–37 may, for certain parameter values, exhibit complex oscillatory behaviour for both changes in population density and allele frequency. In general, polymorphism can only be maintained provided resistance carries some cost such as reduced reproductive performance (i.e. in Equation 36 the b_i for resistant genotypes being greater than that for vulnerable hosts). An example of the maintenance of both resistant and vulnerable hosts within the population is given in Fig. 9. An oscillatory pattern is recorded where the frequency of the resistant allele exhibits complex cycles. The recurrent epidemics of infection in resistant and vulnerable hosts are out of phase owing to the different mortality rates of the host strains and the different rates of disease transmission (Fig. 9).

Continuously overlapping generations

Discrete generation models (difference equations) are more likely to exhibit oscillatory behaviour than their overlapping generation counterparts (differential equations) as a consequence of the instabilities produced by the time delays implicit in such formulations (the generation delay between successive bouts of reproduction; see May (1974)). An alternative approach to that outlined in the previous section is to consider an appropriate description of genetic and population dynamic changes in terms of differential equations.

For a diploid host there will be three genotypes of susceptibles and infectives labelled X_i and Y_i respectively where the subscript i denotes genotype ($1 = AA$, $2 = Aa$, $3 = aa$). The dynamics of the system are described by a generalization of Equations 8 and 9;

$$dX_i/dt = a\theta_i \sum_{i=1}^{3} X_i - b_iX_i - X_i(\sum_{j=1}^{3} \beta_{ij}Y_j) + \gamma_iY_i \tag{38}$$

$$dY_i/dt = X_i(\sum_{j=1}^{3} \beta_{ij}Y_j) - (\alpha_i + b + \gamma_i)Y_i. \tag{39}$$

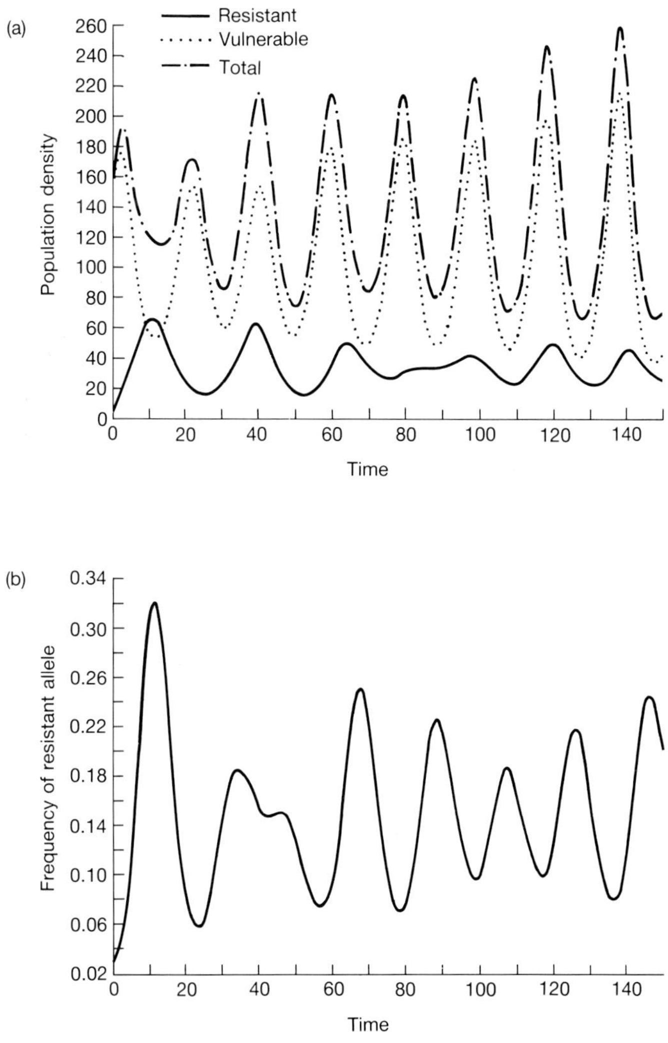

Fig. 9. Frequency- and density-dependent selection — non-overlapping generations.
(a). Changes in the total host density and the densities of vulnerable and resistant genotypes through time as predicted by Equations 33–37 in the main text (Parameter values. $a = 0.5$, $b_1 = 0.25$, $b_2 = 0.2$, $\alpha_1 = 0$, $\alpha_2 = 0.2$, $\beta_{11} = 0.019$, $\beta_{12} = \beta_{21} = 0.001$, $\beta_{22} = 0.01$.)
(b). Records the associated changes in the frequency of the resistant allele, p.

It is here assumed that the resistant and vulnerable hosts differ in their respective natural death rates (b_i), parasite-induced death rates (α_i), recovery rates (γ_i) and transmission rates between and within genotypes (β_{ij}). Random mating is assumed so that newly born susceptibles are apportioned among the three genotypes in the standard proportions $\theta_i = p^2, 2pq, q^2$ for

AA, *Aa* and *aa*, respectively. The gene frequency of the vulnerable allele in susceptible hosts, $\hat{q}$, is

$$\hat{q} = (X_3 + X_2/2)/(X_1 + X_2 + X_3) \qquad (40)$$

Note that Equations 38 and 39 assume that only susceptible hosts reproduce.

The study of Equations 38–40 is a difficult task owing to the large number of parameters (11 in total) and the dimensionality of the system (four equations).

A necessary condition for the persistence of both resistant and vulnerable hosts is $\beta_{11}\,\beta_{22} > \beta_{12}\,\beta_{21}$. This condition is identical to that derived from the simpler model given in Equations 21 and 22 (which represents population dynamic changes in the absence of genetic considerations; see Equation 27). However, the condition is not sufficient since the stability of the polymorphic equilibrium (both resistant and vulnerable genotypes maintained) is dependent on initial conditions. Preliminary numerical results show that the two host genotypes can coexist, or that one always excludes the other, or that either genotype can win depending on the initial conditions. The outcome depends on the relative values of the transmission coefficients, the disease-induced mortality rates and the reproductive rates. An example of stable coexistence is presented in Fig. 10 where damped oscillatory behaviour creates complex changes in gene frequencies before the system settles to a stable polymorphic equilibrium. The full dynamical properties of this system are yet to be determined although related work suggests that stable cyclic behaviour may arise for certain sets of parameters (Beck 1984). In general, however, the more exact description of genetic and population changes provided by the continuous time model (Equations 38 and 39) indicates that sustained oscillations are a rarer phenomenon than suggested by the discrete generation models.

Discussion

Theoretical studies indicate that a range of mechanisms can act to maintain stable polymorphisms within invertebrate host populations subject to selection by infectious disease agents. A summary of these results is presented in Table 2. However, too often in the field of population genetics, the natural situation and the theoretical model bear a disturbingly metaphorical relation to each other. Few data are available, either from laboratory or field studies, to enable critical evaluation of the relative significances of the mechanisms listed in Table 2. However, the theoretical results outlined in this paper raise a series of important biological questions. For example, does resistance confer a selective disadvantage in the absence of the parasite?

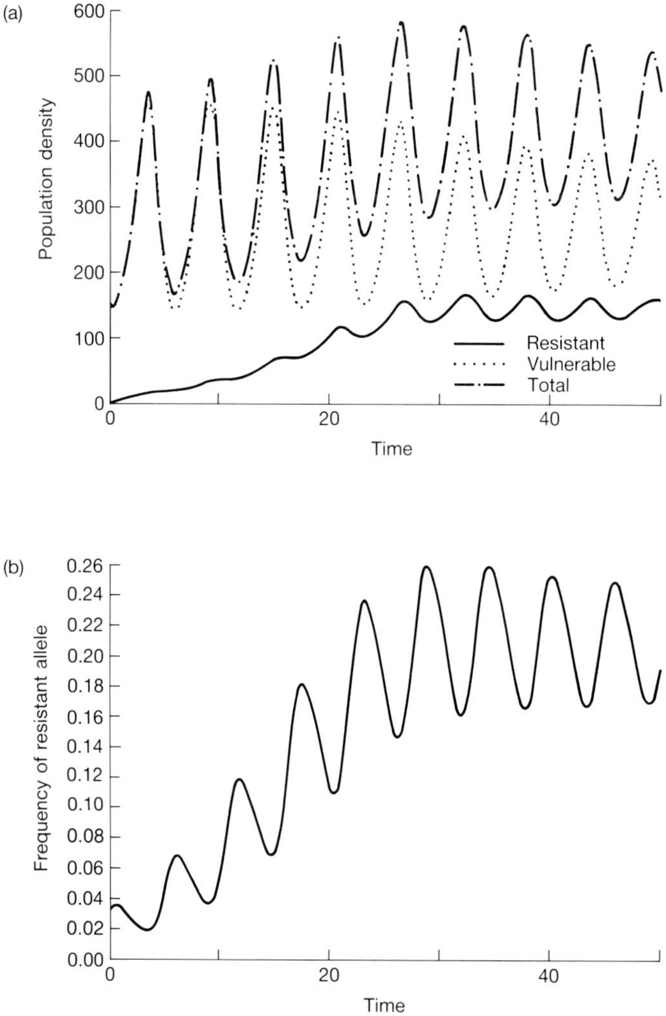

Fig. 10. Frequency- and density-dependent selection — overlapping generations.
(a). Changes in the total host density and the densities of vulnerable and resistant hosts through time as predicted by Equations 38 and 39 in the main text.
(Parameter values: $a = 1.7$, $\beta_{11} = \beta_{22} = 0.01$, $\beta_{12} = \beta_{21} = 0.001$, $\alpha_1 = 0.5$, $\alpha_2 = 1$, $b_1 = 1.5$, $b_2 = 1.0$, $\gamma_1 = \gamma_2 = 0$.)
(b). Records the associated changes in the frequency of the resistant allele p.

Alternatively, do heterozygous genotypes with resistant and vulnerable alleles have a selective advantage over homozygous resistant genotypes in the presence of the parasite? These questions are, in principle, amenable to experimental study. Work on bacteria-phage interactions (Levin & Lenski 1983) and mollusc-digenetic fluke associations (Minchella & Loverde 1983) provides some information on the first of these questions. Resistance appears to carry a fitness cost. As far as heterozygous advantage is concerned, however, very few quantitative experimental studies of invertebrate host-parasite systems have been published at present (see Macdonald 1962; Townson 1971; Minchella & Loverde 1983).

Data from field populations are extremely limited. What is required are studies of longitudinal changes in the proportion of resistant and vulnerable hosts in populations with fluctuating levels of parasite infection. For the 'epidemic' viral and bacterial infections there is clear evidence that the proportion of resistant hosts tends to increase following an epidemic (L. Bailey 1973; Martignoni 1957). With respect to 'endemic' infections (generally protozoan and helminth parasites), Paige & Craig (1975) have reported an interesting pattern of vulnerability of East African strains of *A. aegypti* to *Brugia pahangi* (a filarial nematode). Strains collected from inside houses or from domestic water supplies were largely resistant to infection, strains taken from areas remote from human habitation were mainly or partially vulnerable, and strains collected in areas between the inhabited regions and the remote areas adopted intermediate levels of resistance and vulnerability. This pattern may be interpreted as indicating strong selection against vulnerability in areas where the parasite is endemic. More generally, the highly aggregated frequency distributions of filarial nematodes in their insect vectors probably reflect genetic variability in host resistance to infection. In areas where the prevalence of infection is high in human communities most intermediate hosts are uninfected, while a few harbour high burdens of larval parasites (Fig. 11). Such patterns possibly reflect balanced polymorphisms where resistant and vulnerable hosts coexist in the same population.

Of the mechanisms listed in Table 2 it appears likely that frequency- and density-dependent selection are of greatest importance in maintaining polymorphism within host populations subject to microparasitic infection. These are in general more pathogenic to invertebrates than macroparasites, and they tend to be epidemic in character. Recurrent epidemics induce a fluctuating selective pressure and are therefore more likely to maintain genetic heterogeneity in the host population. More broadly, recent work has suggested that the fluctuating selective pressures induced by epidemics of infectious diseases may, in part, be responsible for the evolution of sexual reproduction (Hamilton 1980; May & Anderson 1983). In the case of stable endemic infections frequency- and density-dependent factors are probably of lesser significance. Under these circumstances heterozygous advantage, migration and sex linkage are likely to play significant roles in the maintainance of genetic diversity. However, these conclusions must be accepted with caution given the scarcity of experimental and field evidence.

In the future there is a need for attention to be focused not solely on the genetic mechanisms that control invertebrate immunity, but also on the

Table 2. Mechanisms that maintain polymorphisms within host populations subject to selection by infectious disease agents.

Mechanisms	Example(s)—comment: Host	Parasite	Reference(s)
1. Heterozygous advantage (over dominance)	*Biomphalaria glabrata*	*Schistosoma mansoni*	Minchella & Loverde (1983)
	Man	*Plasmodium falciparum*	Allison (1954)
2. Migration	(Probably important in many host-parasite associations)		This paper
3. Sex linkage	*Cicadulina mbila*	Maize streak disease virus	Storey (1932)
	Aedes aegypti	*Brugia malayi*	Macdonald (1962)
4. Unequal selection in the different sexes	*Aedes aegypti*	*Plasmodium gallinaceum*	Ward (1963)
	Aedes aegypti	*Brugia malayi*	Macdonald (1962)
5. Frequency-dependent	(Acts to a greater or lesser degree in all host-parasite associations)		Haldane (1949), Gillespie (1975), May & Anderson (1983)
6. Frequency- and density-dependent selection	(Acts to a greater or lesser degree in all host-parasite associations)		Anderson & May (1982), Beck (1984), Kemper (1982)

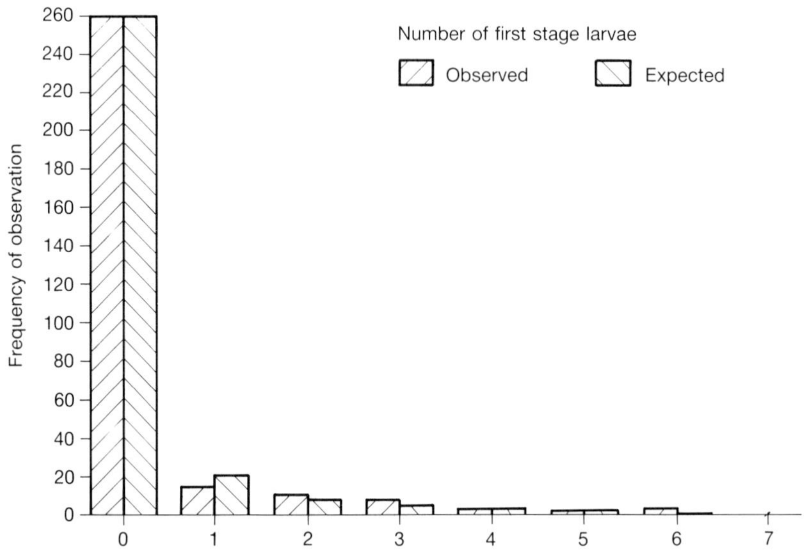

Fig. 11. Frequency distribution of larval filarial worms in a sample population of the mosquito host (data from Cheke, Garms & Kerner 1982). The fit of the negative binomial probability distribution is shown in the graph.

broader population implications of such processes. Geneticists invariably consider changes in gene frequencies without reference to changes in animal abundance, ecologists have tended to study changes in abundance without reference to changes in genetic structure, while invertebrate immunologists have focused on the mechanisms of resistance without reference to population or genetic changes. The integration of these approaches presents a fascinating challenge for future research.

References

Abbey, H. (1952). An examination of the Reed-Frost theory of epidemics. *Hum. Biol.* **24**: 201–33.

Allison, A.C. (1954). Protection afforded by sickle-cell trait against subtertian malarial infection. *Br. med. J.* **1**: 290–4.

Anderson, E.S. (1957). The relations of bacteriophages to bacterial ecology. *Symp. Soc. gen. Microbiol.* 7: 189–217.

Anderson, R.M.(1981). Vertebrate populations, pathogens and the immune system. In *Population and biology*: 249–68. (Symposium Volume of the International Union for the Scientific Study of Population.) (Ed. Keyfitz, N.). Ordina Editions, Liege, Belgium.

Anderson, R.M. & May, R.M. (1978). Regulation and stability of host-parasite population interactions. I. Regulatory processes. *J. Anim. Ecol.* **47**: 219–47.

Anderson, R.M. & May, R.M. (1979). Population biology of infectious diseases: part I. *Nature, Lond.* **280**: 361–7.

Anderson, R.M. & May, R.M. (1981). The population dynamics of microparasites and their invertebrate hosts. *Phil. Trans. R. Soc.* (B) **291**: 451–524.

Anderson, R.M. & May, R.M. (1982). Coevolution of hosts and parasites. *Parasitology* **85**: 411–26.

Bailey, L. (1973). Control of invertebrates by viruses. In *Viruses and invertebrates*: 533–53. (Ed. Gibbs, A.J.). Elsevier, Amsterdam.

Bailey, N.T.J. (1975). *The mathematical theory of infectious diseases and its applications.* Charles Griffin, London.

Bang, F.B. (1975). Phagocytosis in invertebrates. In *Invertebrate immunity*: 137–51. (Eds Maramorosch, K. & Shope, R.E.). Academic Press, New York, San Francisco & London.

Bayne, C.J. (1983). Molluscan immunobiology. In *Biology of Mollusca*: 407–86. (Eds Saleuddin, A.S.M. & Wilbur, K.M.). Academic Press, New York.

Bayne, C.J. & Kime, J.B. (1970). *In vivo* removal of bacteria from the hemolymph of the land snail *Helix pomatia* (Pulmonata: Stylommatophora). *Malac. Rev.* **3**: 103–13.

Beck, K. (1984). Coevolution: mathematical analysis of host-parasite interactions. *J. math. Biol.* **19**: 63–78.

Benz, G. (1962). Untersuchungen über die Pathogenität eines Granulosis-Virus des Graven Larchenwicklers, *Zeiraphera diniana* (Guence). *Agron. Glas.* **34**: 566–74.

Blackwell, J., Freeman, J. & Bradley, D.J. (1980). Influence of H-2 complex on acquired resistance to *Leishmania donovani* infection in mice. *Nature, Lond.* **283**: 72–4.

Bodmer, W.F. (1980). The HLA system and diseases. *J. R. Coll. Phys. Lond.* **14**: 43–50.

Cheers, C., McKenzie, I.F.C., Mandel, T.E. & Yu Yu Chan (1980). A single gene (Lr) controlling natural resistance to murine listeriosis. In *Genetic control of natural resistance to infection and malignancy*: 141–148. (Eds Skamene, E., Kongshavn, P.A.L. & Landy, M.). Academic Press, London.

Cheke, R.A., Garms, R. & Kerner, M. (1982). The fecundity of *Simulium damnosum* s.l. in northern Togo and infections with *Onchocerca* spp. *Ann. trop. Med. Parasit.* **76**: 561–8.

Clarke, B. (1976). The ecological genetics of host-parasite relationships. *Symp. Br. Soc. Parasit.* **14**: 87–103.

Clarke, B. & O'Donald, P. (1964). Frequency-dependent selection. *Heredity* **19**: 201–6.

Crow, J.F. & Kimura, M. (1970). *An introduction to population genetics theory.* Harper and Row, New York.

Curtis, C.F. & Graves. P.M. (1983). Genetic variation in the ability of insects to transmit filarial, trypanosome, and malarial parasites. In *Current topics in vector research* **1**: 32–62. (Ed. Harris, K.F.). Praeger, New York.

Fisher, R.A. (1930). *The genetical theory of natural selection.* Clarendon Press, Oxford.

Ford, E.B. (1975). *Ecological genetics.* Chapman & Hall, London.

Gillespie, J.H. (1975). Natural selection for resistance to epidemics. *Ecology* **56**: 493–5.

Greenbatt, H.C., Rosenstreich, D.L. & Diggs, C.L. (1980). Genetic control of natural resistance to *Trypanosoma rhodesiense* in mice. In *Genetic control of natural resistance to infection and malignancy*: 89–96. (Eds Skamene, E., Kongshavn, P.A.L. & Landy, M.). Academic Press, New York.

Groves, M.G., Rosenstreich, D.L. & Osterman, J.V. (1980). Genetic control of natural resistance to *Rickettsia tsutsugamushi* infection in mice. In *Genetic control of natural resistance to infection and malignancy*: 165–72. (Eds Skamene, E., Kongshavn, P.A.L. & Landy, M.). Academic Press, New York.

Haldane, J.B.S. (1924). A mathematical theory of natural and artificial selection. Part II. The influence of partial self-fertilization, inbreeding, assortative mating and selective fertilization on the composition of Mendelian populations and on natural selection. *Proc. Camb. phil. Soc. biol. Sci.* **1**: 158–63.

Haldane, J.B.S. (1949). Disease and evolution. *Ricerca scient. Suppl.* **19**: 68–76.

Haldane, J.B.S. (1964). Conditions for stable polymorphism at an autosomal locus. *Nature, Lond.* **193**: 1108.

Haldane, J.B.S. & Jayakar, S.D. (1964). Polymorphism due to selection depending on the composition of a population. *J. Genet.* **58**: 318–23.

Hamilton, W.D. (1980). Sex versus non-sex versus parasite. *Oikos* **35**: 282–90.

Holt, R.D. & Pickering, J. (In press). Infectious disease and species coexistence; a model of Lotka-Volterra form. *Am Nat.*

Huff, C.G. (1931). The inheritance of natural immunity to *Plasmodium cathemerium* in two species of *Culex*. *J. prev. Med. (U.S.A.)* **5**: 249–59.

Kemper, J.T. (1982). The evolutionary effect of endemic infectious disease: continuous models for an invariant pathogen. *J. math. Biol.* **15**: 65–77.

Kilama, W.L. & Craig, G.B., Jr. (1969). Monofactorial inheritance of susceptibility to *Plasmodium gallinaceum* in *Aedes aegypti*. *Ann. trop. Med. Parasit.* **63**: 419–32.

Kirchner, H., Engler, H., Zawatzky, R. & Schroder, C.H. (1980). Studies of resistance of mice against Herpes-Simplex virus. In *Genetic control of natural resistance to infection and malignancy*: 267–76. (Eds Skamene, E., Kongshavn, P.A.L. & Landy, M.). Academic Press, London.

Lackie, A.M. (1980). Invertebrate immunity. *Parasitology* **80**: 393–412.

Levin, B.R. & Lenski, R.E. (1983). Coevolution in bacteria and their viruses and plasmids. In *Coevolution*: 99–128. (Eds Futuyma, D.J. & Slatkin, M.) Sinauer Associations Inc., Sunderland, Mass.

Macdonald, W.W. (1962). The genetic basis of susceptibility to infection with semi-periodic *Brugia malayi* in *Aedes aegypti*. *Ann. trop. Med. Parasit.* **56**: 373–82.

Macdonald, W.W. & Ramachandran, C.P. (1965). The influence of the gene f^m (filarial susceptibility, *Brugia malayi*) on the susceptibility of *Aedes aegypti* to seven strains of *Brugia*, *Wuchereria*, and *Dirofilaria*. *Ann. trop. Med. Parasit.* **59**: 64–73.

McGreevy, P.B., McClelland, G.A.H. & Lavoipierre, M.M.J. (1974). Inheritance of susceptibility to *Dirofilaria immitis* infection in *Aedes aegypti*. *Ann. trop. Med. Parasit.* **68**: 97–109.

McKay, D. & Jenkin, C.R. (1969). Immunity in the invertebrates. II. Adaptive immunity in the crayfish (*Parachaeraps bicarinatus*). *Immunology* **17**: 127–37.

McKay, D. & Jenkin, C.R. (1970a). Immunity in the invertebrates. The fate and distribution of bacteria in normal and immunized crayfish (*Parachaeraps bicarinatus*). *Aust. J. exp. Biol. med. Sci.* **48**: 599–607.

McKay, D. & Jenkin, C.R. (1970b). Immunity in the invertebrates. Correlation of the phagocytic activity of haemocytes with resistance to infection in the crayfish (*Parachaeraps bicarinatus*). *Aust. J. exp. Biol. med. Sci.* **48**: 608–17.

Maramorosch, K. & Shope, R.E. (Eds) (1975). *Invertebrate immunity*. Academic Press, New York, San Francisco & London.

Martignoni, M.E. (1957). Contributo alla conoscenza di una granulosi di *Eucosoma griseana* (Hubner) (Tortricidae: Lepidoptera) quale fattori limitante il pullulamento dell'insetto nella Engadina alta. *Mitt. schweiz. ZentAnst. forstl. Versuchsw.* **32**: 371–418.

Martignoni, M.E. & Schmid, P. (1961). Studies on the resistance to virus infection in natural populations of Lepidoptera. *J. Insect Path.* **3**: 62–74.

May, R.M. (1974). *Stability and complexity in model ecosystems* (2nd edn.). Princeton University Press, Princeton.

May, R.M. (Ed.) (1976). *Theoretical ecology. Principles and applications*. Blackwell Scientific Publications, Oxford.

May, R.M. & Anderson, R.M. (1978). Regulation and stability of host-parasite population interactions. II. Destabilizing processes. *J. Anim. Ecol.* **47**: 249–67.

May, R.M. & Anderson, R.M. (1979). Population biology of infectious diseases: part III. *Nature, Lond.* **280**: 455–61.

May, R.M. & Anderson, R.M. (1983). Epidemiology and genetics in the coevolution of parasites and hosts. *Proc. R. Soc. Lond.* (B) **219**: 281–313.

Minchella, D.J. & Loverde, P.T. (1983). Laboratory comparison of the relative success of *Biomphalaria glabrata* stocks which are susceptible and insusceptible to infection with *Schistosoma mansoni*. *Parasitology* **86**: 335–44.

Paige, C.J. & Craig, G.B., Jr. (1975). Variation in filarial susceptibility among East African populations of *Aedes aegypti*. *J. med. Ent. Honolulu* **12**: 485–93.

Rager-Zisman, B., Neighbour, P.A., Grace, J. & Bloom, B.R. (1980). The role of H–Z in resistance and susceptibility to measles virus infection. In *Genetic control of natural resistance to infection and malignancy*: 313–20. (Eds Skamene, E., Kongshavn, P.A.L. & Landy, M.). Academic Press, New York.

Ratcliffe, N.A. & Rowley, A.F. (1981). *Invertebrate blood cells* **1** & **2**. Academic Press, London/New York.

Renwrantz, L.R. & Cheng, T.C. (1977a). Identification of agglutinin receptors on hemocytes of *Helix pomatia*. *J. Invert. Path.* **29**: 88–96.

Renwrantz, L.R. & Cheng, T.C. (1977b). Agglutinin-mediated attachment of erythrocytes in hemocytes of *Helix pomatia*. *J. Invert. Path.* **29**: 97–100.

Richards, C.S. (1970). Genetics of a molluscan vector of schistosomiasis. *Nature, Lond.* **227**: 806–10.

Richards, C.S. (1973). Susceptibility of adult *Biomphalaria glabrata* to *Schistosoma mansoni* infection. *Am. J. trop. Med. Hyg.* **22**: 749–56.

Richards, C.S. (1975). Genetic factors in susceptibility of *Biomphalaria glabrata* for different strains of *Schistosoma mansoni*. *Parasitology* **70**: 231–41.

Richards, C.S. (1977). *Schistosoma mansoni*: susceptibility reversal with age in the snail host *Biomphalaria glabrata*. *Expl. Parasit.* **42**: 165–8.

Richards, C.S. (1984). Influence of snail age on genetic variations in susceptibility of *Biomphalaria glabrata* for infection with *Schistosoma mansoni*. *Malacologia* **25**: 493–502.

Roitt, I. (1984). *Essential immunology*. Blackwell Scientific Publications, Oxford.

Rollinson, D. & Southgate, V.R. (1985). Schistosome and snail populations: genetic variability and parasite transmission. In *Ecology and genetics of host-parasite interactions*: 91–110. (Eds Rollinson, E. & Anderson, R.M.). Academic Press, London.

Roughgarden, J. (1983). The theory of coevolution. In *Coevolution*: 33–64. (Eds Futuyma, D.J. & Slatkin, M.). Sinauer Associations Inc., Sunderland, Mass.

Salt, G. (1970). *The cellular defence reactions of insects*. Cambridge University Press, Cambridge.

Skamene, E., Kongshavn, P.A.L. & Landy, M. (Eds) (1980). *Genetic control of natural resistance to infection and malignancy*. Academic Press, New York.

Storey, H.H. (1932). The inheritance by an insect vector of the ability to transmit a plant virus. *Proc. R. Soc.* (B) **112**: 46–60.

Thomas, V. & Ramachandran, C.P. (1970). Selection of *Culex pipiens fatigans* for vector ability to the rural strain of *Wuchereria bancrofti*—a preliminary report. *Med. J. Malaya.* **24**: 196–9.

Townson, H.T. (1971). Mortality of various genotypes of the mosquito *Aedes aegypti* following the uptake of microfilaria of *Brugia pahangi*. *Ann. trop. Med. Parasit.* **65**: 93–106.

Wakelin, D. (1984). *Immunity to parasites*. Edward Arnold, London.

Ward, R.A. (1963). Genetic aspects of the susceptibility of mosquitoes to malarial infection. *Expl. Parasit.* **13**: 328–41.

Wright, S. (1969). *Evolution and the genetics of populations*. **2**. *The theory of gene frequencies*. University of Chicago Press, Chicago.

Yoshino, T.P. & Granath, W.O. (1985). Surface antigens of *Biomphalaria glabrata* (Gastropoda) hemocytes: functional heterogeneity in cell subpopulations recognised by a monoclonal antibody. *J. Invert. Path.* **45**: 174–86.

Index